BAYESIAN METHODS IN COSMOLOGY

In recent years, cosmologists have advanced from largely qualitative models of the Universe to precision modelling using Bayesian methods in order to determine the properties of the Universe to high accuracy. This timely book is the only comprehensive introduction to the use of Bayesian methods in cosmological studies, and is an essential reference for graduate students and researchers in cosmology, astrophysics and applied statistics.

The first part of the book focuses on methodology, setting the basic foundations and giving a detailed description of techniques. It covers topics including the estimation of parameters, Bayesian model comparison, and separation of signals. The second part explores a diverse range of applications, from the detection of astronomical sources (including through gravitational waves), to cosmic microwave background analysis and the quantification and classification of galaxy properties. Contributions from 24 highly regarded cosmologists and statisticians make this an authoritative guide to the subject.

MICHAEL P. HOBSON is Reader in Astrophysics and Cosmology at the Cavendish Laboratory, University of Cambridge, where he researches theoretical and observational cosmology, Bayesian statistical methods, gravitation and theoretical optics.

ANDREW H. JAFFE is Professor of Astrophysics and Cosmology at Imperial College, London, and a member of the Planck Surveyor Satellite collaboration, which will create the highest-resolution and most sensitive maps of the cosmic microwave background ever produced.

ANDREW R. LIDDLE is Professor of Astrophysics at the University of Sussex. He is the author of over 150 journal articles and four books on cosmology, covering topics from early Universe theory to modelling astrophysical data.

PIA MUKHERJEE is a Postdoctoral Research Fellow in the Astronomy Centre at the University of Sussex, specializing in constraining cosmological models, including dark energy models, from observational data.

DAVID PARKINSON is a Postdoctoral Research Fellow in the Astronomy Centre at the University of Sussex, working in the areas of cosmology and the early Universe.

BAYESIAN METHODS IN COSMOLOGY

MICHAEL P. HOBSON

Cavendish Laboratory, University of Cambridge

ANDREW H. JAFFE

Imperial College, London

ANDREW R. LIDDLE

University of Sussex

PIA MUKHERJEE

University of Sussex

DAVID PARKINSON

University of Sussex

CAMBRIDGE
UNIVERSITY PRESS

CAMBRIDGE
UNIVERSITY PRESS

The Edinburgh Building, Cambridge CB2 8RU, UK

Published in the United States of America by Cambridge University Press, New York

Cambridge University Press is part of the University of Cambridge.

It furthers the University's mission by disseminating knowledge in the pursuit of
education, learning and research at the highest international levels of excellence.

www.cambridge.org
Information on this title: www.cambridge.org/9781107631755

First published 2010
First paperback edition 2013

A catalogue record for this publication is available from the British Library

Library of Congress Cataloguing in Publication data

Bayesian methods in cosmology / [edited by] M. P. Hobson ... [et al.].
p.   cm.
ISBN 978-0-521-88794-6 (Hardback)
1. Cosmology–Statistical methods.   2. Bayesian statistical decision theory.
I. Hobson, M. P. (Michael Paul), 1967–   II. Title.
QB991.S73B34 2009
523.101´519542–dc22

2009035034

ISBN 978-0-521-88794-6 Hardback
ISBN 978-1-107-63175-5 Paperback

Contents

Contributors

Filipe B. Abdalla
Department of Physics and Astronomy,
University College London,
Gower Street,
London WC1E 6BT, UK

Stefano Andreon
INAF–Osservatorio Astronomico di
Brera via Brera 28, 20121 Milano, Italy

M. A. J. Ashdown
Astrophysics Group, Cavendish
Laboratory, JJ Thomson Avenue,
Cambridge CB3 0HE, UK

Manda Banerji
Department of Physics and Astronomy,
University College London,
Gower Street, London WC1E 6BT, UK

Sarah Bridle
Department of Physics and Astronomy,
University College London,
Gower Street, London WC1E 6BT, UK

Neil Cornish
Department of Physics, Montana
State University, Bozeman,
MT 597717, USA

Martin A. Hendry
Department of Physics and Astronomy,
University of Glasgow,
Glasgow G12 8QQ, UK

M. P. Hobson
Astrophysics Group, Cavendish
Laboratory, JJ Thomson Avenue,
Cambridge CB3 0HE, UK

Andrew H. Jaffe
Astrophysics Group, Imperial College
London, Blackett Laboratory,
London SW7 2AZ, UK

Martin Kunz
Astronomy Centre, University of
Sussex, Brighton BN1 9QH, UK

Ofer Lahav
Department of Physics and Astronomy,
University College London, Gower Street,
London WC1E 6BT, UK

Antony Lewis
Institute of Astronomy and Kavli
Institute for Cosmology,
Madingley Road,
Cambridge CB3 0HA, UK

Andrew R. Liddle
Astronomy Centre, University of
Sussex, Brighton BN1 9QH, UK

Thomas J. Loredo
Department of Astronomy,
Cornell University, Ithaca,
NY 14853, USA

Daniel Mortlock
Astrophysics Group, Imperial College
London, Blackett Laboratory, London
SW7 2AZ, UK

Pia Mukherjee
Astronomy Centre, University of
Sussex, Brighton BN1 9QH, UK

David Parkinson
Astronomy Centre, University of
Sussex, Brighton BN1 9QH, UK

Steve Rawlings
Astrophysics, Department of Physics,
Oxford University, Keble Road, Oxford
OX1 3RH, UK

Graça Rocha
California Institute of Technology,
1200 East California Boulevard,
Pasadena, CA 91125, USA

Richard S. Savage
Astronomy Centre, University
of Sussex, Brighton BN1 9QH, UK,
and Systems Biology Centre, University
of Warwick, Coventry CV4 7AL, UK

D. S. Sivia
St John's College, St. Giles,
Oxford OX1 3JP, UK

John Skilling
Maximum Entropy Data Consultants
Ltd, Kenmare, County Kerry, Ireland

V. Stolyarov
Astrophysics Group, Cavendish
Laboratory, JJ Thomson Avenue,
Cambridge CB3 0HE, UK

Roberto Trotta
Astrophysics, Department of Physics,
Oxford University, Keble Road,
Oxford OX1 3RH, UK
and Astrophysics Group, Imperial
College London, Blackett Laboratory,
London SW7 2AZ, UK

Preface

A revolution is underway in cosmology, with largely qualitative models of the Universe being replaced with precision modelling and the determination of Universe's properties to high accuracy. The revolution is driven by three distinct elements – the development of sophisticated cosmological models and the ability to extract accurate predictions from them, the acquisition of large and precise observational datasets constraining those models, and the deployment of advanced statistical techniques to extract the best possible constraints from those data.

This book focuses on the last of these. In their approach to analyzing datasets, cosmologists for the most part lie resolutely within the Bayesian methodology for scientific inference. This approach is characterized by the assignment of probabilities to all quantities of interest, which are then manipulated by a set of rules, amongst which Bayes' theorem plays a central role. Those probabilities are constantly updated in response to new observational data, and at any given instant provide a snapshot of the best current understanding. Full deployment of Bayesian inference has only recently come within the abilities of high-performance computing.

Despite the prevalence of Bayesian methods in the cosmology literature, there is no single source which collects together both a description of the main Bayesian methods and a range of illustrative applications to cosmological problems. That, of course, is the aim of this volume. Its seeds grew from a small conference 'Bayesian Methods in Cosmology', held at the University of Sussex in June 2006 and attended by around 60 people, at which many cosmological applications of Bayesian methods were discussed. CUP editor Vince Higgs, who attended the conference, saw the need for a comprehensive volume covering these topics, and suggested that we put together an edited volume of articles. And here it is!

The book is divided into two part. The first part, 'Methods', concentrates on the formalism, methods and algorithms, with only limited illustrative examples. The focus is very much on those aspects that have proven valuable in cosmological studies, complementing the much more complete treatments of Bayesian inference given in the excellent books by MacKay (2003; see page 35), Gregory (2005; see page 97), and Sivia and Skilling (2006; see page 35), all of which we recommend to the interested reader. The second part, 'Applications', studies a wide range of cosmological applications in detail. Many of the codes used in these applications are publicly available.

Part I

Methods

1

Foundations and algorithms

John Skilling

Why and how – simply – that's what this chapter is about.

1.1 Rational inference

Rational inference is important. By helping us to understand our world, it gives us the predictive power that underlies our technical civilization. We would not function without it. Even so, rational inference only tells us *how* to think. It does not tell us *what* to think. For that, we still need the combination of creativity, insight, artistry and experience that we call intelligence.

In science, perhaps especially in branches such as cosmology, now coming of age, we invent models designed to make sense of data we have collected. It is no accident that these models are formalized in mathematics. Mathematics is far and away our most developed logical language, in which half a page of algebra can make connections and predictions way beyond the precision of informal thought. Indeed, one can hold the view that frameworks of logical connections are, by definition, mathematics. Even here, though, we do not find absolute truth. We have conditional implication: 'If axiom, then theorem' or, equivalently, 'If not theorem, then not axiom'. Neither do we find absolute truth in science.

Our question in science is not 'Is this hypothetical model true?', but '*Is this model better than the alternatives?*'. We could not recognize absolute truth even if we stumbled across it, for how could we tell? Conversely, we cannot recognize absolute falsity. If we believe dogmatically enough in a particular view, then no amount of contradictory data will convince us otherwise, if only because the data could be dismissed as evidence of conspiracy to deceive. Yet even a determined sceptic might be sufficiently charitable to acknowledge that a model with demonstrable ability to predict future effects could have practical value.

Let us, then, avoid the philosophical minefields of belief and truth, and pay attention to what we really need, which is predictive ability. We anticipate the Sun will rise tomorrow, not just because it always has done so far, but because this is predicted by models of stellar structure and planetary dynamics, which accord so well with such a variety of data that perceived failure of the Sun to rise might more likely be hallucination.

Rational assessment of different models is the central subject of Bayesian methods, so called after Revd Thomas Bayes, the eighteenth century clergyman generally associated with the beginnings of formal probability theory. We will find that probability calculus is forced upon us as the only method which lets us learn from data irrespective of their order – surely a required symmetry. We will also discover how to use it properly, with the aid of modern computers and algorithms. Inference was held back for a century by technical inability to do the required sums, but that sad era has closed.

1.2 Foundations

Suppose we are given a choice of basic models, $a = apple$, $b = banana$, $c = cherry$, which purport to explain some data. Such models can be combined, so that $apple$-OR-$banana$, written $a \vee b$, is also meaningful. Data that excluded cherries would, in fact, bring us down from the original $apple$-OR-$banana$-OR-$cherry$ combination to just that choice. With n basic models, there are 2^n possible combinations. They form the elements of a lattice, ranging from the absurdity in which none of the models is allowed, up to the provisional truism in which all of them remain allowed. In inference, we need to be able to navigate these possibilities as we refine our knowledge.

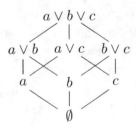

The absurdity $\emptyset$ is introduced merely because analysis is cleaner with it than without it, rather as 0 is often included with the positive integers.

The three core concepts of *measure*, *information* and *probability* all have wider scope than inference alone. They apply to lattices in general, whether or not the lattice fills out all 2^n possibilities. By exposing just the foundation that we need, and no more, we can allow wider application, as well as clarifying the basis so that alternative formulations of these concepts become even less plausible than they may have been before.

1.2.1 Lattices

The critical idea we need is 'partial ordering'. We always have "=": every element equals itself, $x = x$. Sometimes, we have "<", as in $x < y$, meaning that y includes x. In inference, we say that *apple* is included in *apple*-OR-*banana*, because the scope of the latter is wider and includes all of the former, but we would not try to include *apple* within *banana*. We don't need that particular motivation, though. All we need is "<" in the abstract. Ordering is to be transitive,

$$x < y \text{ and } y < z \text{ implies } x < z, \tag{1.1}$$

otherwise it would not make sense.

The other idea we need is 'least upper bound'. The upper bounds to elements x and y are those elements at or including both x and y. If there is a least such bound, we write it as $x \vee y$ and call it the least upper bound:

$$\left\{ \begin{array}{c} x \leq x \vee y \\ y \leq x \vee y \end{array} \right\}, \text{ and } x \vee y \leq u \text{ for all } u \text{ obeying } \left\{ \begin{array}{c} x \leq u \\ y \leq u \end{array} \right\}. \tag{1.2}$$

In inference, the unique least upper bound $x \vee y$ is that element including all the components of x and y, but no more. There, the existence of least upper bound is obvious.

Technically, a *lattice* is a partially ordered set with least upper bound, so that "<" and "$\vee$" are defined. Any pair of elements x and y also has lower bounds, being all those elements at or beneath both. There is a unique greatest lower bound, written $x \wedge y$. (If there were alternatives u and v, then $u \vee v$ could be ambiguously x or y, contradicting uniqueness of their least upper bound.) Mathematicians (Klain & Rota 1997) traditionally define a lattice in terms of $\vee$ and $\wedge$, but our use of < and $\vee$ is equivalent, and (with =) underlies their traditional axioms of reflexivity, antisymmetry, transitivity, idempotency, commutativity, associativity and absorption. Of these, the associativity property

$$(x \vee y) \vee z = x \vee (y \vee z) \tag{1.3}$$

is of particular importance to us.

What we now seek is a numerical *valuation* $v(x)$ on our lattice of models, so that we can rank the possibilities. Remarkably, there is only one way of conforming to lattice structure, and this leads us to measure theory, thence to information and probability. Though modernized following Knuth (2003), the approach dates back to Cox (1946, 1961).

1.2.2 Measure

Addition

For a start, we want valuations to conform to "$\leq$", so we require

$$x \leq y \quad \Longrightarrow \quad v(x) \leq v(y). \tag{1.4}$$

Moreover, whatever our valuations were originally, we can shift them to give a standard value 0 to the ubiquitous absurdity $\emptyset$, so that the range of value becomes $0 = v(\emptyset) \leq v(x)$.

We next assume that if x and y are disjoint, so that $x \wedge y = \emptyset$ and they have nothing in common, then the valuation $v(x \vee y)$ should depend only on $v(x)$ and $v(y)$. Write this relationship as a binary operation $\oplus$,

$$v(x \vee y) = v(x) \oplus v(y) \text{ when } x \wedge y = \emptyset. \tag{1.5}$$

To conform with associativity (1.3), we require

$$\big(v(x) \oplus v(y)\big) \oplus v(z) = v(x) \oplus \big(v(y) \oplus v(z)\big). \tag{1.6}$$

This has to hold for arbitrary values $v(x)$, $v(y)$, $v(z)$, and the *associativity theorem* (Azcél 2003) then tells us that there must be some invertable function F of our valuations v such that

$$F\big(v(x \vee y)\big) = F\big(v(x)\big) + F\big(v(y)\big). \tag{1.7}$$

That being the case, we are free to discard the original valuations v and use $m = F(v)$ instead, for which $\vee$ is simple addition: for disjoint x and y we have the *sum rule*

$$\boxed{m(x \vee y) = m(x) + m(y)} \tag{1.8}$$

In other words, valuation can without loss of generality be taken to be what mathematicians call a '*measure*'. They traditionally define measures from the outset as additive over infinite sets, but offer little justification. Mathematicians just do it. Physicists want to know why. Here we see that there's no alternative, and although we start finite we can extend to arbitrarily many elements; it is the same structure. This is *why* measure theory works – it is because of associativity – and we physicists don't have to worry about the infinite.

Assignment

As for actual numerical values, we can build them upwards by addition – except for foundation elements that are not equal to any least upper bound of different elements. Those values alone cannot be determined by the sum rule.

Thus, in the inference example, we can value an apple, a banana and a cherry arbitrarily, but there is a scale on which combinations add. On that scale, if an apple costs 3¢ and a cherry costs 4¢, then their combination costs $3 + 4 = 7$¢, not $3^2 + 4^2 = 25$ or other non-linear construction. Associativity underlies money. Personal assignments may be on a different scale. In economics, for example, personal benefit is sometimes held to be logarithmic in money, $m = \log(\$)$, to reflect the asymmetry between devastating downside risk and comforting upside reward. On that scale, money combines non-linearly, as $\log \$(x \vee y) = \log \$(x) + \log \$(y)$. That's permitted, the point being that there *is* a scale on which one's numbers add. Thus quantification is intrinsically linear – because of associativity.

In inference, $\vee$ behaves as logical OR and $\wedge$ as logical AND, obeying the extra property of distributivity:

$$
\begin{aligned}
(x \text{ OR } y) \text{ AND } z &= (x \text{ AND } z) \text{ OR } (y \text{ AND } z), \\
(x \text{ AND } y) \text{ OR } z &= (x \text{ OR } z) \text{ AND } (y \text{ OR } z).
\end{aligned}
\tag{1.9}
$$

Equivalently, they behave as set union and set intersection of the foundation elements, which can therefore be assigned arbitrary values. In other applications, $\vee$ and $\wedge$ might not be distributive, and the foundation assignments become restricted by the non-equality of combinations that would otherwise be identical. But their calculus would still be additive.

Multiplication

As well as by addition, measures can also combine by multiplication. Here, we consider a direct product of lattices. For example, one lattice might have playing-card foundation elements ($\spadesuit$, $\heartsuit$, $\clubsuit$, $\diamondsuit$) while the other has music-key foundations ($\flat$, $\natural$, $\sharp$). The direct-product lattice treats both together, here with 12 foundation elements like $\heartsuit \times \natural$ and 2^{12} elements overall:

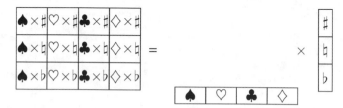

We now assume that the measure $m(x \times y)$ should depend only on $m(x)$ and $m(y)$. Write this relationship as a binary operation

$$
m(x \times y) = m(x) \otimes m(y).
\tag{1.10}
$$

Now the direct-product operator is associative, $(x \times y) \times z = x \times (y \times z)$,

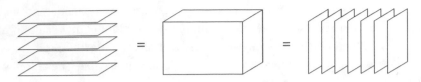

so

$$\big(m(x) \otimes m(y)\big) \otimes m(z) = m(x) \otimes \big(m(y) \otimes m(z)\big). \qquad (1.11)$$

This has to hold for arbitrary values $m(x)$, $m(y)$, $m(z)$, and the associativity theorem then tells us that there must be some invertable function Φ of the measures m such that

$$\Phi(m(x \times y)) = \Phi(m(x)) + \Phi(m(y)). \qquad (1.12)$$

We cannot now re-grade to $\Phi(m)$ and ignore m because we have already fixed the behaviour of m to be additive. What we can do is require consistency with that behaviour by requiring the sum rule (1.8) to hold for composite elements, $m(x \times t) + m(y \times t) = m((x \vee y) \times t)$ for any t. The *context theorem* (Knuth and Skilling, in preparation) then shows that Φ has to be logarithmic, so that

$$m(x \times y) = m(x)\, m(y). \qquad (1.13)$$

While $\oplus$ is addition, $\otimes$ is multiplication. Combination is intrinsically multiplicative – because of associativity. There is no alternative.

Commutativity

Technically, we have not used the commutative property $x \vee y = y \vee x$ of a lattice. However, the sum rule automatically generates values that are equal, $v(x \vee y) = v(y \vee x)$. So real valuations cannot capture non-commutative behaviour. Quantum mechanics is an example, where states lack ordering so do not form a lattice, and the calculus is complex. Inference is not an example. There, *apple*–OR–*banana* is the same as *banana*–OR–*apple* so $\vee$ is commutative for us, and we are allowed to use the real values that we need.

1.2.3 Information

Different measures can legitimately be assigned to the same foundation elements, as when different individuals value apples, bananas and cherries differently. The difference between source measure μ and destination measure m can be quantified, consistently with lattice structure, as '*information*' $H(m \mid \mu)$.

One way of deriving the form of H is as a variational potential, in which destination m is obtained at the extremal (minimum, actually) of H, subject to whatever constraints require the change from μ. Suppose the playing-card example above has source measure μ, with destination m obtained by some constraint on card suits. Independently, the music-key example has source measure ν, with destination n obtained by some constraint on music keys.

Equivalently, we must be able to analyze the problems jointly. Measures multiply, so the joint element 'card suit i and music key j' has source measure $\mu_i \nu_j$ and destination $m_i n_j$. The latter is to be obtained at the extremal of $H(m_i n_j \mid \mu_i \nu_j)$, under one constraint acting on i and another on j. Temporarily suppressing the fixed source $\mu\nu$, the variational equation for the destination measure is

$$H'(m_i n_j) = \lambda_1(i) + \lambda_2(j), \tag{1.14}$$

where the λ's are the Lagrange multipliers of the i and j constraints. Writing $x = m_i$ and $y = n_j$, and differentiating $\partial^2/\partial x \partial y$, the right-hand side is annihilated, leaving

$$xy H'''(xy) + H''(xy) = 0, \tag{1.15}$$

whose solution is

$$H(z) = A - Bz + Cz \log z. \tag{1.16}$$

Setting $C = 1$ an as arbitrary scale (positive to ensure a minimum), $B = 1$ to place that minimum correctly at $m = \mu$, and $A = \mu$ to make the minimum zero, we reach (Skilling 1988)

$$\boxed{H(m \mid \mu) = \mu - m + m \log \frac{m}{\mu}} \tag{1.17}$$

This obeys (1.14), so the potential we seek exists, and is required to be of this unique form. The difference between measures plays a deep rôle in Bayesian analysis.

1.2.4 Probability

Acquiring data involves a reduction of possibilities. Some outcomes that might have happened, did not. In terms of the lattice of possibilities, the all-encompassing top element moves down. To deal with this, we seek a *bi*-valuation $p(x \mid t)$, in which the context t of model x can shrink. Within any fixed context, p is to be a measure, being non-negative and obeying the sum rule. But we want to change the numbers when the context changes.

To find the dependence on context, take ordered elements $x \le y \le z \le t$. As before, we require conformity with lattice ordering, here

$$x \le y \le z \quad \Longrightarrow \quad p(x \mid z) \le p(x \mid y) \tag{1.18}$$

so that a wider context dilutes the numerical value. Ordering such as $x \leq z$ can be carried out in two steps, $x \leq y$ and $y \leq z$. Our bi-valuation should conform to this, meaning that we require a "$\odot$" operator combining the two steps into one:

$$p(x \mid z) = p(x \mid y) \odot p(y \mid z). \tag{1.19}$$

Extending this to three steps and considering passage $p(x \mid t)$ from x to t, via y and z, gives another associativity relationship,

$$\big(p(x \mid y) \odot p(y \mid z)\big) \odot p(z \mid t) = p(x \mid y) \odot \big(p(y \mid z) \odot p(z \mid t)\big), \tag{1.20}$$

representing $(((x \leq y) \leq z) \leq t) = (x \leq (y \leq (z \leq t)))$. As before, this induces some invertible function Φ of our valuations p such that

$$\Phi\big(p(x \mid z)\big) = \Phi\big(p(x \mid y)\big) + \Phi\big(p(y \mid z)\big). \tag{1.21}$$

Again, we require consistency with the sum rule $p(x \vee y \mid t) = p(x \mid t) + p(y \mid t)$ for arbitrary context t. A variant of the context theorem (Knuth and Skilling, in preparation) then shows that Φ has to be logarithmic as before, so $\odot$ was multiplication. Specifically, we recognize p as *probability*, hereafter "pr",

$$
\left.
\begin{array}{lll}
0 = \mathrm{pr}(\emptyset) \leq \mathrm{pr}(x) \leq \mathrm{pr}(t) = 1 & \text{Range} \\
\mathrm{pr}(x \vee y) = \mathrm{pr}(x) + \mathrm{pr}(y) & \text{Sum rule for disjoint } x, y \\
\mathrm{pr}(x \wedge y) = \mathrm{pr}(x \mid y)\,\mathrm{pr}(y) & \text{Product rule}
\end{array}
\right\} \ \| \ t \tag{1.22}
$$

(The "$\| \ t$" notation means that all probabilities are conditional on t, and avoids proliferation of "$\mid t$" without introducing ambiguity.)

Just as measure theory was forced for valuations, so probability theory is forced for bi-valuations. We need not be distracted by claimed alternatives because they conflict with very general requirements. It is all very simple. There's only this one calculus for numerical bi-valuations on a lattice. If, say, we seek a calculus for conditional beliefs, then this has to be it. But the calculus itself is abstract and motive-free. We don't have to subscribe to an undefined idea like 'belief' in order to use it. In fact, the reverse holds. It is probability, with its defined properties, that would underpin belief, not the other way round.

Most simply of all, probability calculus can be subsumed in the single definition of probability as a ratio definition of measures:

$$\boxed{\ \mathrm{pr}(x \mid t) = \frac{m(x \wedge t)}{m(t)}\ } \tag{1.23}$$

This is the original discredited frequentist definition, as the ratio of number of successes to number of trials, now retrieved at an abstract level, which bypasses the catastrophic difficulties of literal frequentism when faced with isolated non-reproducible situations. The calculus of probability is no more than the calculus of proportions.

1.3 Inference

Henceforward, in accordance with traditional accounts, we take all foundation elements to be disjoint, and work in terms of these. The OR operator $\vee$ can be replaced by the summation to which it reduces, while the AND operator $\wedge$ can be written as the traditional comma. It is also usual to use I for context, and allow the discrete choice x to be continuous θ. The rules of probability calculus then reduce to

$$
\left.
\begin{array}{ll}
\mathrm{pr}(\theta) \geq 0 & \text{Positivity} \\
\int \mathrm{pr}(\theta)\, \mathrm{d}\theta = 1 & \text{Sum rule} \\
\mathrm{pr}(\phi, \theta) = \mathrm{pr}(\phi \mid \theta)\, \mathrm{pr}(\theta) & \text{Product rule}
\end{array}
\right\} \quad \| \, I \qquad (1.24)
$$

1.3.1 Bayes' theorem

In inference, we need to consider both parameter(s) θ and data D, all in the overarching context I of all possibilities we are currently considering. By the product law, the joint probability of model and data factorizes:

$$
\begin{array}{ccccccl}
\mathrm{pr}(\theta)\,\mathrm{pr}(D \mid \theta) & = & \mathrm{pr}(\theta, D) & = & \mathrm{pr}(D)\,\mathrm{pr}(\theta \mid D) & & \| \, I \\
\text{Prior} \times \text{Likelihood} & = & \text{Joint} & = & \text{Evidence} \times \text{Posterior} & & \\
\pi(\theta)\,\mathcal{L}(\theta) & = & \cdots\cdots & = & E\,\mathcal{P}(\theta) & & \\
\text{Inputs} & & \Longrightarrow & & \text{Outputs} & &
\end{array}
\qquad (1.25)
$$

On the left lies the prior probability $\pi(\theta) = \mathrm{pr}(\theta \mid I)$, representing how we originally distributed the parameters' unit mass of probability. This assignment has provoked legendary argumentation, and we discuss it below. Also on the left is the likelihood $\mathcal{L}(\theta) = \mathrm{pr}(D \mid \theta)$, representing the probability distribution of the data for each allowed input θ. This is less controversial. The instrument acquiring the data can usually be calibrated with known inputs θ to find how often it produces specific outputs D, which effectively fixes the likelihood to any desired precision. If there remain any unknown calibration parameters in the likelihood, they can be incorporated in θ as extra parameters to be determined, leading to extra computation but no difficulty of principle.

On the far right is the posterior $\mathcal{P}(\theta) = \mathrm{pr}(\theta \mid D, I)$, representing our inferred distribution of probability among the models, after using the data. The difference between prior and posterior is the information (1.17)

$$
H(\mathcal{P} \mid \pi) = \int \mathcal{P}(\theta) \log\left(\mathcal{P}(\theta)/\pi(\theta)\right) \mathrm{d}\theta \qquad (1.26)
$$

gleaned about θ. Also on the right is the evidence $E = \mathrm{pr}(D \mid I)$, representing how well our original assignments managed to predict the data. E is also known as 'prior predictive' (how it is often used), 'marginal likelihood' (how it is often computed), and various similar terms. However, there ought to be a simple moniker (what it is) for this key quantity in Bayesian analysis, and 'evidence' (not to be confused with dataset) is that name (MacKay 2003).

Of course, the terminology is for convenience only. It is not hard and fast. A posterior to a first analyst may become a prior to a second with new data. Evidence values become likelihoods if the context is widened, so that I becomes merely a provisional model within a wider analysis, and so on. There is really just one quantity, *probability*.

The two outputs, evidence and posterior, can be disentangled by noting that the posterior, being a probability, sums to 1. Here, then, is the complete calculus of inference:

$$
\begin{array}{ll}
\displaystyle\int \pi(\theta)\,\mathrm{d}\theta = 1 & \text{Prior} \\[2mm]
\displaystyle E = \int \pi(\theta)\,\mathcal{L}(\theta)\,\mathrm{d}\theta & \text{Evidence} \\[2mm]
\displaystyle \mathcal{P}(\theta) = \frac{\pi(\theta)\,\mathcal{L}(\theta)}{E} & \text{Bayes' theorem}
\end{array}
\tag{1.27}
$$

Bayes' theorem shows how the prior is modulated into posterior through the likelihood/evidence ratio. The same ratio $\mathcal{L}/E = \mathcal{P}/\pi$ shows up in the information H, alternatively known as the negative entropy.

1.3.2 Prior probability

Before using the data, we need to assign a distribution of prior probability. Probability calculus tells us how to manipulate probability values, but not what they should be in the first place. Neither does the world tell us. The only restriction is that, by the sum rule, all the possibilities must add to 1. Beyond that, we are free to invent any model we want. In that sense, anything goes. It is a challenge. However, the world does give its opinion through data. Better hypotheses predict the data better, through having high values of evidence. That and that alone is what we get, and it is all we need for understanding and for technology. The quest for certainty is mistaken and naive.

Guidelines have been developed for assigning priors, but *beware the dangers!* The first step is to decide on a model – which parameters are to be used to predict the data? The range of these parameters defines what is known as the 'hypothesis space', over which unit mass of prior probability is to be distributed.

Informal

No matter how sophisticated the methodology, there is in the end no escape from an informal assessment of what is judged reasonable in the light of whatever background knowledge is available. Your author proceeds by contemplating perhaps ten points, each representing 10% of the prior, and assigning plausible θ to them. This introspection gives a rough range and indication of shape for the prior, which is then assigned some algebraic form conforming to these. Instead of putting prior mass onto θ, this procedure puts θ onto prior mass, which seems more sympathetic to the basic equations. It is also more sympathetic to the computational requirements, because the prior is uniform by definition when prior mass is the underlying coordinate. Either way, the prior doesn't have to be 'right' in some undefinable sense – it just has to be reasonable.

For a location parameter, an informal centre c and width w might suggest a Cauchy distribution,

$$\pi(\theta) = \frac{w/3.14159\ldots}{(\theta - c)^2 + w^2}, \quad (-\infty < \theta < \infty), \tag{1.28}$$

which comfortably tolerates quite wide excursions from the guessed centre if the data demand them. Note that we don't need to interpret c and w as moments, and indeed we may be wiser not to. Probability calculus requires normalization, but not the existence of mean and standard deviation. To some extent, moments are a holdover from the days of manual paper-and-pencil calculation.

A necessarily positive intensity parameter of plausible magnitude a might be assigned either a truncated Cauchy or an exponential distribution,

$$\pi(\theta) = a^{-1}e^{-\theta/a}, \quad (\theta > 0). \tag{1.29}$$

If magnitude was accompanied by width, then a Gamma distribution

$$\pi(\theta) = \frac{\theta^{-1+\mu}e^{-\theta/\lambda}}{\Gamma(\mu)\lambda^\mu}, \quad (\theta > 0) \tag{1.30}$$

might be appropriate. Whatever the choice, the prior has to be normalized because it is a probability.

Symmetry

Sometimes, our knowledge of some or all of the states is invariant to exchange. The classic example is a six-sided die, for which θ can be 1 or 2 or 3 or 4 or 5 or 6. Given this knowledge and nothing more, we can only assign equal probability to each: $\pi(1) = \pi(2) = \cdots = \pi(6) = \frac{1}{6}$. If we did anything else, say $\pi(1) > \pi(2)$, then we could exchange the labels 1 and 2 and reach a different assignment with $\pi(2) > \pi(1)$ on the basis of a null change to our prior knowledge. So the same state of knowledge would be coded two different ways, which is unlikely to be helpful. Prior assignments should conform to any symmetry in our prior knowledge. This does *not* mean that the object being investigated need be symmetric. Indeed, data may well tell us it is not.

Symmetry arguments can be over-played. Here, the classic example is a location parameter θ for which the hypothesis space is unbounded, $-\infty < \theta < \infty$. Given this, and nothing more, one's prior knowledge would be invariant to offset of origin, implying $\pi(\theta) = $ constant. After acquiring data, the posterior distribution $\mathcal{P} = \pi\mathcal{L}/E$ would be independent of whatever constant was chosen. $\mathcal{P}$ would be proportional to the likelihood, which would plausibly prohibit infinite values. With non-zero constant, the prior would become un-normalized (the dreaded 'improper prior'), but otherwise all might be well.

Actually, no. The posterior is the lesser half of Bayesian inference. The evidence comes first. As the allowed range W of θ increases indefinitely, the prior $\pi = 1/W$ decreases indefinitely, and so does the evidence. This means that the model with $W \to \infty$ loses by an infinite factor when compared with any prior that includes even the slightest knowledge of the expected range. In the limit, the posterior becomes $\mathcal{P}(\theta) = 0 \times \mathcal{L}(\theta) / 0$, and total ignorance is seen to be total stupidity. Moreover, the improper prior extending arbitrarily far fails the sanity check of informal assessment. Are you *really* almost certain that $|\theta| > 10^{100}$?

Approximate invariance to small offsets of origin suggests that the prior should be smooth, but that's as far as the argument should go.

Maximum entropy

Sometimes, informal background knowledge is accompanied by 'testable' constraints in the form of known means $\langle Q \rangle = \int Q(\theta)p(\theta)\,\mathrm{d}\theta$. Here, the informal prior π can be modified to a revised p by minimizing the information $H(p \mid \pi)$. 'Entropy' is just the traditional word for the negative of information, so that maximum entropy just means minimum information. Note that maximum entropy is a method of assignment, not inference. It assigns a single p to be used in later inference. It does not infer a probabilistic distribution of plausible p's.

The variational equation is

$$\delta\left(\int p(\theta) \log \frac{p(\theta)}{\pi(\theta)}\,\mathrm{d}\theta + \lambda_0 \int p(\theta)\,\mathrm{d}\theta + \lambda_1 \int Q(\theta)p(\theta)\,\mathrm{d}\theta + \cdots \right) = 0, \quad (1.31)$$

where the first term is H, λ_0 is the Lagrange multiplier for normalization, λ_1 is the multiplier for the constraint, and so on for as many constraints as are given. The solution is

$$p(\theta) = \pi(\theta) \exp(-1 - \lambda_0 - \lambda_1 Q(\theta) - \cdots), \quad (1.32)$$

where the λ's fit the constraints to their required values.

One standard example is a location parameter subject to first and second moments:

$$\mu = \int \theta \, p(\theta) \, d\theta, \qquad \mu^2 + \sigma^2 = \int \theta^2 p(\theta) \, d\theta. \qquad (1.33)$$

The original π becomes modified by a Gaussian,

$$p(\theta) = \pi(\theta) \exp(-1 - \lambda_0 - \lambda_1 \theta - \lambda_2 \theta^2). \qquad (1.34)$$

If the original informal knowledge was weaker than the new constraints, so that π was effectively constant within a few σ of μ, the Gaussian modification would dominate, leading to the standard normal (or Gaussian) distribution:

$$p(\theta) = \frac{\exp\left[-(\theta - \mu)^2 / 2\sigma^2\right]}{\sqrt{2\pi\sigma^2}}, \qquad (\pi = 3.14159\ldots). \qquad (1.35)$$

Another standard example is an intensity parameter subject to mean value μ. Here, the original π becomes modified by an exponential,

$$p(\theta) = \pi(\theta) \exp(-1 - \lambda_0 - \lambda_1 \theta), \qquad (\theta > 0). \qquad (1.36)$$

Again, if the original knowledge was appropriately weaker than the new constraints, so that π was effectively constant for small or moderate θ, this result becomes of standard exponential form:

$$p(\theta) = \mu^{-1} \exp(-\theta/\mu), \qquad (\theta > 0). \qquad (1.37)$$

On the other hand, if the original knowledge was weak but different, perhaps effectively uniform over θ^2 so that $\pi \propto \theta$, the result

$$p(\theta) = (\theta/\mu^2) \exp(-\theta/\mu), \qquad (\theta > 0) \qquad (1.38)$$

would be different too. Maximum entropy refines prior knowledge, but does not replace it.

Continuous problems

Here, we want to infer a measure m_i defined over so many states i that we may as well use continuum notation $m(x)$. Such a model might represent a spectrum or image, whose intensity is distributed across frequency or spatial coordinate(s) x. In practice, x is digitized into cells i, with the continuum limit merely meaning that macroscopic results stop changing when the digitization gets arbitrarily fine.

We commonly want cell boundaries to be invisible. This means that, in the absence of data saying otherwise, we have the same expectation for the intensity accumulated in a domain Δx whether the domain is treated as a whole, or as the sum of two or more subdivisions. In symbols, with cell $k = i \cup j$ decomposed into i and j (known as 'stick-breaking'), we require

$$\pi(m_k) = \iint \delta(m_k - m_i - m_j)\, \pi(m_i, m_j)\, \mathrm{d}m_i\, \mathrm{d}m_j \tag{1.39}$$

so that the intensities add correctly as $m_k = m_i + m_j$. Additionally, we often suppose that each cell is to behave independently,

$$\pi(m_i, m_j) = \pi(m_i)\pi(m_j), \tag{1.40}$$

so that there is no prior expectation of internal correlation.

These conditions are actually quite restrictive on the form of prior. As an example of a prior that does not work, take the candidate $\pi(m_k) = \delta(m_k-1)+\delta(m_k-2)$ with two-point support (1 and 2) for the values. This cannot be subdivided at all, let alone infinitely. In any subdivision into symmetric halves, each half would need at least two points of support, because giving each only one would be insufficient. But the combination would then cover at least three points, which is too many: QED. Another prior that does not work is $\pi(m) \propto \exp(-H(m))$, proposed in the hope that maximum entropy assignment of a single m might be promoted to a distribution of m's. That cannot be subdivided either.

In technical parlance, a prior that behaves consistently on all scales right down to the infinitesimal limit is called a 'process', and the property of such consistency is called 'infinite divisibility' (Steutel 1979). One prior that does work, common in physics, is the Poisson process. Each small cell i is usually empty, but has small probability λ_i of receiving a quantum, and negligible chance of more than one. By construction, this works in the infinitesimal limit. Occupancies r (usually 0, occasionally 1, negligibly more) of small cells are distributed as

$$\pi(r_1, r_2, \ldots, r_n) = \prod_{i=1}^{n} \big((1 - \lambda_i)\delta(r_i - 0) + \lambda_i\delta(r_i - 1)\big). \tag{1.41}$$

If the quanta are allowed to have individually variable intensity, say exponential

$$\pi(m \mid \text{quantum}) = \mathrm{e}^{-m} \tag{1.42}$$

in suitable units, the intensity pattern among the small cells is

$$\pi(m_1, m_2, \ldots, m_n) = \prod_{i=1}^{n} \big((1 - \lambda_i)\delta(m_i) + \lambda_i\mathrm{e}^{-m_i}\big), \tag{1.43}$$

with each small cell having a small chance of holding a macroscopic intensity. If micro-cells are combined, their Poisson rates λ add, so that $\lambda(x)$ is itself a measure on x. In fact, there is an exact macroscopic formula,

$$\pi(m) = \mathrm{e}^{-\lambda}\big(\delta(m) + \mathrm{e}^{-m}\sqrt{\lambda/m}\, I_1(2\sqrt{\lambda m})\big), \tag{1.44}$$

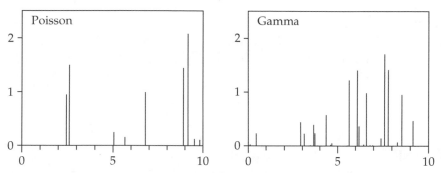

Fig. 1.1. (Left) Sample of Poisson process averaging 10 spikes of mean intensity 1. (Right) Sample of Gamma process of same mean and variance.

for the Poisson model. It is parameterized by the unit of quantum intensity (here 1) and by the production measure λ, and is not too difficult to program in terms of its constituent quanta.

Another prior that works, more popular among statisticians, is the Gamma process:

$$\pi(m_1, m_2, \ldots, m_n) = \prod_{i=1}^{n} \frac{m_i^{-1+\lambda_i}}{\Gamma(\lambda_i)} \, e^{-m_i}, \qquad (1.45)$$

where, as before, $\lambda(x)$ is a measure over x. Although the formula is arguably simpler algebraically, it is less interpretable and more expensive to program because every micro-cell enters the prior (instead of a limited number of quanta). Not that random samples look very different. As calculated at high resolution, both give spiky results (Figure 1.1). The difference is that the Gamma process produces a lot of extremely low-level grass which the Poisson process cuts away. If the Gamma measure m is normalized, the formula reduces to the Dirichlet process (Ferguson 1973)

$$\pi(p_1, p_2, \ldots, p_n) = \delta\big(1 - \textstyle\sum p\big) \, \Gamma\big(\textstyle\sum \lambda\big) \prod_{i=1}^{n} \frac{p_i^{-1+\lambda_i}}{\Gamma(\lambda_i)} \qquad (1.46)$$

for inferring a probability distribution p.

Geometry

When two measures m and $m + \delta m$ are close, their information H becomes approximately symmetric,

$$H(m + \delta m \mid m) \approx \sum_{j=1}^{n} \frac{(\delta m^j)^2}{m^j}, \qquad (1.47)$$

and behaves as a distance-squared in parameter space, here digitized to n points for notational clarity, and with coordinates written as contravariant superscripts because the space is about to become Riemannian. The metric describing this distance is diagonal,

$$g_{jk} = \begin{cases} 1/m^j & \text{if } j = k, \\ 0 & \text{otherwise,} \end{cases} \tag{1.48}$$

and, locally, $H = (\mathrm{d}s)^2 = \sum g_{jk}\,\mathrm{d}m^j\,\mathrm{d}m^k$. Measures within a small fixed distance ϵ of m fill an ellipsoid of 'radius' ϵ and volume proportional to $(\det g)^{-1/2}$. By supposing that ϵ-ellipsoids all contain the same mass, regardless of central location, the m's induce their own natural density proportional to $(\det g)^{1/2}$. In the absence of any better guidance, one might try to use this as the prior on m.

For example, suppose parameter space has just two intensities (a 2-cell image, perhaps), over which we seek a prior $\pi(m^1, m^2)$. The proposal is

$$(\det g)^{1/2} = \sqrt{\det\begin{pmatrix} 1/m^1 & 0 \\ 0 & 1/m^2 \end{pmatrix}} = \frac{1}{\sqrt{m^1 m^2}}. \tag{1.49}$$

An immediate objection is that this expression is not normalizable, leading to an improper prior for m. However, the components of m could represent proportions p and $1 - p$ of some fixed total, and the proposed prior for p (the *shape* of the 2-cell image) would then be

$$\pi(p) = \frac{1/3.14159\ldots}{\sqrt{p(1 - p)}}, \tag{1.50}$$

which is normalized, and might well be acceptable.

The generalization to a larger number n of proportions p is a Dirichlet distribution (1.46), but with all indices λ equal to $\frac{1}{2}$. This would not be acceptable for inference about an image digitized to arbitrarily many cells. The reason is that macroscopic structure is washed out. For example, the total proportion in any $n/2$ cells is almost certain to be very close to the uniform, featureless $\frac{1}{2}$. Dirichlet indices λ need to be a measure, thereby getting individually smaller as cells are subdivided. The geometrical indices of $\frac{1}{2}$ do not get smaller. Here, the geometric proposal fails to be infinitely divisible.

The geometrical formulation can be generalized to a parameterized subspace $m^i = m(i \mid \theta)$ restricted to the range of r parameters $\theta^1, \ldots, \theta^r$. The information between neighbours becomes

$$H(\theta + \mathrm{d}\theta \mid \theta) = \sum_{i=1}^{n} \frac{(\mathrm{d}m^i)^2}{m^i} = \sum_{i=1}^{n} \frac{1}{m^i}\left(\sum_{j=1}^{r} \frac{\partial m^i}{\partial \theta^j}\,\mathrm{d}\theta^j\right)\left(\sum_{k=1}^{r} \frac{\partial m^i}{\partial \theta^k}\,\mathrm{d}\theta^k\right), \tag{1.51}$$

in which the coefficient of $\mathrm{d}\theta^j \, \mathrm{d}\theta^k$, namely

$$
g_{jk} = \begin{cases} \displaystyle\sum_{i=1}^{n} \frac{1}{m(i \mid \theta)} \frac{\partial m(i \mid \theta)}{\partial \theta^j} \frac{\partial m(i \mid \theta)}{\partial \theta^k} & \text{(discrete)} \\[2ex] \displaystyle\int \frac{\mathrm{d}x}{m(x \mid \theta)} \frac{\partial m(x \mid \theta)}{\partial \theta^j} \frac{\partial m(x \mid \theta)}{\partial \theta^k} & \text{(continuous)}, \end{cases} \tag{1.52}
$$

defines the 'Fisher metric' with respect to changes in θ. Thus, if the parameters θ are being used to define a measure or probability distribution m, that relationship induces a metric g, and thence a 'Fisher density' $(\det g)^{1/2}$ which might be usable as a prior on θ. After all, this density is derived from the coordinate-invariant H, so represents a coordinate-invariant mass. That sounds attractive.

There has even been a proposal to automate the whole process by taking the induced measure $m(x \mid \theta)$ to be the likelihood function $\mathcal{L}(x \mid \theta)$ that characterizes the experiment (er, *which* experiment that we haven't built yet?) that eventually observes data $x = D$. However, that idea doesn't work. A sufficient reason is that the prior on θ would then depend on the entire form of likelihood function, integrated over all anticipated x. After the data were found to be D and D alone, posterior inference would still depend on all the other data values that might have been observed but were not. That's a recipe for disaster, because with the falsity in play one can 'prove' *anything*.

Even discarding the likelihood idea, the geometric proposal can fail to be normalized. As a simple example, take the family

$$
p(x \mid \theta) = \frac{\mathrm{e}^{-\theta \cos x}}{2\pi I_0(\theta)} \tag{1.53}
$$

of maximum entropy probability distributions on the unit circle $0 \le x < 2\pi$, parameterized by θ determined by a constraint on the mean $\langle \cos x \rangle$. Their Fisher density evaluates to

$$
(\det g)^{1/2} = \sqrt{\mathrm{d}^2 \log I_0 \, / \, \mathrm{d}\theta^2}, \tag{1.54}
$$

which behaves acceptably for small θ but not for large, where its asymptotic $|\theta|^{-1}$ behaviour cannot be normalized. If used as a prior for θ, it would put all its mass at infinite θ, as if $x = 0$ and $x = \pi$ were the only two points on the circle that mattered.

Geometry leaves no place for prior knowledge or judgment, so it represents deep ignorance. Yet that ignorance can easily be too extreme, and contradict reasonable expectation. The fact is that we never are as completely ignorant as the geometrical approach supposes. A geometrical prior can be acceptable, but one should always use the sanity check of informal assessment, and over-ride the proposal if necessary.

The general need to employ informal assignment, or at least assessment, indicates that the long search for general 'objective' priors is doomed. Inference is inherently subjective, through its dependence on incompletely formalized background knowledge. Like it or not, that's the way it is. The world assesses our priors through their evidence values, requiring us to question and observe. It does not give divine instruction.

1.4 Algorithms

Practical Bayesian inference requires us to compute the evidence E for the current model. If E is outclassed by alternative models, there is no point in proceeding further. Otherwise, and insofar as the model is reasonable, it is worth proceeding to compute the posterior for more detailed inference about parameters and properties. But the primary task is to calculate E:

$$E = \int \mathcal{L}(\theta)\pi(\theta)\,\mathrm{d}\theta. \tag{1.55}$$

The also-useful $H = \left(\int \mathcal{L}(\theta) \log \mathcal{L}(\theta)\,\pi(\theta)\,\mathrm{d}\theta - \log E \right) / E$ can be accumulated in parallel. Actually, the computer has less work to do if part of the likelihood can be transferred analytically to the prior, so that $E = \int (\mathcal{L}/w)(w\pi)\,\mathrm{d}\theta$ for some useful weighting function w. This trick is known as 'importance sampling', and can be borne in mind.

Unfortunately, there is a general difficulty. In problems of even moderate complexity, θ involves more than just a few parameters, and we are faced with a high-dimensional integral. More fundamentally, if θ is encoded into ν bits of computer memory, the evaluation of E involves a sum over 2^ν possibilities. If storage ν is considered as in class aleph-null, then possibilities 2^ν are in class aleph-one. (Beyond that again is the domain of possible questions, in 2^{2^ν} class aleph-two (Knuth 2005).)

We do not have aleph-one resources, so cannot evaluate E by definitive exhaustive summation. In practice, we can only explore a limited number of points θ. This means that the estimation of E is itself *an inference problem*, as opposed to classical numerical analysis. The result will be uncertain, and we ought to know what that uncertainty is.

Moreover, the information H scales with the size of the problem, and is the logarithm of the prior-to-posterior compression ratio. In other words, the bulk of the posterior occupies an exponentially small fraction $\mathcal{O}(\mathrm{e}^{-H})$ of the prior domain. That means it is hard to find. It also means that, unable to do better, we are forced towards random 'Monte Carlo' sampling methods to find our points θ. There are two basic approaches to this, nested sampling and simulated annealing, each with strengths and weaknesses. Each has a kernel program that codes the general strategy of evaluation, to which is attached an exploration program involving the likelihood function appropriate to a particular application. The kernels are really very simple. Only the specific exploration code needs to reflect the complexity of an awkward likelihood.

1.4.1 Nested sampling

Nested sampling (Skilling 2004b; Sivia & Skilling 2006) is a twenty-first-century algorithm, already being used in cosmology (Mukherjee, Parkinson & Liddle 2006).

Theory

We can think of ordering the 2^ν states by decreasing likelihood value, accumulating prior mass as we do so. As a technicality to ensure unambiguous ordering when $\mathcal{L}$ values coincide, we can append a 'key' $\kappa(\theta)$ to the least significant bits of the computer's coding of $\mathcal{L}$, chosen so that key values don't repeat and all $\mathcal{L}$'s become different. This ordering defines a strictly decreasing function

$$\mathcal{L}(X) = \text{'enclosing likelihood' function of prior mass } X, \qquad (1.56)$$

with its strictly decreasing inverse (Figure 1.2)

$$X(\mathcal{L}) = \int_{\mathcal{L}(\theta)>\mathcal{L}} \pi(\theta)\,\mathrm{d}\theta = \text{'enclosed prior' of likelihood } \mathcal{L}. \qquad (1.57)$$

Here your author adopts the computer-science practice of overloading function names according to argument type; $\mathcal{L}(\theta)$ and $\mathcal{L}(X)$ are technically different functions, but the same likelihood.

Our task is to find the value of what is now a one-dimensional sum along the states, ordered-by-likelihood as they now are, over unit prior mass:

$$E = \int_0^1 \mathcal{L}(X)\,\mathrm{d}X. \qquad (1.58)$$

The integral for E is now very well behaved (Figure 1.2), with a decreasing positive integrand $\mathcal{L}$ over unit range of X. But we only have resources for a limited number of θ, from which we must *infer* the evidence and consequential properties. Let these states be $\theta_1, \theta_2, \ldots, \theta_n$ with $\mathcal{L}_i = \mathcal{L}(\theta_i)$ and $X_i = X(\mathcal{L}_i)$, and let them be ordered as

$$0 < X_n < \cdots < X_2 < X_1 < X_0 = 1. \qquad (1.59)$$

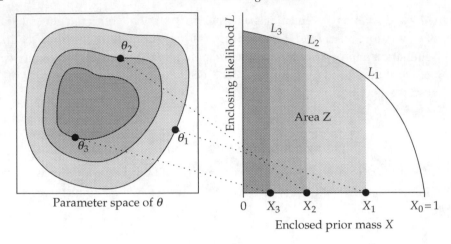

Fig. 1.2. Nested likelihood contours sort to enclosed prior mass X.

In any interval, $\mathcal{L}$ is constrained by the end-points, $\mathcal{L}_{i+1} \geq \mathcal{L}(X) \geq \mathcal{L}_i$ in $X_{i+1} \leq X \leq X_i$, so E is not far from its trapezoid-rule quadrature estimate

$$\widehat{E} = \sum_{i=1}^{n} \frac{X_{i-1} - X_{i+1}}{2} \mathcal{L}_i, \qquad (\mathcal{L}_0 = 0). \tag{1.60}$$

Around this is a $\mathcal{O}(n^{-1})$ quadrature error, and what will become an exponentially small $\mathcal{O}(X_n)$ truncation error. Progress is systematically inward within the nested likelihood contours, hence the name *nested sampling*.

Kernel

If, in practice, we could assign X and compute its $\mathcal{L}(X)$, the evaluation of E would reduce to the above elementary numerical analysis. But we cannot. What we *can* do is assign X *at random*, as follows. Run the algorithm with N 'particles', where N can be as low as 1 though larger N gives improved accuracy and power. Each particle has known location θ, and known likelihood $\mathcal{L}(\theta)$. At iterate i, suppose that all N particles are uniformly distributed in X, though constrained within $X < X_i$, equivalently $\mathcal{L} > \mathcal{L}_i$. At the initial iterate $i = 0$, this just means that the N particles are independent samples from the prior, with $X_0 = 1$ being no constraint at all because all the prior mass is accessible.

Let $\mathcal{L}_i$ be the outermost (lowest) of the N likelihood values current at iterate $i - 1$. The $N - 1$ other particles are uniformly distributed over the prior, except that they are now known to have yet larger likelihood $\mathcal{L} > \mathcal{L}_i$, equivalently $X < X_i$.

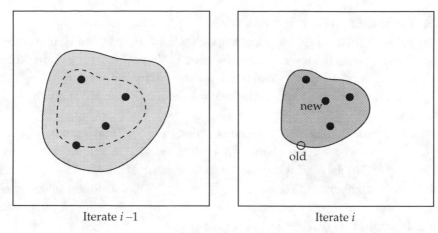

Fig. 1.3. One nested sampling iterate with $N = 4$ particles.

For iterate i, discard this outermost particle and replace it with a new one within this same constraint $\mathcal{L} > \mathcal{L}_i$.

$$\boxed{\mathrm{pr}(\theta) \propto \pi(\theta) \text{ in } \mathcal{L}(\theta) > \mathcal{L}_i} \qquad (1.61)$$

(How to do this is discussed below.) We again have N particles uniformly distributed in X, but now they are constrained within $X < X_i$, equivalently $\mathcal{L} > \mathcal{L}_i$, and this completes an iterate (Figure 1.3).

Uncertainty

If we could actually do the $\theta \to X$ ordering, we would know exactly what X was, but we don't. We are stuck with the partial knowledge that X_i was the highest of N values, each uniformly distributed between 0 and X_{i-1}. This means that we can claim no more than

$$X_i = t_i X_{i-1} \text{ where } \mathrm{pr}(t_i) = N t_i^{N-1} \text{ in } 0 < t_i < 1, \qquad (1.62)$$

$N t^{N-1}$ being the distribution of the highest of N random numbers less than 1. Equivalently,

$$\log X_i = \log X_{i-1} - \tau_i/N \text{ where } \tau = -N \log t \text{ has } \mathrm{pr}(\tau) = \mathrm{e}^{-\tau} \text{ in } \tau > 0. \qquad (1.63)$$

However, we do know $\mathcal{L}_i = \mathcal{L}(\theta_i)$ definitively.

The upshot is that we have a method which generates a correctly ordered sequence of $(\mathcal{L}, X)$ pairs, with $\mathcal{L}$ known definitively and X known statistically. This suffices to infer E *statistically*. Take 100 or so independent guesses about what the compression sequence of τ's might have been by sampling each from

(1.63). We know no better. Each sequence defines its corresponding X's, which can be substituted in (1.60) to yield an estimate of $\widehat{E}$, preferably presented as $\log \widehat{E}$. These 100 estimates are samples from $\mathrm{pr}(\log \widehat{E})$, so they define the required result – conveniently summarized as mean and standard deviation. The result is uncertain, and we do know what that uncertainty is. Nested sampling is well formulated.

As iterates proceed, the inward trend is geometrical. After r iterates, $T = \tau_1 + \tau_2 + \cdots + \tau_r = -N \log X_r$ is distributed as $\mathrm{pr}(T) = T^{r-1} \mathrm{e}^{-T}/(r-1)!$; basically $T = r \pm \sqrt{r}$. The bulk of the posterior, where most of the evidence integral is found, is near compression factor $X = \mathrm{e}^{-H}$, so it is reached after about $NH \pm \sqrt{NH}$ iterates, and usually left behind quite soon after. Even though the posterior volume is small, it is reached in linear time because of the geometrical nature of the algorithm. Hence nested sampling is feasible. The computational cost scales as size-squared, one factor being due to the $\mathcal{O}(H)$ number of steps and the other factor being the intrinsic cost of an iterate.

If the minor overhead of taking 100 samples of the τ's is deemed too awkward, it is often enough to set $-N \log X_r$ to its mean value r, thereby fixing $X_r = \mathrm{e}^{-r/N}$, and then evaluate a single central $\widehat{E}$ from (1.60). This is subject to uncertainty $\pm \sqrt{H/N}$ through the $\pm \sqrt{NH}$ uncertainty in number of iterates, each of which compresses by about $1/N$ in the logarithm.

Exploring the prior

Nested sampling requires a particle to explore randomly with respect to the prior, within a hard constraint on likelihood value. Such exploration is easiest to program if carried out in coordinates over which the prior is flat. Your author goes further, and places a d-parameter problem in the unit hypercube $\rho \in (0,1)^d$ within which the prior is flat, $\mathrm{pr}(\rho) = 1$. To ensure that all sample points are a-priori-equivalent, and to avoid possible singularity at the edges, the controlling random variables can conveniently be 32-bit unsigned integers $u \in [0, 2^{32}-1]$, with wraparound topology. In terms of these, $\rho_i = 2^{-32}(u + \frac{1}{2})$, which stays far enough away from 0 and 1 for functions like logarithm to avoid singularity. With 4 billion allowed values, the 32 bits give enough choice for practical purposes, and unsigned integers are automatically wraparound in most computers.

The desired parameters θ are then appropriate functions of ρ. For example, an intensity distributed as $\mathrm{pr}(\theta) \propto \exp(-\theta/q)$ would be constructed as $\theta = -q \log \rho$. A Cauchy distribution $\mathrm{pr}(\theta) \propto 1/(\theta^2 + w^2)$ would be constructed as $\theta = w \tan((\rho - \frac{1}{2})\pi)$, and so on. This ρ-to-θ approach is opposite to many presentations, which start with θ and consider ρ, if at all, as the cumulant $\int^\theta \mathrm{pr}(\theta')\,\mathrm{d}\theta'$. However, ρ-to-θ is more in keeping with the philosophy of defining the prior by placing typical θ on top of prior mass ρ, and it is easier to program too.

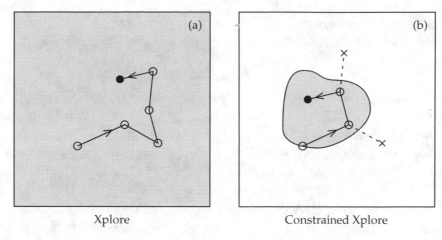

Fig. 1.4. Xplore explores the prior (a), constrained by likelihood (b).

Sampling from the prior alone is easy – for each coordinate, just call a random-number generator for a 32-bit integer u, scale it to ρ and transform to θ. Moving within the prior is also easy. To perturb a coordinate u within some limit k, generate a random integer between $-k$ and k, and add it to u. If this is done for only one, or a few, of the coordinates, this exploration strategy is 'Gibbs sampling' (Gelman *et al.* 1995). If it is done for all coordinates, the exploration can take any direction. One of your author's favourite tricks (Skilling 2004a) is to raster through the d-dimensional 2^{32}-cube with a space-filling Hilbert curve, so that each point is labelled by a $32d$-bit integer in a way that substantially preserves neighbourhood relationships. Exploration is then 1-dimensional, with movement coded by extended-precision addition and subtraction. Whatever strategy is adopted, we suppose that we have an algorithm 'Xplore' that can explore the prior (Figure 1.4a).

What matters is that Xplore is symmetric. The probability of reaching v from u should equal the probability of reaching u from v:

$$\mathrm{pr}(v \mid u) = \mathrm{pr}(u \mid v) \tag{1.64}$$

– a condition known as 'detailed balance'. In the very long term ergodic limit, this would ensure equal numbers of transitions between u and v when u and v are equally populated. This uniform equilibrium state models the uniform prior that we seek. Of course, we never wait that long, and our random-number generator is likely to be a definitive algorithm, so discussion of ergodicity and equilibration are really just metaphor. The rationale is that our predictive belief about the occupancies in our single system equilibrates in the same way that an ensemble of very many systems would. Probability and relative frequency obey the same laws, so we can use frequentist arguments as metaphor without believing in frequentism.

Regarding the random-number generator, we agree to be ignorant of its inner workings, so that it is (to us) unpredictable except in terms of probability.

Exploring the constraint

Finding a point within nested sampling's hard constraint on likelihood would become very difficult if carried out directly, because the constraint volume shrinks to become exponentially small. In practice, we are forced to start with some location obeying the constraint, and re-equilibrate it sufficiently to obtain the effectively new point that we seek. Using past history is technically known as a Markov chain, so our algorithms are 'Markov chain Monte Carlo' (MCMC).

At each iterate, there are N points already available, $N-1$ of them in the interior plus the about-to-be-discarded outer point on the constraint boundary. Any of these can be used as a source location to be re-equilibrated to the destination, though it seems preferable (if $N > 1$) to start at one of the survivors within the strict interior. So the first step is to replace the discarded outer point with a *copy* of one of the others.

The next step is to equilibrate the new point, which can be done with the same `Xplore` algorithm as before – but with the restriction (Figure 1.4b) that any transition to a prohibited destination is rejected, leaving the source in place.

$$\boxed{\text{Accept } \theta \text{ if and only if } \mathcal{L}(\theta) > \mathcal{L}_{\text{constraint}}} \qquad (1.65)$$

Transitions between interior points remain in detailed balance, transitions from interior to exterior are blocked, and no transitions can start from the exterior, so the scheme is fully balanced and equilibrates towards the constrained prior, as required. As always in MCMC, the step-size of a proposed perturbation has to be large enough to make useful progress in 'forgetting' the source location, but not so large that too many expensive transitions are rejected.

It is the user's responsibility to acquire a new location, a task that can become more difficult if the enclosing likelihood has awkward shape in high dimension. Note, though, that copying has the very useful side effect that any point becoming stuck in a local false likelihood maximum will find itself trapped as the likelihood constraint becomes more severe, and when it becomes the outer point it gets discarded. This means that false maxima can be left behind without need for `Xplore` to propose an unlikely transition from false to favoured location.

Posterior

At the end of a nested-sampling run, it has accumulated a sequence of samples θ with known likelihoods $\mathcal{L}(\theta)$. Perhaps 100 guesses of the corresponding sequence of X's have been computed, each of which has given its evidence estimate (1.60) and thereby tried to weight θ_i by $\widehat{w}_i = \frac{1}{2}(\widehat{X}_{i-1} - \widehat{X}_{i+1})\mathcal{L}_i/\widehat{E}$. The average $\overline{w}_i$ of these weights gives the best estimate of the posterior as a weighted sum

$$\widehat{\mathcal{P}}(\theta) = \sum \overline{w}_i \delta(\theta - \theta_i) \tag{1.66}$$

of the samples. Already randomly selected around the likelihood contours, this list gives the most faithful posterior distribution that is available. It can be used for estimating any property $Q(\theta)$.

The effective number $\mathcal{N}$ of independent samples is given by the information content of the weights through

$$\log \mathcal{N} = -\sum \overline{w}_i \log \overline{w}_i \tag{1.67}$$

and is usually close to $\mathcal{N} \approx Ne$. If this is too few to sample the posterior adequately, then typical θ taken from the sequence can be used to seed fuller exploration, using the Metropolis–Hastings balancing described next.

1.4.2 Simulated annealing

Theory

Nested sampling took a direct view of the enclosing likelihood function $\mathcal{L}(X)$. The traditional method of simulated annealing (Kirkpatrick, Gelatt & Vecchi 1983) takes an indirect view, introducing the likelihood gradually through a fractional power β. At β, the annealed likelihood is $\mathcal{L}^\beta$, the evidence is $E(\beta) = \int \mathcal{L}^\beta \, dX$, and the posterior is $\mathcal{L}^\beta dX / E(\beta)$. At the start, $\beta = 0$ switches the likelihood off, leaving just the prior with $E(0) = 1$. At the finish, $\beta = 1$ gives the likelihood its proper weight, with required evidence $E = E(1)$. The term 'simulated annealing' marks a conscious analogy with thermal physics, with $\mathcal{L} = \exp(-\text{energy})$ and $\beta = 1/\text{temperature}$. The identity

$$\frac{d \log E}{d\beta} = \frac{dE/d\beta}{E} = \frac{\int \mathcal{L}^\beta \log \mathcal{L} \, dX}{\int \mathcal{L}^\beta \, dX} \equiv \langle \log \mathcal{L} \rangle_\beta \tag{1.68}$$

expresses the E differential in terms of the current log-likelihood average, and integrates to the 'thermodynamic integration' formula

$$\log E = \int_0^1 \langle \log \mathcal{L} \rangle_\beta \, d\beta. \tag{1.69}$$

Kernel

Simulated annealing evaluates E with an ordered sequence of coolness,

$$0 = \beta_{-1} = \beta_0 \leq \beta_1 \leq \beta_2 \leq \cdots \leq \beta_n = \beta_{n+1} = \cdots = 1, \tag{1.70}$$

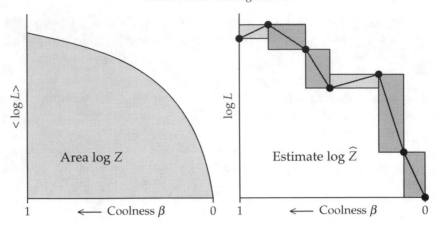

Fig. 1.5. Thermodynamic integral (left) and trapezoid estimate (right).

known as the 'annealing schedule' and locates a particle at each station according to

$$\mathrm{pr}(\theta) \propto \mathcal{L}^{\beta}(\theta)\pi(\theta) \qquad (1.71)$$

Equivalently, X_i is taken from $\mathcal{L}^{\beta_i}(X)$.

Either of the two quadrature forms,

$$\log \widehat{E}^- = \sum_{i=0}^{n}(\beta_{i+1} - \beta_i)\log \mathcal{L}_i\,, \qquad \log \widehat{E}^+ = \sum_{i=0}^{n}(\beta_i - \beta_{i-1})\log \mathcal{L}_i, \quad (1.72)$$

could be used as an estimate of $\log E$, with the central trapezoid estimate

$$\log \widehat{E} = \tfrac{1}{2}(\log \widehat{E}^- + \log \widehat{E}^+) = \sum_{i=0}^{n} \Delta\beta_i \log \mathcal{L}_i\,, \qquad \Delta\beta_i = \frac{\beta_{i+1} - \beta_{i-1}}{2}, \qquad (1.73)$$

preferred. As the annealing schedule is made denser, these estimates plausibly converge to the correct $\log E$, as the sum approaches the integral and the statistical
variation of the terms averages out (Figure 1.5).

Uncertainty

The uncertainty is estimated by taking $\xi = \log X$ as the abscissa and observing
that, at coolness β, ξ is sampled from

$$\mathrm{pr}(\xi)\,\mathrm{d}\xi \propto \mathcal{L}^{\beta}(X)\mathrm{d}X = \mathcal{L}^{\beta}(X)X\,\mathrm{d}\xi = \mathrm{e}^{\beta\lambda+\xi}\,\mathrm{d}\xi, \qquad (1.74)$$

where $\lambda = \log \mathcal{L}$. The maximum (if there is one!) occurs where $\mathrm{d}\xi/\mathrm{d}\lambda = -\beta$.
Around this, second-order expansion of the exponent yields a Gaussian of variance

$$\text{var}(\xi) = \left(-\frac{d^2(\beta\lambda + \xi)}{d\xi^2}\right)^{-1} = \left(-\beta\frac{d^2\lambda}{d\xi^2}\right)^{-1} = \beta^2\frac{d\lambda}{d\beta}, \qquad (1.75)$$

where the last expression is obtained by eliminating $d\xi$. This is the variability to be expected for a sample obtained at coolness β, which needs to be mirrored in the uncertainty of our inference from that sample. Since $d\xi = -\beta\,d\lambda$,

$$\text{var}(\lambda) = \frac{\text{var}(\xi)}{\beta^2} = \frac{d\lambda}{d\beta} \approx \frac{\Delta\lambda}{\Delta\beta}, \qquad (1.76)$$

and the variance of the estimate (1.73) of $\log\widehat{E} = \sum(\Delta\beta)\lambda$ becomes

$$\text{var}(\log\widehat{E}) = \sum(\Delta\beta_i)^2\text{var}(\lambda_i) = \sum\Delta\beta_i\,\Delta\lambda_i = \log\widehat{E}^+ - \log\widehat{E}^-. \quad (1.77)$$

So the uncertainty variance that we seek is very simply estimated by the difference of the two quadrature forms (1.72). In Figure 1.5, this difference is shaded dark for positive contributions, and light for negative contributions.

Schedule

For minimum uncertainty, the annealing schedule should be chosen so that the terms in (1.77) balance, with $\Delta\beta_i \propto 1/\sqrt{\text{var}(\lambda_i)}$. Performance of the algorithm is only weakly dependent on the schedule details, and it suffices to estimate the variance from the spread of λ's in recent iterates. As it happens, $\beta^2\text{var}(\lambda) = R/2$, where R is the locally effective dimensionality of the likelihood function – an exact relation if $\mathcal{L}(\theta)$ is Gaussian (*aka* rank-R multivariate normal) within a flat prior, and an informative diagnostic in general.

As with most statistical methods, accuracy is improved by $\sqrt{N}$ by taking N times more samples. Accordingly, the preferred annealing schedule is to cool at each step by something like

$$\Delta\beta_i = \frac{1}{N\sqrt{\text{var}(\lambda)}}, \qquad (1.78)$$

in which the variance of λ derives from about $10N$ previous values. Here, the factor 10 guards against accidental or local coalescence of values leading to a wrongly enhanced cooling step. Setting the schedule is something of a black art.

Exploration

Again, it is necessary to use the Markov chain idea and start from a known location when seeking a new point. This is another black art, involving choice between the most recent point and some previous point that might afford better escape from a false likelihood maximum. Theory does not give firm guidance here. Once a source has been selected, equilibration to the new destination can be performed by modulating Xplore's exploration of the prior according to the Metropolis–Hastings criterion:

$$\text{Accept } \theta_{\text{proposed}} \text{ if and only if } \frac{\mathcal{L}^{\beta}(\theta_{\text{proposed}})}{\mathcal{L}^{\beta}(\theta_{\text{source}})} > t \sim \texttt{Uniform}(0,1) \qquad (1.79)$$

When a transition links two points u and v having likelihoods $\mathcal{L}(u) < \mathcal{L}(v)$, passage from less likely u to more likely v is unimpeded, whereas the reverse is restricted by the average factor $\mathcal{L}^{\beta}(u)/\mathcal{L}^{\beta}(v)$. At equilibrium, u is thus underpopulated relative to v, consistent with occupation density proportional to $\mathcal{L}^{\beta}$, as required by simulated annealing.

$$\text{Less likely } u \quad \overset{\text{free}}{\underset{\text{restricted}}{\longrightarrow}} \quad \text{More likely } v$$

Posterior

To explore the posterior after cooling to $\beta = 1$, all one need do is stop cooling by holding β fixed. The Xplore procedure modulated by Metropolis–Hastings will allow a particle to diffuse faithfully to the posterior for as long as desired.

1.4.3 Comparison

The relative costs (measured by number of steps) and uncertainties of nested sampling and simulated annealing, programmed using the same N for an R-dimensional Gaussian (rank-R multivariate normal) likelihood, are roughly

$$\frac{\#\text{steps(nested)}}{\#\text{steps(anneal)}} \approx 0.2\sqrt{R}, \qquad \frac{\text{var} \log E(\text{nested})}{\text{var} \log E(\text{anneal})} \approx 2.0\sqrt{R} \qquad (1.80)$$

for large R. At least for this simplest of applications, nested sampling is slower and less accurate than simulated annealing. It is less accurate because its estimate of X is subject to Brownian-motion accumulation of uncertainty, whereas simulated annealing's schedule of β is known. It is slower because its particles are constrained by the hard likelihood constraint to occupy a range $\Delta \log X \sim 1$, whereas simulated annealing's soft constraint allows the particles to spread out over $\Delta \log X \sim \sqrt{R}$ (Figure 1.6).

On the other hand, nested sampling has the advantage of being more robust. It systematically tracks inwards, generating a complete map of the likelihood function $\mathcal{L}(X)$. Nothing gets lost. Simulated annealing depends on tracking the maximum of $\beta \log \mathcal{L} + \log X$ as the inverse slope β increases. That requires $\log \mathcal{L}$ to be

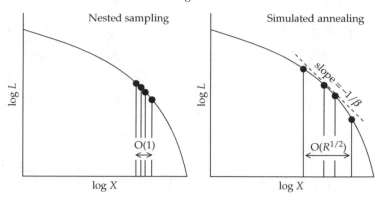

Fig. 1.6. $N = 4$ samples, in nested sampling and simulated annealing.

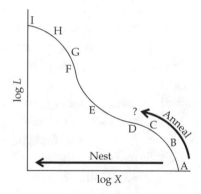

Fig. 1.7. Nested sampling and simulated annealing for a partly convex likelihood function.

a concave $(\frown)$ function of $\log X$. Many simple problems, such as the archetypical Gaussian, are of this gratifyingly simple form.

Other problems, specifically those involving phase changes, are not of this form. In fact, many of today's difficult problems *do* involve phase changes. Simulated annealing is unable to deal with these. Consider the likelihood function plotted in Figure 1.7. Nested sampling would start near A at $\log X = 0$ and move systematically inwards along $\log X$, plotting out the entire $\mathcal{L}(X)$ and enabling computation of the evidence.

Simulated annealing would start the same, but move inwards by tracking the tangent slope along ABC... At or before the point of inflection D, though, particles will discover the unstable convex $(\smile)$ region DEF in which the curvature is the wrong way round. In seeking to increase their likelihood-volume $\beta \log \mathcal{L} + \log X$, the particles will crash inward towards HI... If that situation were recognized, and

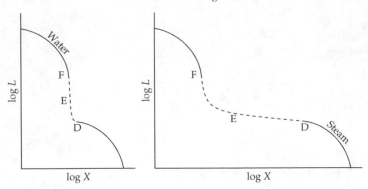

Fig. 1.8. Annealing is unable to distinguish very different problems!

if it were deemed to be overshoot, annealing could be reversed to track IHG...
However, the convex region remains unstable and at or before F particles would
crash back out to B or beyond. During a crash, the evidence value is lost, and with
it knowledge of the relative importance of the outer phase (steam?) near B and the
inner phase (water?) near H.

Relating the phases demands connecting them across the convex region, which
simulated annealing can only do if allowed exponential resources. For example,
Figure 1.8 shows likelihood functions for which the annealing integrands $\langle \log \mathcal{L} \rangle_\beta$
– computed by inevitably local exploration – behave *identically* with coolness β.
The functions differ only in their inaccessible convex region DEF, but that difference suffices to change $\log E$ enough to reverse the dominant phases. To put it
mildly, that could be important.

Simulated annealing could warn of impending catastrophe because the relevant
curvature,

$$-\frac{\mathrm{d}^2 \log \mathcal{L}}{\mathrm{d} \log X^2} = \beta^2 \, \mathrm{var}(\lambda) = \frac{R}{2}, \qquad (1.81)$$

is already being tracked in order to fix the cooling schedule. The warning signal
would be a dangerously small effective dimension R as the coolness lost control
of the likelihood. The estimated dimension could even be negative, because the
formula (1.77) does not have positive-definite form. But simulated annealing could
not avoid the catastrophe. Avoidance requires a more robust strategy, for which
nested sampling is available.

1.5 Concluding remarks

The foundations of Bayesian inference are solid. Recent work has deepened our understanding of the measure/information/probability nexus, without lending support

to any alternative fuzzy or quantum logics. Quite the reverse. The applications have been widened, but not the calculus. The message is that generalizing the uniquely correct is a bad idea.

Another message is that Bayesian analysis is prescriptive as to method, but not as to content. Within a model, we are required to assign a prior probability distribution before we can reach posterior conclusions, but that assignment is essentially a matter of educated intelligence.

Nature judges models through our observations, which yield evidence values allowing us to discriminate. This 'evidence' is the prime output of Bayesian inference, and precedes any evaluation of posterior distribution. However, it is not on an absolute scale. There is no magic level at which a hypothetical model can be 'rejected with 95% confidence' in old-fashioned 'orthodox' style. All we can do is compare models (A and B, say) through their evidence ratios known as Bayes factors,

$$\mathrm{BayesFactor}(\mathrm{A}, \mathrm{B}) = \frac{E_\mathrm{A}}{E_\mathrm{B}}. \tag{1.82}$$

We may feel that model A is unsatisfactory, but we are stuck with it until we can do better. We cannot stand outside our current horizon, in some mythic state of holistic totality. On the contrary, in problems of significant size we often don't manage to fit the data well enough to satisfy an orthodox χ^2 significance test. Some stray effect that we haven't thought about or bothered with often upsets the fine detail of a fit, but that doesn't make us reject the model. The point about a good model is not that it explains *all* of a large dataset, but that it explains *most* of it better than the alternatives. We get nothing more than that. The adult view of inference is that we *need* nothing more.

In practice, unfortunately, the stray effect that we haven't thought about can corrupt the parameter values that we seek, so that we reach wrong values with over-stated precision. In principle, we should take the effort to model the stray effect. Often, though, we just allow ourselves to scale up the experimental noise until the data are more-or-less satisfied by the model we are actually interested in. The scale parameter is just another number to be determined in the analysis and, unsurprisingly, that factor tends to make χ^2 very acceptable. Real life often demands such compromise with perfection.

Computing evidence values requires good algorithms, which nowadays usually means some variant of Markov chain Monte Carlo. Before good algorithms were available, Bayesian inference was desperately handicapped, but now we can do it properly. Presented and compared here are two central methods: nested sampling for evaluating $E = \int_0^1 \mathcal{L} \, \mathrm{d}X$ and simulated annealing for evaluating $\log E = \int_0^1 \langle \log \mathcal{L} \rangle_\beta \, \mathrm{d}\beta$. Of the two, nested sampling is more direct and robust.

For easy problems, simulated annealing is likely to be somewhat quicker and more accurate, although nested sampling can increase its accuracy by any factor $\sqrt{N}$ by using N times more resources. For difficult problems, simulated annealing can fail completely, unless given impractical exponential resources. So, if computational resources are not a constraint, nested sampling may be preferred. When properly formulated, both methods yield estimates of the uncertainty that accompanies their Monte Carlo sampling, and arrive at

$$\log E = \text{estimate} \pm \text{uncertainty}, \tag{1.83}$$

which usually represents a nice single-humped probability distribution. Your author recommends, as a courtesy to future workers who may have alternative models, that presentations of Bayesian results should always be accompanied by this sort of statement of the evidence, including the relevant inverse-data units.

The uncertainties matter. Suppose models A and B yield

$$\log_e \frac{E_A}{(\text{km/s})^{-200}} = -1200 \pm 30\,, \qquad \log_e \frac{E_B}{(\text{km/s})^{-200}} = -1260 \pm 40\,, \tag{1.84}$$

when explaining 200 velocity measurements. Naive interpretation might indicate that model A was preferred by the hugely convincing Bayes factor e^{60}. Actually, though, the difference of 60 ± 50 is only a 1.2-sigma effect, from which we properly conclude that

$$\text{pr}(\text{A better than B}) = 0.885\,. \tag{1.85}$$

Yes, A is preferred, but only at an odds ratio of 8:1, not e^{60}:1. There's a 12% chance that the algorithm's random sampling led to the wrong conclusion, and that's what matters. More computation would be needed to gain the accuracy needed to settle the issue. Even then, we are likely to be content with uncertainties considerably above ± 1 in the logarithm, corresponding to orders of magnitude of uncertainty in E if we were foolish enough to try to undo the logarithm.

Parasitic upon the computation of $\log E$ is the generation of random samples of θ which define the posterior $\mathcal{P}(\theta)$ and any desired properties. With nested sampling, the E computation itself yields a weighted sequence of θ fit for purpose. With simulated annealing, a separate exploration phase needs to follow the 'burn-in' cooling phase. Either way, the posterior is available. But, contrary to orthodox presentation, the evidence comes first.

References

Aczél, J. (2003). In G. J. Erickson and Y. Zhai, eds., *AIP Conf. Proc.*, **707**, 195.

Cox, R. T. (1946). *Am. J. Phys.*, **14**, 1.

Cox, R. T. (1961). *The Algebra of Probable Inference*. Baltimore, MD: Johns Hopkins Press.

Ferguson, T. S. (1973). *Ann. Stat.*, **1**, 209.

Gelman, A., Carlin, J. B., Stern, J. S. and Rubin, D. B. (1995). *Bayesian Data Analysis.* London: Chapman and Hall, Chapter 11.

Kirkpatrick, S., Gelatt, C. D. and Vecchi, M. P. (1983). *Science*, **220**, 671.

Klain, D. A. and Rota, G.-C. (1997). *Introduction to Geometric Probability.* Cambridge: Cambridge University Press.

Knuth, K. H. (2003). In G. J. Erickson and Y. Zhai, eds., *AIP Conf. Proc.*, **707**, 204.

Knuth, K. H. (2005). *Neurocomputing*, **67**, 245.

Knuth, K. H. and Skilling, J., in preparation.

MacKay, D. J. C. (2003). *Information Theory, Inference, and Learning Algorithms.* Cambridge: Cambridge University Press.

Mukherjee, P., Parkinson, D. and Liddle, A. R. (2006). *Astrophys. J. Lett.*, **638**, L51.

Sivia, D. S. and Skilling, J. (2006). *Data Analysis: A Bayesian Tutorial*, 2nd edn. Oxford: Oxford University Press, Chapter 9.

Skilling, J. (1988). In G. J. Erickson and C. R. Smith, eds., *Maximum-Entropy and Bayesian Methods in Science and Engineering.* Dordrecht: Kluwer, p. 173.

Skilling, J. (2004a). In G. J. Erickson and Y. Zhai, eds., *AIP Conf. Proc.*, **707**, 381, 388.

Skilling, J. (2004b). In R. Fischer, R. Preuss and U. von Toussaint, eds., *AIP Conf. Proc.*, **735**, 395.

Steutel, F. W. (1979). *Scand. J. Stat.*, **6**, 57.

2

Simple applications of Bayesian methods

D. S. Sivia and S. G. Rawlings

Having seen how the need for rational inference leads to the Bayesian approach for data analysis, we illustrate its use with a couple of simplified cosmological examples. While real problems require analytical approximations or Monte Carlo computation for the sums to be evaluated, toy ones can be made simple enough to be done with brute force. The latter are helpful for learning the basic principles of Bayesian analysis, which can otherwise become confused with the details of the practical algorithm used to implement them.

2.1 Introduction

In science, as in everyday life, we are constantly faced with the task of having to draw inferences from incomplete and imperfect information. Laplace (1812, 1814), perhaps more than anybody, developed probability theory as a tool for reasoning quantitatively in such situations where arguments cannot be made with certainty; in his view, it was 'nothing but common sense reduced to calculation'. Although this approach to probability theory lost favour soon after his death, giving way to a frequency interpretation and the related birth of statistics (Jaynes 2003), it has experienced a renaissance since the late twentieth century. This has been driven, in practical terms, by the rapid evolution of computer hardware and the advent of larger-scale problems. Theoretical progress has also been made with the discovery of new rationales (Skilling 2010), but most scientists are drawn to Laplace's viewpoint instinctively.

In the past few years, several introductory texts have become available on the Bayesian (or Laplacian) approach to data analysis written from the perspective of the physical sciences (Sivia 1996; MacKay 2003; Gregory 2005). This chapter aims to fulfill a similar tutorial need in cosmology. To set the scene, a brief orientation on the pertinent astrophysics is given in Section 2.2. This is followed by two examples, in Sections 2.3 and 2.4, chosen to illustrate some of the issues that might be faced by a theorist and an experimentalist respectively, simplified to highlight the principles underlying the analysis. The chapter concludes with a summary and discussion of the salient points in Section 2.5.

2.2 Essentials of modern cosmology

On suitably large scales, and to some level of approximation, the Universe obeys the cosmological principle of isotropy and homogeneity: it looks the same in every direction, and would appear so to (hypothetical) fundamental observers (FOs) in other galaxies. Together with its expansion, as measured by the Hubble parameter $H = \dot{a}/a$, where a is the separation of two FOs, this implies a remarkable relationship between the local geometry of the Universe and the energy densities of its contents. Adopting Einstein's general relativity, this is encapsulated succinctly in Friedmann's equation, which describes the temporal, or redshift z, dependence of H on all the Ω parameters defined at the current epoch ($z = 0$) and H_0:

$$H^2 = H_0^2 \left[\Omega_k (1+z)^2 + \Omega_{\mathrm{M}} (1+z)^3 + \Omega_{\mathrm{DE}} (1+z)^{3+3w} + \Omega_{\mathrm{R}} (1+z)^4 \right]. \quad (2.1)$$

The local geometry is determined by $\Omega_k = -k\,c^2/H_0^2 a_0^2$, with $k = -1, 0, 1$ representing the hyperbolic, Euclidean and spherical cases respectively; the hyperbolic and spherical geometries have a constant curvature, associated with the length a_0, and Ω_k and k have opposite signs. The contents of the Universe are denoted by a set of density parameters $\Omega_j = 8\pi G \rho_j/3H_0^2$, where $j = \mathrm{M}, \mathrm{DE}, \mathrm{R}$ and G is Newton's gravitational constant. Matter, baryonic plus cold dark matter, has total density ρ_{M}; the mysterious dark energy has an equation-of-state parameter (modelled as a constant) $w = p_{\mathrm{DE}}/\rho_{\mathrm{DE}} c^2$, where p_{DE} and ρ_{DE} are its pressure and density respectively; and relativistic material, photons and hot dark matter, has total density ρ_{R}. The connection between the local geometry and the contents of the Universe becomes clear when Eq. (2.1) is evaluated at $z = 0$; namely,

$$\Omega_k + \Omega_{\mathrm{M}} + \Omega_{\mathrm{DE}} + \Omega_{\mathrm{R}} = 1. \quad (2.2)$$

2.2.1 Standard rulers and candles

The most straightforward way of probing the geometry and contents of the Universe is with a 'standard (cosmic) ruler'. Its fixed size means that the apparent length observed at various redshifts gives direct information on everything influencing the separation of light rays as they propagate through an expanding and, perhaps, inherently curved Universe. If the standard ruler is of an intrinsic co-moving size

s, and oriented perpendicular to the line-of-sight, then its transverse angular size, $\theta_{\rm obs}$, at a redshift of z is given by

$$\theta_{\rm obs} = \frac{s}{D_{\rm M}}, \qquad (2.3)$$

in radians, where

$$D_{\rm M} = \begin{cases} \frac{c}{H_0} \frac{1}{\sqrt{\Omega_k}} \sinh\left(\sqrt{\Omega_k}\, \frac{H_0 D_{\rm C}}{c}\right) & \text{for } k=-1,\ \Omega_k>0, \\ D_{\rm C} & \text{for } k=\ \ 0,\ \Omega_k=0, \\ \frac{c}{H_0} \frac{1}{\sqrt{|\Omega_k|}} \sin\left(\sqrt{|\Omega_k|}\, \frac{H_0 D_{\rm C}}{c}\right) & \text{for } k=+1,\ \Omega_k<0, \end{cases} \qquad (2.4)$$

and $D_{\rm C}$ is the integral of c/H with respect to redshift from 0 to z. Neglecting the contribution from relativistic material, which is reasonable when dealing with the Universe after recombination,

$$D_{\rm C} = \frac{c}{H_0} \int_0^z \left[\Omega_k(1+\zeta)^2 + \Omega_{\rm M}(1+\zeta)^3 + \Omega_{\rm DE}(1+\zeta)^{3+3w}\right]^{-1/2} \mathrm{d}\zeta. \qquad (2.5)$$

Note that, regardless of the effects of the local geometry, any sources of energy density close to the (ruler's) light rays curve space-time in accordance with Einstein's general relativity, so that Eqs. (2.3), (2.4) and (2.5) are only true along an average path; detailed gravitational lensing calculations are needed for any specific line-of-sight.

The quantity $D_{\rm C}$, the co-moving distance, represents the current radial separation between the observer and the standard ruler, assuming that neither has a significant 'peculiar' velocity. This can be thought of in two ways: either, the sum of distances measured simultaneously by a chain of FOs between the observer and the ruler at the epoch corresponding to $z=0$; or, the line integral of the distance travelled by a photon moving from the ruler to the observer, where each element of the integral is weighted by $(1+z)$ to account for the cosmic expansion since the photon passed that point (FO) along its path. We see from Eq. (2.3) that a measurement of $\theta_{\rm obs}$ and a 'knowledge' of s together determine $D_{\rm M}$ and hence, via Eqs. (2.4) and (2.5), have the potential to constrain the cosmological parameters strongly.

The search for a suitable cosmic ruler, or something of 'known' physical length, has been in progress since the birth of observational cosmology. It came to dramatic fruition around the year 2000, when cosmic microwave background (CMB) experiments reached sufficient sensitivity and fidelity to map out the power in the anisotropies of the CMB as a function of angular scale. A distinctive peak in this distribution, at an angular scale just below 1 degree, and its harmonics at smaller angles were seen, just as had been predicted by theoretical cosmologists.

These peaks reflect the so-called baryon acoustic oscillations (BAOs). The relevant physics here is that when the Universe is young and hot, the photon–baryon fluid is prevented from freely falling into dark-matter potential wells (and down from potential hills) by radiation pressure, resulting in acoustic oscillations, or sound waves. These oscillations cease when the expansion of the Universe causes the temperature to drop sufficiently for the electrons and atoms to combine, and thus for the photons to be released, and ultimately be observed as the CMB, with physical compressions and rarefactions corresponding to hot and cold spots on the CMB. Since these processes occur around a known universal temperature ($\sim$3000 K at $z \sim 1100$), and hence, for a particular choice of cosmological parameters, at an equivalent age of the Universe ($\sim$4 $\times$ 10^5 yr), and because the speed of sound in a plasma is finite ($\lesssim c$) then no compressions (or rarefactions) can form on physical scales larger than $\sim$130 kpc. Allowing for cosmic expansion from redshift $z \sim 1100$, this yields a standard ruler of co-moving size $s \approx 150$ Mpc in the form of the fundamental BAO, a scale that is also seen in the separation of the harmonics.

The best current measurements of θ_{obs} from BAOs seen in projection in the CMB are around $0.8°$ (Hinshaw *et al.* 2007), whence Eq. (2.3) implies $D_{\mathrm{M}} \sim 11$ Gpc. With reasonable values of $\Omega_{\mathrm{M}} = 0.3$ and $H_0 = 70$ km s^{-1} Mpc^{-1}, and Eqs. (2.4) and (2.5), this rules out $\Omega_{\mathrm{DE}} = 0$, as the implied curvature would then predict $\theta_{\mathrm{obs}} \approx 0.4°$ for $s \approx 150$ Mpc. The CMB data suggest that the geometry is close to spatially flat ($k = 0$) and $\Omega_{\mathrm{DE}} \approx 0.7$.

The BAO standard ruler has now also been detected in large galaxy redshift surveys (Percival *et al.* 2007) at low redshift ($z \sim 0.3$), showing that, on large scales, galaxies trace the matter distribution seen at much earlier times in the CMB. The details of this link comprise the complex subject of 'galaxy bias' (e.g., Dekel & Lahav 1999). Such methods also probe the ruler scale both perpendicular and parallel to the line-of-sight, with the latter type of observation giving, for a ruler at redshift z of length $\Delta D_{\mathrm{C}}/\Delta z$, a direct measurement of $c/H(z)$. The cosmological principle requires identical results for rulers parallel and perpendicular to the line-of-sight, the idea behind the cosmological test of Alcock and Paczynski (1979), and which on relatively small scales yields corrections for 'velocity–space distortion' – this effect is caused by the gravity-driven peculiar velocities of galaxies, and provides crucial independent information on galaxy bias.

Somewhat remarkably, these CMB- and galaxy-based experiments were not the first to provide evidence for dark energy. Ostriker and Steinhardt (1995) argued that the preponderance of data (such as direct and indirect measurements of $\Omega_{\mathrm{M}} \lesssim 1$; the relatively low Hubble Constant; and the overall level of CMB anisotropy) argued for a flat Universe with non-zero cosmological constant. Jaffe (1996) applied Bayesian model selection (Chapter 4) to a subset of this data and reached a similar conclusion. Most importantly, astronomers studying distant type-Ia supernovae (e.g., Perlmutter *et al.* 1999) had previously noted that they were too dim, or too far away at a given z, if treated as good 'standard candles' in either a flat $\Omega_M = 1$ or a curved $\Omega_M = 0.3$ ($\Omega_{DE} = 0$) cosmology, meaning that a flat $\Omega_M = 0.3$ ($\Omega_{DE} = 0.7$) cosmology is preferred. A standard candle assumes that the intrinsic luminosity of an object is 'known', or in practice calibratable, and provides a measurement of the luminosity distance $D_L = (1+z) D_C$.

It is important to appreciate the different nature of the constraint that supernovae provide on the content and geometry of the Universe. Current supernovae standard candle measurements typically probe the fairly local ($z \lesssim 1$) Universe where, for $w < -1/3$, the same D_C is obtained from positively correlated values of Ω_M and Ω_{DE} (to see this either substitute $\Omega_k = 1 - \Omega_M - \Omega_{DE}$ into Eq. (2.5), or note equivalently that the time derivative of the Friedmann equation yields an expression in which dark energy provides acceleration, gravity provides deceleration and the effects of curvature differentiate away). On the other hand, the CMB standard ruler method probes such a high redshift that $D_C \gg c/H_0$ so, from Eq. (2.4), angular size measurements become dominated by curvature (and hence $\Omega_M + \Omega_{DE}$), and the same D_M results from negatively correlated values of Ω_M and Ω_{DE}. The combination of the different constraints imposed on the parameters, such as Ω_M and Ω_{DE}, by a variety of experiments is a good way of breaking degeneracies inherent in a single type of measurement. Note that much of the decorrelating power comes from observations at different redshifts because: (i) the $(1+z)$ scalings of Eq. (2.1) changes the mix of components with redshift; and (ii) Eq. (2.4) shows the strong effects of curvature on photon paths originating at large distances ($D_C \gg c/H_0$). Note also that at a given redshift there is no difference between a standard candle method and a standard ruler method applied to a ruler parallel to the line-of-sight as both probe D_C and hence, via Eq. (2.5), $H(z)$.

In practice, both the measurements and the inference related to the size and luminosities of standard rulers and candles are fraught with difficulties. Even for a 'clean' BAO ruler, the physics determining how the length scale gets imprinted on the CMB and in the galaxy distributions is sufficiently complicated that theoretical cosmologists should always worry about subtle model dependencies in the absolute ruler length. Experimental cosmologists must also measure angular wavelengths of 'oscillatory' features against non-trivial backgrounds in the presence of noise which may have its own preferred angular scales.

2.2.2 Motivation

The discussion of cosmology above is brief, but designed to motivate the simple examples used in Sections 2.3 and 2.4; more comprehensive treatments can be found in standard text books (e.g., Peacock 1999). Our intention was to choose instructive illustrations that can be related to potentially fundamental experiments over the coming decades.

Cosmologists expect that 'standard ruler' experiments will eventually provide decoupled observations of the key parameters. For simplicity, we will set most at fiducial values in Section 2.3 and focus on w and Ω_k. Detection of $w \neq -1$, presumably then mappable as a time and space varying quantity, would indicate that dark energy is not simply Einstein's cosmological constant and would dictate the need for new physics; there is already much speculation (e.g., Peebles & Ratra 2003). Similarly, $\Omega_k \gg 10^{-5}$ would imply much more power on super-Hubble-length scales than is predicted by cosmic inflation, perhaps casting doubt on this popular theory for the early Universe, or perhaps supporting fascinating anthropic arguments about the origin of dark energy (Knox 2006).

It should be noted that modern cosmology features other fundamental cosmological parameters, such as the amplitude and shape of the primordial power spectrum of scalar perturbations and the ratio of tensor-to-scalar perturbations. Although their measurement is crucial to a full understanding of our Universe, they will not be considered further here, and we shall also largely ignore many critical 'nuisance' parameters, such as galaxy bias and the optical depth to the epoch of recombination.

2.3 Theorists and pre-processed data

From the discussion of Section 2.2, and with reference to Eq. (2.1) in particular, our theoretical model of cosmology is defined by the estimates of the parameters Ω_k, Ω_{M}, Ω_{DE}, Ω_{R} and w. In the Bayesian context, our state of knowledge about their values is encapsulated in the conditional probability density function (PDF) $\mathrm{prob}(\Omega_k, \Omega_{\mathrm{M}}, \Omega_{\mathrm{DE}}, \Omega_{\mathrm{R}}, w | \mathrm{Data}, I)$, where I represents all the information and assumptions relevant to the problem other than the data at hand. Since this is a five-dimensional entity, it is awkward to display graphically. To aid visualization, let us focus on just two of the cosmological parameters, Ω_k and w, and consider $\mathrm{prob}(\Omega_k, w | \mathrm{Data}, I)$. This PDF is related to the former through marginalization, or an integration over the omitted parameters,

$$\mathrm{prob}(\Omega_k, w | \mathrm{Data}, I) = \tag{2.6}$$
$$\iiint \mathrm{prob}(\Omega_k, \Omega_{\mathrm{M}}, \Omega_{\mathrm{DE}}, \Omega_{\mathrm{R}}, w | \mathrm{Data}, I) \, \mathrm{d}\Omega_{\mathrm{M}} \, \mathrm{d}\Omega_{\mathrm{DE}} \, \mathrm{d}\Omega_{\mathrm{R}},$$

and indicates what we can infer about Ω_k and w when the uncertainties in Ω_{M}, Ω_{DE} and Ω_{R} are taken into account. To keep this example simple, let us avoid any complications related to the evaluation of the triple integral by rewriting the integrand as

$$\text{prob}(\Omega_k, w | \Omega_M, \Omega_{DE}, \Omega_R, \text{Data}, I) \times \text{prob}(\Omega_M, \Omega_{DE}, \Omega_R | \text{Data}, I),$$

using the product rule of probability, and removing uncertainty about the values of Ω_M, Ω_{DE} and Ω_R with the assignment

$$\text{prob}(\Omega_M, \Omega_{DE}, \Omega_R | \text{Data}, I) = \text{prob}(\Omega_M, \Omega_{DE}, \Omega_R | I)$$
$$= \delta(\Omega_M - 0.3) \, \delta(\Omega_R) \, \delta(1 - \Omega_k + \Omega_M + \Omega_{DE} + \Omega_R).$$

Whereas the third δ-function follows from the theoretical constraint of Eq. (2.2), the other two represent very strong assumptions imposed on this analysis through the conditioning information I. The marginalization of Eq. (2.6) then yields

$$\text{prob}(\Omega_k, w | \text{Data}, I)$$
$$= \text{prob}(\Omega_k, w | \Omega_M = 0.3, \Omega_{DE} = 0.7 - \Omega_k, \Omega_R = 0, \text{Data}, I),$$

as expected, where Ω_M and Ω_R are fixed at their fiducial values.

Suppose we learn that CMB observations indicate an angular size of $\theta_{obs} = 0.8°$, to within 1%, at $z \sim 1100$ for a cosmic ruler of co-moving size $s \approx 150 \, \text{Mpc}$; what can we infer about the values of Ω_k and w in the light of this datum? This is defined by $\text{prob}(\Omega_k, w | \text{Data}, I)$, of course, which can be related to other PDFs through Bayes' theorem:

$$\text{prob}(\Omega_k, w | \text{Data}, I) = \frac{\text{prob}(\text{Data} | \Omega_k, w, I) \times \text{prob}(\Omega_k, w | I)}{\text{prob}(\text{Data} | I)}, \qquad (2.7)$$

where $\Omega_M = 0.3$, $\Omega_{DE} = 0.7 - \Omega_k$ and $\Omega_R = 0$ have been subsumed into the general conditioning symbol I to reduce algebraic cluttering. The various terms in Bayes' theorem are often given specific names. The left-hand side is the posterior PDF that we seek, and encodes our state of knowledge about the values of Ω_k and w after the analysis of the current data. By contrast, $\text{prob}(\Omega_k, w | I)$ is called the prior PDF since it represents our ignorance (or knowledge) regarding them before a consideration of the new measurements. The experimental results enter the analysis through the likelihood function, $\text{prob}(\text{Data} | \Omega_k, w, I)$, which quantifies the significance of the mismatch between the data and their theoretically predicted values from a given Ω_k and w. The denominator, $\text{prob}(\text{Data} | I)$, is an uninteresting normalization constant from the perspective of an inference about Ω_k and w given a specific I,

$$\text{prob}(\text{Data}|I) = \iint \text{prob}(\text{Data}, \Omega_k, w|I) \, d\Omega_k \, dw \qquad (2.8)$$

$$= \iint \text{prob}(\text{Data}|\Omega_k, w, I) \, \text{prob}(\Omega_k, w|I) \, d\Omega_k \, dw \,,$$

which ensures that the integral of the posterior PDF with respect to Ω_k and w in Eq. (2.7) is unity, but it plays a central role when the relative merit of alternative analysis assumptions needs to be assessed. For this reason, scientists sometimes refer to it as the 'evidence' for a model; conventionally trained (Bayesian) statisticians know it by a variety of other names, such as the prior predictive and the marginal likelihood.

Although the nomenclature above is used widely, and intended to be helpful, it can be misconstrued since people are led to regard the prior as being subjective, whereas the likelihood is seen as objective. In reality, every term is just a conditional PDF and, as such, on a par with each other. The power of Bayes' theorem comes from its ability to turn things around with respect to the conditioning statement and, thereby, allow us to express the PDF of interest in terms of others that we are more comfortable in assigning; indeed, the latter dictates how the rules of probability are used in general. Since all probabilities quantify a state of knowledge from the Bayesian point of view, and so are conditional on the information at hand, the assignment of a PDF, whether it be a prior or a likelihood function, depends on a careful consideration of the relevant contents of I (including assumptions).

So, what should we assign for $\text{prob}(\Omega_k, w|I)$? That depends on what is known about Ω_k and w prior to the analysis of the current data. If the answer is 'not much', then this ought to be reflected with a very broad PDF; a sharply peaked one would indicate that we already had a pretty good idea about the values of Ω_k and w. Since a PDF that is invariant over a suitably large range is both simple and broad, we could legislate that gross ignorance be encoded through the assignment of a uniform prior; subtle information about the nature of the parameters can then be incorporated through a judicious choice of the exact hypothesis space in which the calculation is done. For example, a uniform probability in x makes a slightly different assumption about this parameter than invariance with respect to $\log(x)$. The former PDF, which remains unchanged if the origin is translated, is suitable if absolute differences in x are deemed significant; the latter, which implicitly assumes that x is positive and is invariant under a change of units, is appropriate if relative changes $(\Delta x/x)$ are considered important. With this in mind, the definitions of Ω_k and w, and the roles they play in Eq. (2.1), suggest that it would be reasonable to have a prior which was uniform in $\log(|\Omega_k|)$ and w; to be properly normalized, however, finite ranges of validity also need to be specified. Thus,

$$\text{prob}(\Omega_k, w|I) = \text{prob}(\Omega_k|I) \times \text{prob}(w|I) \,, \qquad (2.9)$$

which assumes logical independence in our knowledge of Ω_k and w as far as I is concerned, with

$$\mathrm{prob}(\Omega_k|I) = \begin{cases} \gamma & \text{for } k=0, \ \Omega_k=0\,, \\ \frac{A\,(1-\gamma)}{|\Omega_k|} & \text{for } k=\pm 1, \ \alpha<|\Omega_k|<\beta\,, \\ 0 & \text{otherwise}\,, \end{cases} \qquad (2.10)$$

where $A = \left[\,2\ln(\beta/\alpha)\,\right]^{-1}$ and $1/|\Omega_k|$ transforms to invariance with respect to $\log(|\Omega_k|)$, and

$$\mathrm{prob}(w|I) = \begin{cases} B & \text{for } w_{\min}<w<w_{\max}\,, \\ 0 & \text{otherwise}\,, \end{cases} \qquad (2.11)$$

where $B = (w_{\max}-w_{\min})^{-1}$. The choice of the bounds, α, β, $w_{\min}$ and $w_{\max}$, and the probability γ that $\Omega_k=0$, are dependent on I as well. Based only on the three distinct possibilities for k, we might set $\gamma=1/3$; α represents the smallest value of $|\Omega_k|$ considered to be significantly different from zero, whereas β may be estimated from the fact that a non-zero Ω_k would have been obvious by now if its magnitude had been large enough. Similar arguments are required for the step-size to straddle $w=-1$.

What should we assign for the likelihood function, $\mathrm{prob}(\mathrm{Data}|\Omega_k,w,I)$? If the measurement of the angular size of a standard ruler at redshift z yields the datum D, which is ascribed a fractional uncertainty ϵ, then most people would be inclined to say that

$$\mathrm{prob}(D|\Omega_k,w,I) = \frac{1}{\sigma\sqrt{2\pi}}\,\exp\left\{-\frac{\left[D-\theta(z)\right]^2}{2\,\sigma^2}\right\}, \qquad (2.12)$$

where $\sigma = \epsilon D$ and $\theta(z)$ is the value of θ_{obs} at z predicted by Eqs. (2.3)–(2.5) for the given Ω_k and w; a knowledge of s, H_0, c and ϵ, as well as $\Omega_{\mathrm{M}} = 0.3$, $\Omega_{\mathrm{R}} = 0$ and $\Omega_{\mathrm{DE}} = 0.7-\Omega_k$, is assumed in I. Although the Gaussian PDF is ubiquitous, it is important to remember that Eq. (2.12) is merely an attempt to quantify the information inherent in I. According to Jaynes' principle of maximum entropy (MaxEnt), it is best regarded as the PDF which satisfies constraints on the mean and variance of the mismatch between a measurement and its prediction,

$$\left\langle D-\theta(z)\right\rangle = 0 \quad \text{and} \quad \left\langle\left[D-\theta(z)\right]^2\right\rangle = \epsilon D\,,$$

but is otherwise maximally non-committal about D (Jaynes 2003). Does the available datum correspond to expectations of this kind? Perhaps the statement of the uncertainty in fractional terms should be interpreted as

$$\left\langle \left[\frac{D - \theta(z)}{D} \right]^2 \right\rangle = \epsilon \,,$$

whence MaxEnt would deem a log-normal PDF (Sivia & Skilling 2006) more appropriate? Such questions rarely have clear-cut answers. This indicates how the perceived objectivity of the likelihood over a prior is largely an illusion, resulting from familiarity, and it is really down to the analysis assumptions and background information in both cases. A toy problem, known as 'Peelle's pertinent puzzle', is used to illustrate this point explicitly in Sivia (2006).

Having assigned the PDFs of Eqs. (2.9)–(2.12), the ingredients required for the calculation of the posterior, and the evidence of the related model, are in place. Our inference about Ω_k and w is appreciated most easily through the computation and graphical display of $\mathrm{prob}(\Omega_k, w \,|\, D, I)$ on a regular two-dimensional grid. The digitization of the w-axis is straightforward: with N points spanning the prior range,

$$w_j = w_{\min} + \left[\frac{w_{\max} - w_{\min}}{N-1} \right] (j-1) \,, \tag{2.13}$$

for $j = 1, 2, \ldots, N$. To accommodate a uniform prior in $\log(|\Omega_k|)$ for $k = \pm 1$, and to handle the case of $k = 0$ correctly, it is best to let the index i for the Ω_k-axis run from $-M$ to $+M$:

$$(\Omega_k)_i = \begin{cases} \Upsilon_i & \text{for } k = -1, \ i = 1, 2, \ldots, M \,, \\ 0 & \text{for } k = 0, \ i = 0 \,, \\ -\Upsilon_i & \text{for } k = +1, \ i = -M, \ldots, -2, -1 \,, \end{cases} \tag{2.14}$$

$$\text{where} \quad \Upsilon_i = \alpha \left(\frac{\beta}{\alpha} \right)^{\frac{|i|-1}{M-1}} \,. \tag{2.15}$$

Taking M and N to be sufficiently large that integrals with respect to Ω_k and w are approximated well by a simple summation over the discrete mesh, the prior of Eqs. (2.9)–(2.11) is equivalent to

$$\mathrm{prob}(i, j \,|\, I) = \begin{cases} \frac{\gamma}{N} & \text{for } i = 0 \,, \\ \frac{1-\gamma}{2MN} & \text{otherwise} \,. \end{cases} \tag{2.16}$$

In conjunction with the corresponding likelihood, it is computationally prudent to evaluate the logarithmic sum

$$L_{ij} = \ln \left[\mathrm{prob}(D \,|\, i, j, I) \right] + \ln \left[\mathrm{prob}(i, j \,|\, I) \right] \,, \tag{2.17}$$

where $\mathrm{prob}(D|i,j,I)$ is the value of Eq. (2.12) when $\Omega_k = (\Omega_k)_i$ and $w = w_j$ on the two-dimensional grid of points defined by i and j. If the greatest value of L_{ij} is $L_{\max}$, then the posterior PDF and the evidence for the analysis model can be ascertained safely through

$$\mathrm{prob}(i,j|D,I) = \frac{e^{L_{ij}-L_{\max}}}{Z}, \qquad (2.18)$$

where

$$Z = \sum_{j=1}^{N}\sum_{i=-M}^{M} e^{L_{ij}-L_{\max}} = \mathrm{prob}(D|I) \times e^{-L_{\max}}. \qquad (2.19)$$

Let us see a few examples.

In accordance with the simplifying assumptions made above, we will remove uncertainty about s, H_0, Ω_M and Ω_R in the illustrations by fixing them at their fiducial values ($s = 150\,\mathrm{Mpc}$, $H_0 = 70\,\mathrm{km\,s^{-1}\,Mpc^{-1}}$, $\Omega_M = 0.3$ and $\Omega_R = 0$) and take them, and the relevant z, to be known exactly. Given the datum $D = 0.629°$ to within 1% at a redshift of 1000, which is θ_{obs} from Eqs. (2.3)–(2.5) for $\Omega_k = 0$ and $w = -1$, the resultant posterior PDF is shown with a greyscale and contours in Figure 2.1(a); the related prior had $\alpha = 10^{-6}$, $\beta = 10^{-1}$, $\gamma = 1/3$, $w_{\min} = -3$ and $w_{\max} = 0$, and the computation was done on a grid with $M = N = 200$. While this two-dimensional function captures all that can be inferred simultaneously about the values of Ω_k and w in the light of D and I, our principal interest may just be in one of them. The associated information is conveyed through the marginal posteriors,

$$\mathrm{prob}(w|D,I) \equiv \mathrm{prob}(j|D,I) = \sum_{i=-M}^{M}\mathrm{prob}(i,j|D,I), \qquad (2.20)$$

and the equivalent expression for $\mathrm{prob}(\Omega_k|D,I)$, which are plotted in Figure 2.1(c) and (d). Also shown, with a dotted line, are the marginals corresponding to a fractional error of 0.2%; to aid comparison, the PDFs have been scaled vertically to have unit maxima. Although the five-fold improvement in the quality of the measurement is reflected in the greater reliability of the estimate of w, there is no significant gain for Ω_k! Indeed, the insensitivity of Ω_k (between α and β) to the datum (for either ϵ) can also be seen from the fact that the posterior probability of a flat universal geometry,

$$\mathrm{prob}(k=0|D,I) = \sum_{j=1}^{N}\mathrm{prob}(i=0,j|D,I), \qquad (2.21)$$

changes imperceptibly from the prior value of a third. (We note that if we had definite prior knowledge of w, for example, if we knew it to have the cosmological constant value $w = -1$, we would be able to measure Ω_k somewhat better, as expected from the discussion of Section 2.2.)

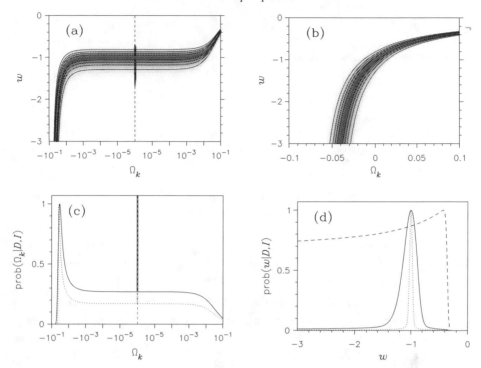

Fig. 2.1. (a) The posterior PDF, $\mathrm{prob}(\Omega_k, w\,|\,D, I)$, of Eqs. (2.13)–(2.19) for $D = 0.629°$ at $z = 1000$ and $\epsilon = 1\%$; the darker regions indicate areas of higher probability, and the vertical dashed line marks the discontinuity at $k = 0$. (b) The corresponding posterior for a uniform prior in Ω_k, between ± 0.1. (c) The marginal posterior PDF for Ω_k resulting from (a), scaled to a maximum of unity ($k \neq 0$); the dotted line shows the equivalent curve for $\epsilon = 0.2\%$. (d) The same as (c) but for w, with an additional dashed line indicating the marginal yielded by (b).

Posterior PDFs are frequently specified with the means, standard deviations and covariances of their parameters. While this is often the only practical way of summarizing the inference, its usefulness is limited when the posterior is not approximated well by a multivariate Gaussian. It would paint an incomplete and potentially misleading picture in the case of Figure 2.1, for example, even allowing for the unusual δ-function at $\Omega_k = 0$ in (a), as non-linear correlations cannot be captured with a covariance (or a Fisher) matrix.

If the preceding numerical experiment is repeated for a datum corresponding to low redshift, $D = 2.60°$ at $z = 1$ with $\epsilon = 1\%$, then the resultant posterior PDF for Ω_k and w will be as shown in Figure 2.2. The fractional error on a measurement of this type is currently closer to 10%, so that $\mathrm{prob}(\Omega_k, w\,|\,D, I)$ would be about ten times more spread out in practice. The marginal posterior PDF for w is also plotted in Figure 2.2, with the case of $\epsilon = 10\%$ indicated with a dotted line; to aid

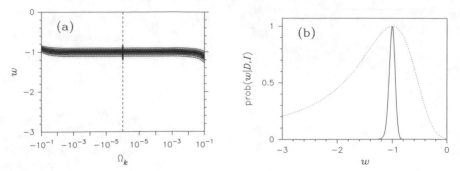

Fig. 2.2. (a) The posterior PDF for Ω_k and w, with $D=2.60°$ at $z=1$ and $\epsilon=1\%$. (b) The marginal for w, with the dotted line being for $\epsilon=10\%$.

comparison, both $\mathrm{prob}(w\,|\,D, I)$ have been scaled vertically to have unit maxima. The marginal posterior for Ω_k has not been displayed since it is very broad and shows little change from the prior.

What would be gained inferentially if both high- and low-redshift data were available? We can assess this by combining the two measurements that have just been considered:

$$\mathrm{Data} \equiv \left\{ \, D_1=0.629° \left(z_1=1000, \epsilon_1\right), \ D_2=2.60° \left(z_2=1, \epsilon_2\right) \right\}.$$

The analysis proceeds as before, except that the previous datum D is replaced by the vector Data, with the likelihood function

$$\mathrm{prob}(\mathrm{Data}\,|\,\Omega_k, w, I) = \prod_{q=1}^{N_D} \mathrm{prob}(D_q\,|\,\Omega_k, w, I)\,, \qquad (2.22)$$

where $q=1, 2, \ldots, N_D$ and $\mathrm{prob}(D_q\,|\,\Omega_k, w, I)$ is given by Eq. (2.12) with $z=z_q$, $D=D_q$ and $\sigma=\epsilon_q D_q$; $N_D=2$ in our case. The decomposition of the likelihood into a product of the probabilities for the individual data assumes independence, implicit in I, where the noise in one measurement is not correlated with that of any other. The resultant posterior PDF, $\mathrm{prob}(\Omega_k, w\,|\,\mathrm{Data}, I)$, with $\epsilon_1 = \epsilon_2 = 1\%$ is shown in Figure 2.3(a). It resembles Figure 2.2(a) more than Figure 2.1(a), because D_2 imposes a much tighter constraint on w; this can also be seen for the corresponding marginals for w in Figures 2.1(d), 2.2(b) and 2.3(b). The two data contain complementary information on Ω_k for $|\Omega_k| \gtrsim 10^{-3}$, which gives rise to a rounded fall off at the extremities. With $\epsilon_2 = 10\%$, Figure 2.3(a) becomes almost identical to Figure 2.1(a). Incidentally, the posterior probability of a flat universal geometry rises from the prior value of $1/3$ to 39.3% with $\epsilon_1 = \epsilon_2 = 1\%$ but only increases to 35.3% when $\epsilon_2 = 10\%$.

The number of unknowns in this example was intentionally kept to just two, to aid visualization and understanding, by imposing very strong constraints in the

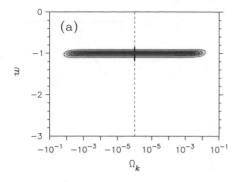
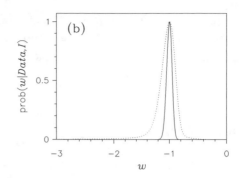

Fig. 2.3. (a) The resultant posterior PDF, $\mathrm{prob}(\Omega_k, w\,|\,\mathrm{Data}, I)$, when the two measurements that yield Figs. 2.1(a) and 2.2(a) separately are used together. (b) The corresponding marginal for w ($\epsilon_1 = \epsilon_2 = 1\%$), scaled to a maximum of unity, with the case of $\epsilon_2 = 10\%$ shown with a dotted line.

conditioning information. Uncertainties with regard to s, z, H_0, Ω_{M}, and so on, need to be taken into account through marginalization, as in Eq. (2.6), with the assignment of appropriate PDFs to quantify our limited knowledge of their values:

$$\mathrm{prob}(\mathrm{Data}|\Omega_k, w, I) = \iint \cdots \int \mathrm{prob}(\mathrm{Data}, w, s, \ldots | \Omega_k, w, I)\,\mathrm{d}s \ldots,$$

where the integrand can be decomposed into a product of the previous likelihood function, $\mathrm{prob}(\mathrm{Data}|\Omega_k, w, s, \ldots, I)$, where $s, \ldots$ are given (formerly subsumed in I), and the prior PDF, $\mathrm{prob}(s, \ldots | I)$, which will not have δ-function character in general. The marginalization results in a broadening of the posterior for Ω_k and w. It rapidly becomes impractical to deal with such problems through an explicit computation of the probabilities on a regular grid in the space of many parameters, and so we are forced to turn to analytical approximations and Monte Carlo methods to carry out the calculations; the latter are, therefore, the focus of much of this volume.

2.4 Experimentalists and raw measurements

The previous section was intended to be a simple illustration of data analysis from the perspective of a theoretical cosmologist. In reality, θ_{obs} is not an actual measurement but the distillation of the pertinent information contained in raw observations. We now model this situation in an elementary way to highlight some of the issues involved in the intermediate step. The example is designed to bear a passing resemblance to problems associated with measuring BAOs in the galaxy power spectrum, where the gravitationally-driven non-linear growth of structures boosts the power preferentially on small scales; the noise can be quite a complicated function of scale, with the precise details depending on the specific experiment. Indeed, in practice the angle θ_{obs} is not measured directly at all, but rather the cosmological parameters' impact on the galaxy power spectrum or correlation function is calculated, which is in turn compared to the observed data in a similar fashion to the case for the CMB power spectrum discussed in various contexts elsewhere in this volume.

For simplicity, suppose that the data are a cosine function of $1/\theta$, of unit amplitude and zero phase, superimposed upon a quadratic background:

$$F(\theta) = \cos\left(\frac{2\pi\theta_{\text{obs}}}{\theta}\right) + \frac{b}{\theta^2}, \qquad (2.23)$$

where the wavelength, θ_{obs}^{-1}, is of interest but there is one nuisance parameter, b. Given a set of measurements, $\{d(\theta_l)\}$ for $l = 1, 2, \ldots, N_d$, such as those in Figure 2.4(a), what can we say about θ_{obs}?

This type of problem is usually addressed by a least-squares fit of the model to the measurements, whereby the best estimates of θ_{obs} and b are taken to be the values which minimize the quadratic mismatch statistic

$$\chi^2 = \sum_{l=1}^{N_d} \left[F(\theta_l) - d(\theta_l)\right]^2. \qquad (2.24)$$

The procedure can be interpreted in Bayesian terms as representing the maximum of the posterior PDF for θ_{obs} and b with a uniform prior (in θ_{obs} and b), and a likelihood function which assumes that every datum is subject to independent and additive Gaussian noise of the same standard deviation, σ:

$$\text{prob}(\{d(\theta_l)\}|\theta_{\text{obs}}, b, \sigma, I) = \left(\sigma\sqrt{2\pi}\right)^{-N_d} \exp\left[-\frac{\chi^2}{2\sigma^2}\right], \qquad (2.25)$$

which basically follows from Eqs. (2.12) and (2.22). Although σ has no impact on the optimal θ_{obs} and b, its value does affect our assessment of their reliability. Without a good estimate of the noise level, σ has to be marginalized out just like any other nuisance parameter:

$$\text{prob}(\{d(\theta_l)\}|\theta_{\text{obs}}, b, I) = \int \text{prob}(\{d(\theta_l)\}|\theta_{\text{obs}}, b, \sigma, I)\, \text{prob}(\sigma|I)\, d\sigma$$

$$\approx \frac{\left(2\pi e/N_d\right)^{-N_d/2}}{\ln(\sigma_{\text{max}}/\sigma_{\text{min}})} \left(\chi^2\right)^{-N_d/2}, \qquad (2.26)$$

where the prior, $\text{prob}(\sigma|I)$, has been taken to be uniform with respect to $\log(\sigma)$, over a large but finite range, and the integral simplified under the assumption that $N_d \gg 1$. The marginal posterior PDF for θ_{obs} that results from Eqs. (2.23), (2.24) and (2.26), given the data of Figure 2.4(a) and a uniform prior for θ_{obs} and b, is shown in Figure 2.4(b).

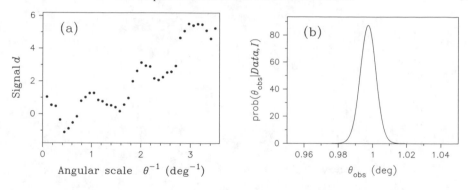

Fig. 2.4. (a) Measurements pertaining to Eq. (2.23), with Gaussian noise of uniform level σ. (b) The corresponding marginal posterior PDF for θ_{obs}, with a uniform prior and a likelihood given by Eqs. (2.23), (2.24) and (2.26).

How should the inference of θ_{obs} from the observations $\{d\,(\theta_l)\}$ be propagated to the analysis of the preceding section? Well, our state of knowledge about Ω_k and w given the 'real' data is encapsulated in the PDF $\mathrm{prob}(\Omega_k, w\,|\,\{d\,(\theta_l)\}\,, I)$. This was equated with $\mathrm{prob}(\Omega_k, w\,|\,D, I)$, as per Figure 2.1, on the understanding that D (and ϵ) contained the same information about θ_{obs} as $\{d\,(\theta_l)\}$. For example, $\theta_{\mathrm{obs}} = 0.997° \pm 0.005°$ constitutes a useful summary of the inference from the data of Figure 2.4(a) because the posterior of Figure 2.4(b) is described well by a Gaussian. The formal link can made through Bayes' theorem of Eq. (2.7), with Data interpreted as $\{d\,(\theta_l)\}$, by using marginalization and the product rule of probability to express the likelihood function as

$$\mathrm{prob}(\{d\,(\theta_l)\}\,|\,\Omega_k, w, I) = \int \mathrm{prob}(\{d\,(\theta_l)\}\,|\,\theta_{\mathrm{obs}}, I)\,\mathrm{prob}(\theta_{\mathrm{obs}}\,|\,\Omega_k, w, I)\,\mathrm{d}\theta_{\mathrm{obs}},$$

where the conditioning on Ω_k and w has been dropped from the PDF in which θ_{obs} is given (without uncertainty), since it is multiplied by another that is non-zero only when they are consistent:

$$\mathrm{prob}(\theta_{\mathrm{obs}}\,|\,\Omega_k, w, I) = \delta(\Theta(\Omega_k, w) - \theta_{\mathrm{obs}}),$$

where $\Theta(\Omega_k, w)$ is the value of θ_{obs} predicted by Eqs. (2.3)–(2.5) for the given Ω_k and w; a knowledge of z, s, and so on, is assumed in I. The integration over the δ-functions confirms the intuitive result that the relevant likelihood in the inference of Ω_k and w is $\mathrm{prob}(\{d\,(\theta_l)\}\,|\,\Theta(\Omega_k, w), I)$.

The data in Figure 2.4(a) were analyzed on the basis that they were subject to independent and additive Gaussian noise of constant magnitude. This was known to be the case since they were generated in a computer under those conditions (with $\sigma = 0.25$); reassuringly, the angular wavelength recovered was in line with

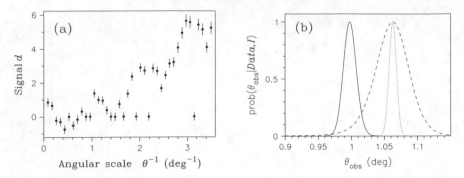

Fig. 2.5. (a) Measurements with Gaussian noise and a few spurious zeros. (b) Marginal posterior PDFs, scaled to unit maxima, corresponding to noise models h_1 (dotted), h_2 (dashed) and h_3 (solid), explained in the text.

the $\theta_{\mathrm{obs}} = 1$ used for the simulation. In real life this would only be an assumption, not least because part of the expected mismatch between the measurements and predictions would stem from an inadequacy of the functional model used to fit the structure in the data, such as Eq. (2.23), and we must never forget that the inference is conditional upon it.

Consider the measurements in Figure 2.5(a), where each datum $d\,(\theta_l)$ is accompanied by an estimate of its error bar, σ_l. Although the previous cosine function and quadratic background are visible, so too are glitches where the signal drops to zero occasionally. If the data were analyzed blindly using least squares, with the definition of Eq. (2.24) generalized to the sum of the squares of the normalized residuals, $\chi^2 = \sum R_l^2$ where

$$R_l = \frac{F(\theta_l) - d\,(\theta_l)}{\sigma_l}, \qquad (2.27)$$

then the resultant marginal posterior PDF, $\mathrm{prob}(\theta_{\mathrm{obs}}|\{d\,(\theta_l)\}, h_1, I)$ where h_1 explicitly denotes the corresponding noise hypothesis, would be as shown by the dotted line in Figure 2.5(b). It is not very surprising that the simulation value of $\theta_{\mathrm{obs}} = 1$ lies well outside the 'credible' region, since h_1 makes no allowance for 'spurious measurements'. While the validity of h_1 would be questionable due to the least-squares fit to the data being too poor ($\chi^2_{\mathrm{min}} \gg N_d$), it is down to us to come up with a better alternative.

A common course of action involves the postulate, h_2, that all the stated error-bars ought to be scaled by an unknown positive constant, σ, so that $\sigma_l \to \sigma\,\sigma_l$; it can be regarded as a 'safety valve' designed to compensate crudely for unaccounted sources of uncertainty. A marginalization over σ leads to the likelihood function of Eq. (2.26), with χ^2 generalized as before and $\ln(\sigma_{\mathrm{max}}/\sigma_{\mathrm{min}})$ multiplied by $\sigma_1\,\sigma_2 \ldots \sigma_{N_d}$, and gives rise to the dashed posterior PDF in Figure 2.5(b),

$\text{prob}(\theta_{\text{obs}}|\{d(\theta_l)\}, h_2, I)$. Even though the 'optimal' estimate of θ_{obs} remains unchanged, the posterior conditional on h_2 is sufficiently broad to encompass the simulation value within the realm of plausibility.

Noise models that intrinsically offer protection against erratic data have likelihood functions which fall off much more slowly in the tails, when $R_l \gg 1$, than a Gaussian. For example,

$$\text{prob}(\{d(\theta_l)\}|\{\sigma_l\}, \theta_{\text{obs}}, b, h_3, I) = \prod_{l=1}^{N_d} \frac{1}{\sigma_l\sqrt{2\pi}} \left(\frac{1 - e^{-R_l^2/2}}{R_l^2}\right), \qquad (2.28)$$

derived in Chapter 8 of Sivia & Skilling (2006), which assumes that the σ_l represent only the lower bounds for the (Gaussian) error bars to be associated with the respective measurements. The posterior PDF for θ_{obs} conditional on h_3 is shown with a solid line in Figure 2.5(b). Since the data of Figure 2.5(a) were generated in a computer, we know that, of the three noise models considered, h_3 yields the inference which is most faithful to the underlying reality. Could this have been surmised without being privy to the simulation program?

The support for a proposition, be it a model for the object of interest or the noise properties of the related data, is quantified by its posterior probability. In the context of h_1, h_2 and h_3 above, we need to evaluate $\text{prob}(h_r|\{d(\theta_l)\}, I)$ for $r = 1, 2, 3$. Using Bayes' theorem, this can be expressed in terms of the prior and evidence probabilities for h_r,

$$\text{prob}(h_r|\{d(\theta_l)\}, I) = \frac{\text{prob}(\{d(\theta_l)\}|h_r, I) \times \text{prob}(h_r|I)}{\text{prob}(\{d(\theta_l)\}|I)}, \qquad (2.29)$$

where, for a total of N_R noise models deemed reasonable given I,

$$\text{prob}(\{d(\theta_l)\}|I) = \sum_{r=1}^{N_R} \text{prob}(\{d(\theta_l)\}|h_r, I)\,\text{prob}(h_r|I). \qquad (2.30)$$

Although we might assign $\text{prob}(h_r|I) = 1/N_R$ in the absence of cogent information, only a few h_r possibilities are considered in practice. This means that most of the evidence values needed for the normalization factor of Eq. (2.30) are not available, and even N_R is usually left unspecified. We can still ascertain the relative merit of alternative h_r possibilities, however, as the denominator of Eq. (2.29) cancels out when calculating

$$\frac{\text{prob}(h_r|\{d(\theta_l)\}, I)}{\text{prob}(h_{r'}|\{d(\theta_l)\}, I)} = \frac{\text{prob}(h_r|I)}{\text{prob}(h_{r'}|I)} \times \frac{\text{prob}(\{d(\theta_l)\}|h_r, I)}{\text{prob}(\{d(\theta_l)\}|h_{r'}, I)},$$

so that the posterior odds are equal to the ratio of the evidences for r and r' to within the prior odds. While an evidence value is meaningless in isolation, it is crucial for a comparison with alternatives. For the case of h_1, h_2 and h_3 in Figure 2.5(b), the

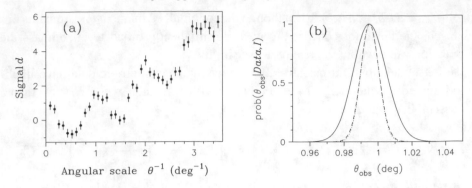

Fig. 2.6. (a) Noisy measurements that are free from spurious glitches. (b) Marginal posterior PDFs, scaled to unit maxima, corresponding to noise models h_1 (dotted), h_2 (dashed) and h_3 (solid), explained in the text.

natural logarithm of their evidences are -333.0, -64.5 ($\sigma_{\min} = 0.01$ and $\sigma_{\max} = 100$) and -45.2 respectively; for equal priors, this makes h_3 favoured by a factor of 10^8 over h_2 and 10^{125} with respect to h_1.

If the previous analysis is repeated for the data of Figure 2.6(a), which do not suffer from spurious zeros, the corresponding (logarithmic) evidence values for h_1, h_2 and h_3 are -8.4, -10.6 and -24.1. As shown in Figure 2.6(b), the posterior PDFs conditional on h_1 and h_2 are virtually identical; the probabilistic evidence prefers the former by a small margin (10^1) since the latter has an extra degree of freedom ($\sigma \neq 1$) that is not warranted by the data. The posterior for θ_{obs} subject to h_3 is about twice as broad as the other two, which is not surprising given its pessimistic stance on the error bars; the evidence now weighs against it by a factor of 10^7 relative to h_1.

Incidentally, the evidence returns a non-committal verdict on the alternative choice of priors for Ω_k in Figure 2.1. Logarithmic values of 5.9 for (a) and 5.6 for (b) give a posterior ratio of essentially unity based on the single datum.

2.5 Concluding remarks

The defining feature of the Bayesian approach to data analysis is not the use of Bayes' theorem or priors per se, but the interpretation of probability as something that quantifies a state of knowledge (or ignorance). The corollary of this viewpoint is that all analyses are understood to be conditional on the information, simplifications and assumptions that go into them. The loss of absoluteness may be uncomfortable, but it is honest.

A Bayesian analysis can seem troublesome, as even the simplest case involves awkward questions before a calculation can be performed: How is the hypothesis to be formulated? What are the alternatives? What is known a priori? How are the modelling and measurement uncertainties characterized? ... No wonder a conventional least-squares procedure, with its associated measure of goodness-of-fit, is so appealing! On reflection, however, the apparent objectivity of orthodox methods starts to look illusory and the difficult issues appear relevant. For the data of Figure 2.5(a), for example, a much higher than expected value of χ^2 warns us that something is awry; but what? Is the description of Eq. (2.23) inadequate, or is the implicit assumption of Eq. (2.25) inappropriate? Without trying specific alternatives, it is hard to tell. An acceptance that the problem might lie with the noise model raises some interesting points: (i) the likelihood function is never really 'known'; (ii) the introduction of noise uncertainties, such as those leading to Eqs. (2.26) and (2.28), makes goodness-of-fit tests impossible in general; (iii) since the hypothesis and data appear on the wrong sides of the probability conditioning symbol for an inference, the use of the likelihood cannot be justified logically without acknowledging the role of the prior.

As Steve Gull once remarked, data analysis is simply a dialogue with the data. The awkward theoretical issues that come to the fore when trying to implement the Bayesian approach are akin to us asking a question only to hear the response, 'Well, that depends on what you mean by ...', requiring qualification. Likewise, on receiving an unsatisfactory answer we might be moved to ask another question by saying, 'No, what I meant to ask was ...' It is a dynamic process, just like a conversation.

In the 1980s, a straw poll of working cosmologists would probably have revealed a strong belief in the then 'preferred' cosmological parameters: $\Omega_M = 1$, $\Omega_{DE} = 0$ and $\Omega_k = 0$. Approaching 2010, we find ourselves driven, largely by observational evidence and statistical inference, towards a radically new consensus: $\Omega_M \sim 0.3$, $\Omega_{DE} \sim 0.7$, $w \sim -1$ and $|\Omega_k| \sim 10^{-5}$. Bayesian methods of analysis will undoubtedly play a major role during the next change in beliefs, potentially aiding a paradigm shift in the way we view the Universe.

References

Alcock, C. and Paczynski B. (1979). *Nature*, **281**, 358.

Dekel, A. and Lahav, O. (1999). *Astrophys. J.*, **520**, 24.

Gregory, P. (2005). *Bayesian Logical Data Analysis for the Physical Sciences*. Cambridge: Cambridge University Press.

Hinshaw, G. *et al.* (2007). *Astrophys. J. Supp.*, **170**, 288.

Jaffe, A. J. (1996). *Astrophys. J.*, **471**, 24.

Jaynes, E. T. (2003). In G. L. Bretthorst, ed., *Probability Theory: The Logic of Science*. Cambridge: Cambridge University Press.

Knox, L. (2006). *Phys. Rev. D*, **73**, 023503.

de Laplace, P. S. (1812). *Théorie Analytique des Probabilités*. Paris: Courcier Imprimeur.

de Laplace, P. S. (1814). *Essai Philosophique sur les Probabilités*. Paris: Courcier Imprimeur.

MacKay, D. J. C. (2003). *Information Theory, Inference, and Learning Algorithms*. Cambridge: Cambridge University Press.

Ostriker, J. P. and Steinhardt, P. J. (1995). *Nature*, **377**, 600.

Peacock, J. A. (1999). *Cosmological Physics*. Cambridge: Cambridge University Press.

Peebles, P. J. E. and Ratra, B. (2003). *Rev. Mod. Phys.*, **75**, 559.

Percival, W. J. *et al.* (2007). *Mon. Not. Roy. Astron. Soc.*, **381**, 1053.

Perlmutter, S. *et al.* (1999). *Astrophys. J.*, **517**, 565.

Sivia, D. S. and Skilling, J. (2006). *Data Analysis: A Bayesian Tutorial*. Oxford: Oxford University Press.

Sivia, D. S. (2006). In P. Ciarlini *et al.*, eds., *Advanced Mathematical and Computational Tools in Metrology VII*. Singapore: World Scientific Press, p. 108.

Skilling, J. (2010). Chapter 1 in this volume.

3

Parameter estimation using Monte Carlo sampling

Antony Lewis and Sarah Bridle

In this chapter we assume that you have a method for calculating the likelihood of some data from a parameterized model. Using some prior on the parameters, Bayes' theorem then gives the probability of the parameters given the data and model. A common goal in cosmology is then to find estimates of the parameters and their error bars. This is relatively simple when the number of parameters is small, but when there are more than about five parameters it is often useful to use a sampling method. Therefore in this chapter we focus mainly on finding parameter uncertainties using Monte Carlo methods.

3.1 Why do sampling?

We suppose that you have (i) some data, d, (ii) a model, M, (iii) a set θ of unknown parameters of the model, and (iv) a method for calculating the probability of these parameters from the data $P(\theta|d, M)$. For convenience we mostly shall leave the dependence on d and M implicit, and thus write $P(\theta) = P(\theta|d, M)$.

An example we will consider throughout this chapter is the estimation of cosmological parameters from cosmic microwave background (CMB) data. For example, you could consider that (i) the data is the CMB power spectrum, (ii) the cosmological model is a Big Bang inflation model with cold dark matter and a cosmological constant, and (iii) the unknown parameters are cosmological parameters such as the matter density and expansion rate of the Universe. We discuss methods for obtaining the probability of the parameters from this data in Section 3.2.

From the posterior probability distribution we can answer statistical questions about functions of the parameters, $P(f(\theta)|P(\theta))$. For example, the expected value of a function of parameters $f(\theta)$ is given exactly by

$$\langle f(\theta) \rangle = \int d\theta \, f(\theta) P(\theta). \tag{3.1}$$

In the simplest case, we could use this to find the expectation value of a cosmological parameter $f(\theta) = \theta = \Omega_\mathrm{m}$, conditional on all the other cosmological parameters being known; this would require calculating the one-dimensional probability distribution $P(\Omega_\mathrm{m})$ as a function of Ω_m and performing the one-dimensional integration above. In the more realistic case where we do not know the values of the other cosmological parameters, then the posterior must be evaluated in the full multidimensional parameter space to perform the above integral for the expectation value of Ω_m. This full posterior probability distribution could also be used to obtain the expectation value of other parameters which are not included in the original parameter set, such as $\Omega_\mathrm{m} h$, by carrying out the above integral.

In principle, this integral can be calculated numerically to machine precision (or to the precision with which we can calculate $P(\theta)$). However, if there are n parameters then the posterior is an n-dimensional scalar-valued function. The simplest computational method would sum over an evenly spaced grid in parameter space. This grid needs to cover a large enough volume of parameter space to capture all the regions in which $f(\theta)P(\theta)$ is significantly non-zero, with some width w_i in each direction. The grid resolution will depend on the smoothness of the function and probability distribution, with some resolution Δ_i required in each dimension. Therefore the number of grid points is $\prod_1^n (w_i/\Delta_i) \sim (w_1/\Delta_1)^n$, where the latter applies if all dimensions contain a similar amount of structure. Note the exponential scaling with the number of dimensions, which makes direct integration numerically prohibitive in large dimensions.

Instead we can try to radically compress $P(\theta)$ into a small manageable collection of numbers by sampling. The probability of taking a sample from a given position in parameter space θ is proportional to the probability $P(\theta)$ at that position in parameter space. A set of n_s samples $\{\theta_i\}$ specifies n_s positions in parameter space and therefore consists of $n \times n_\mathrm{s}$ numbers. The number density of samples should then be proportional to the probability distribution itself.

From the set of samples $\{\theta_i\}$ we can then try to infer properties of the full distribution. So instead of directly calculating $P(f(\theta)|P(\theta))$ we now instead calculate $P(f(\theta)|\{\theta_i\})$. For general properties of $f(\theta)$ this is difficult; however, if we are primarily interested in a set of expectation values the central limit theorem comes to the rescue. As long as the distribution of $f(\theta_i)$ has finite variance, the central limit theorem states that the distribution over different sets of samples of any sum of numbers $\sum_i f(\theta_i)$ will tend to a normal distribution for large numbers of independent samples. For example, from $\{\theta_i\}$ we can estimate $\langle f(\theta) \rangle$ using

$$\hat{E}_f \equiv \frac{1}{n_\mathrm{s}} \sum_{i=1}^{n_\mathrm{s}} f(\theta_i). \tag{3.2}$$

The expected value of the $\hat{E}_f$ estimator is just the true expectation value, $\langle \hat{E}_f \rangle = \langle f(\theta) \rangle$. For large numbers of samples the central limit theorem tells us that $P(\hat{E}_f)$ tends to the normal distribution $N(\langle f(\theta) \rangle, \sigma_E^2)$, where $\sigma_E = \sigma_f / \sqrt{n_s}$ and σ_f^2 is the true variance of $f(\theta)$. This is true even if the distribution of $f(\theta)$ is not Gaussian, so we can reliably infer expectation values from samples even for general distributions.

Compressing the posterior distribution to a set of samples is very convenient when calculating low-dimensional statistical properties of the full distribution, such as the posterior expectation value of parameters or some function of them. Expectation values can be calculated very quickly from a fixed number $n_s \gg 1$ of samples independent of the dimensionality of θ. This is useful since often the time to generate a fixed number of independent samples has a low scaling with dimension. Sampling therefore potentially saves an exponentially large fraction of the time when computing expectation values compared to integrating the posterior over the full n-dimensional parameter space.

3.2 How do I get the samples?

Compressing the posterior into samples is a useful trick. But given some distribution $P(\theta)$, how do we actually generate the samples?

3.2.1 Direct sampling methods

We first describe two sampling methods that work well in low-dimensional parameter spaces: 'inverse transform sampling' and 'rejection sampling'. Both methods are 'direct', in the sense that they generate exactly independent samples, rather than more complicated methods that often draw independent samples only asymptotically.

'Inverse transform sampling' is generally applicable for one-dimensional probability distributions $P(\theta)$ and is sometimes more efficient than rejection sampling, because all computed samples are used. We consider the simplest case, which uses the fact that it is easy to sample uniformly from values of the cumulative distribution $Q(\theta)$, a monotonic function ranging from zero to one,

$$Q(\theta) = \int_{-\infty}^{\theta} \mathrm{d}\theta' P(\theta').$$ (3.3)

Given a sample x_i from the uniform distribution between zero and one, we can convert it into a sample θ_i from $P(\theta)$ by finding θ_i for which $Q(\theta_i) = x_i$.

This method may be inefficient, especially if $Q(\theta)$ or its inverse cannot be computed analytically. As a simple example that works well, consider sampling from the normalized positive exponential distribution $P(\theta) = \mathrm{e}^{-\theta/2}/2$ for $0 \leq \theta < \infty$, which has a cumulative distribution $Q(\theta) = 1 - \mathrm{e}^{-\theta/2}$: generate uniform samples x_i (with $0 \leq x_i < 1$) and then compute the sample parameter value $\theta_i = -2\ln(1 - x_i)$.

'Rejection sampling' is sometimes more efficient if a suitable 'umbrella' or 'envelope' distribution $U(\theta)$ can be found that is easy to sample from, where $U(\theta) \geq P(\theta)$ for all points in parameter space θ (where U can be scaled so that this is true and hence is not normalized). Samples from U will occur with a density that differs from P by a factor $\propto U(\theta)/P(\theta)$. Rejection sampling works by generating a sample θ from $U(\theta)$, and keeping it with probability $P(\theta)/U(\theta)$ or rejecting it otherwise. Non-rejected samples will then be independent samples from $P(\theta)$.

A simple example is sampling uniformly within the unit circle. Given a standard random number generator for the interval $[-1, +1]$, we can easily generate a point (x, y) that lies within an enclosing square, $-1 \leq x \leq 1$, $-1 \leq y \leq 1$ (the umbrella distribution). To generate a sample from within the unit circle we then simply keep the point if $r^2 = x^2 + y^2 < 1$, otherwise we reject it and try generating another point. The method will generate a valid point a fraction $\pi/4$ of the time, so rejection sampling is a fairly efficient sampling method in this case.

Alternatively, we may wish to sample from a probability distribution $P(\theta)$ which is a function of a single parameter θ, and normalized to have a maximum value of unity, for example $P(\Omega_\mathrm{m}) = \exp(-0.5(\Omega_\mathrm{m} - 0.258)^2/0.03^2)$. It is helpful to define a range of our parameter, e.g., $0 < \Omega_\mathrm{m} < 0.5$. We could then define the umbrella distribution U to be unity within the allowed parameter range and zero outside. Samples from U are obtained by drawing a trial sample θ_t, which is simply a uniform random number from the parameter range. We now keep the trial sample at a rate proportional to $P(\theta_\mathrm{t})$. This can be achieved by generating another random number for each trial sample, which is uniform in the range zero to unity. The trial sample is kept if the random number is less than $P(\theta_\mathrm{t})$; otherwise it is removed from the final list.

An efficient way to sample from a Gaussian distribution is to use a combination of rejection sampling and inverse transform sampling, the Box–Muller algorithm. The idea is to draw a sample uniformly from within the unit circle (x_{ci}, y_{ci}) and then scale it to produce a sample from a 2-D Gaussian distribution (x_{Gi}, y_{Gi}). Samples from the unit circle have a uniform distribution in polar angle ϕ_{ci}, but a probability distribution in the radial r_{ci} direction $P(r_{ci}) \propto r_{ci}$. Therefore $R_{ci} \equiv r_{ci}^2$ is uniformly distributed between zero and one, following the rules for mapping probability distributions when changing variables. A point from a 2-D Gaussian distribution is also uniformly distributed in angle ϕ_{Gi} but has $P(r_{Gi}, \phi_{Gi}) \propto r_{Gi}\mathrm{e}^{-r_{Gi}^2/2}$, and therefore $P(R_{Gi}) \propto \mathrm{e}^{-R_{Gi}/2}$. A sample from the unit circle can be transformed into a sample from a 2-D Gaussian using the inverse transform sampling result we derived earlier, $R_{Gi} = -2\ln(R_{ci})$ (where $R_{ci} = 1 - x_i$ since $0 < R_{ci} \leq 1$ and $0 \leq x_i < 1$). The new components are thus $(x_{Gi}, y_{Gi}) = (x_{ci}, y_{ci})\sqrt{-2\ln(r_{ci}^2)}/r_{ci}$. Each component is an independent sample from a 1-D unit Gaussian distribution, giving two independent Gaussian random numbers of unit variance.

Samples from Gaussian distributions are often useful in cosmological applications, and a general n-dimensional distribution can easily be sampled from, if its covariance is known, by using the Box–Muller algorithm to sample in each independent eigenvector. More efficient methods for generating Gaussian variates, such as the Ziggurat algorithm,[1] are also based on rejection sampling. These methods essentially cover the distribution in a series of rectangular boxes (that are easy to sample from, and edges can be computed analytically), so that the probability of rejection is very low. For a comparison of methods for sampling from a Gaussian distribution, see Thomas *et al.* (2007).

3.2.2 Problems with large dimensions

Rejection sampling and inverse transform sampling often work well in low dimensions. However, as the number of dimensions increases we have to be careful. As an example consider trying to sample uniformly from within a unit n-sphere, generalizing the method described before for sampling uniformly within a unit circle. Rejection sampling will certainly work in principle: we just generate n uniform numbers $x_i \in [-1, +1]$ and test whether $r^2 = \sum_{i=1}^{n} x_i^2 < 1$. The problem comes when we calculate the acceptance probability, given by the ratio of the volume of a unit n-sphere to the volume of an n-cube of size 2. The cube has volume 2^n, and the volume of the sphere is

$$V_n = \int d\Omega_n \int_0^\infty dr\, r^{n-1} = \frac{1}{n}\int d\Omega_n. \tag{3.4}$$

The surface area of a unit n-sphere $\int d\Omega_n$ can be calculated readily from known properties of Gaussian distributions:

$$\int d^n\theta \frac{e^{-\theta^2/2}}{(2\pi)^{n/2}} = (2\pi)^{-n/2}\int d\Omega_n \int_0^\infty dr\, r^{n-1}e^{-r^2/2} = 1 \tag{3.5}$$

$$\implies \int d\Omega_n = \frac{2\pi^{n/2}}{\Gamma(n/2)}, \tag{3.6}$$

so that

$$V_n = \frac{2\pi^{n/2}}{n\Gamma(n/2)} \implies \frac{V_n}{2^n} = \left(\frac{\sqrt{\pi}}{2}\right)^n \frac{1}{\Gamma(1+n/2)}. \tag{3.7}$$

[1] See, e.g., http://en.wikipedia.org/wiki/Ziggurat_algorithm.

As n becomes large this ratio becomes exponentially small: for $n = 2$ the acceptance probability is fine at $\pi/4$, but for $n = 10$ it is only ~ 0.025 and for $n = 100$ it is a prohibitive $\sim 10^{-70}$. Clearly rejection sampling from an n-cube is not a good way to sample uniformly from a unit n-sphere when n is large. In this case we can use the symmetries of the distribution to find a better method. For example, we could generate a vector θ in a random direction by generating n Gaussian random numbers, generate a unit vector $\theta/|\theta|$, then re-scale so that $P(r) \propto r^{n-1}$ for $r \le 0 < 1$. This alternative method only scales linearly with the number of dimensions, but it is only possible because of the spherical symmetry of the target distribution.

For a general distribution it will not normally be possible to find a fast direct sampling method. Rejection sampling in high dimensions either requires an umbrella distribution that is very close to the true distribution, or it requires an exponentially large number of tries, and is therefore not usually possible. The problem is that in high dimensions most of the probability tends to lie in an exponentially tiny fraction of the space; if the umbrella distribution is a factor of w broader than the target in all dimensions, the relevant volume ratio that determines the acceptance probability is $\sim 1/w^n$. For this reason, sampling methods for high-dimensional problems usually use some kind of local sampling method, setting up a random walk through parameter space so that only regions with fairly high probability are explored.

3.2.3 *Markov chain sampling*

The most common methods for sampling from general distributions in high dimensions are based on Markov chains. The idea is to construct a rule for choosing a sequence of points in parameter space such that after a long time the probability of the current position being θ_i is proportional to $P(\theta_i)$. If the rule for moving from θ_i to θ_{i+1} depends only on θ_i (and not on the previous history), then the sequence of points is called a Markov chain. Some of these points may later be considered 'samples' if certain conditions are met (see below).

We can define a transition probability $T(\theta_i, \theta_{i+1})$ that determines the probability of the chain moving to θ_{i+1} given that it is currently at position θ_i. In order for the chain to have the correct asymptotic distribution, the probability of arriving at a given point must be equal to the probability at that point,

$$P(\theta_j) = \int \mathrm{d}\theta_i P(\theta_i) T(\theta_i, \theta_j). \tag{3.8}$$

It can be seen by substitution that one way to ensure this is to use a (normalized) transition probability satisfying detailed balance

$$P(\theta_{i+1}) T(\theta_{i+1}, \theta_i) = P(\theta_i) T(\theta_i, \theta_{i+1}), \tag{3.9}$$

so that 'the probability of being here and going there is the same as the probability of being there and coming here'. There are numerous ways to choose T so that detailed balance holds, so we shall discuss only some of the most popular. The only general constraint on T is that every point in the parameter space must be accessible within a finite number of transitions. Excellent references available online include MacKay (2003) and Neal (1993), and there are also several books: Gilks, Richardson and Spiegelhalter (1996), Liu (2001), Gelman *et al.* (2003), and Gamerman and Lopes (2006).

It important to emphasize that although asymptotically the position at any given time follows the correct distribution, $P(\theta_{i+1})$ and $P(\theta_i)$ are certainly not independent. Running a chain until it equilibrates and then stopping generally gives *one* independent sample from the distribution. To generate more than one independent sample it would then be necessary to wait for the chain to move about parameter space until it has lost all memory of the previous independent sample. The efficiency of the method depends on how long this wait is; i.e., how efficiently the chain moves about parameter space.

3.2.4 Metropolis–Hastings algorithm

The Metropolis–Hastings algorithm (Metropolis *et al.* 1953; Hastings 1970) is one of the simplest and most popular ways to set up a Markov chain satisfying detailed balance. It uses an arbitrary *proposal density* distribution $q(\theta_i, \theta_{i+1})$ to propose a new point θ_{i+1} given the chain is currently at θ_i. We can choose a proposal density that is easy to sample from. The proposed new point is then accepted with probability

$$\alpha(\theta_i, \theta_{i+1}) = \min\left\{1, \frac{P(\theta_{i+1})q(\theta_{i+1}, \theta_i)}{P(\theta_i)q(\theta_i, \theta_{i+1})}\right\}. \tag{3.10}$$

The total transition probability is therefore

$$T(\theta_i, \theta_{i+1}) = \alpha(\theta_i, \theta_{i+1})q(\theta_i, \theta_{i+1}). \tag{3.11}$$

It is straightforward to show by substitution that this satisfies the detailed balance condition. Note that when the proposal is rejected, $\theta_{i+1} = \theta_i$, so there are two (or more) samples at exactly the same position. Equivalently these can be combined into a 'weighted sample' with weight two (see the later section on importance sampling). Of course these samples are not independent.

Often it is convenient to choose a symmetric proposal density, $q(\theta_{i+1}, \theta_i) = q(\theta_i, \theta_{i+1})$, in which case the proposal density cancels out of the acceptance probability. This is the original Metropolis algorithm, which says 'move to the new position if the probability there is higher than at the current position, otherwise

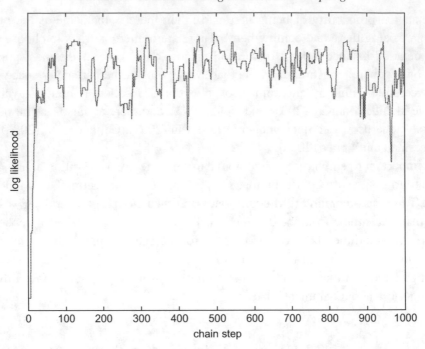

Fig. 3.1. The log-likelihood value of the first 1000 points in a Markov chain sampling a typical CMB posterior. The likelihood grows rapidly initially until the chain equilibrates around good parameter values. Horizontal lines correspond to more than one sample at exactly the same parameter values following rejection of a proposed move.

move to the new position with a probability equal to the ratio of the probabilities at the new and current positions'. Following convention, we will refer to Metropolis and Metropolis–Hastings as Metropolis methods hereafter. The likelihood values at the start of a typical chain run are shown in Figure 3.1.

Note that it is important that the proposal distribution q is only a function of the current position, otherwise the chain is not Markov; in particular, q cannot be chosen based on details of the past history of the chain. However, during an initial 'burn-in' phase where samples are not kept, it is not necessary for the chain to be Markov as long as it is Markov when samples are being collected. It is therefore often useful to use the early movements of the chain to refine the proposal density. After this burn-in phase the proposal density must be fixed, or it must at least only be updated in such a way that it asymptotically converges to a fixed distribution.

At best, the time to produce a fixed number of independent samples using Metropolis sampling methods can scale *linearly* with the number of dimensions. For example, to sample from an n-dimensional Gaussian $P(\theta) \propto e^{-\theta^2/2}$ we could propose a change to each component of θ using a Gaussian of width $1/\sqrt{n}$. The acceptance probability is then $\mathcal{O}(1)$ (in fact, a width of $\sim 2.4/\sqrt{n}$ is optimal in this particular case, with acceptance rate ~ 0.234 (Gelman, Roberts & Gilks 1996; Roberts, Gelman & Gilks 1997)). In each component, a random walk with step-size $1/\sqrt{n}$ will take $\sim n$ steps to cover the unit width of the distribution. So we can expect to generate an independent sample about every $\mathcal{O}(n)$ steps of the chain. If this is the case, Metropolis sampling to compress the posterior into a set of samples can be very efficient for moderate numbers of dimensions. In realistic situations the scaling will be worse, both because a priori we don't know the best proposal distribution, and because the distribution may have a more complicated shape.

The choice of proposal density can have a large effect on how the algorithm performs in practice. In general, it is best to have a proposal density that is of similar shape to the posterior, since this ensures that large changes are proposed to parameters along the degeneracy directions. Fortunately, with cosmological data, we have a reasonable idea of what the posterior might look like, and so choosing a sensible proposal density is often not difficult. However, it is a limitation of the method that it is not at all robust to bad choices of proposal density: if the distribution is too narrow it takes a long time to random walk around the space; if it is too broad then the acceptance probability becomes very low. For the method to perform well, the acceptance probability should be around unity. In many cases a good choice of proposal density gives an acceptance rate in the range $0.2-0.3$ (Gelman, Roberts & Gilks 1996; Roberts, Gelman & Gilks 1997; Dunkley *et al.* 2005).

One common way to choose the proposal density is to make a preliminary short chain with some simple proposal, calculate the covariance matrix from the samples generated, and then use the (possibly scaled) covariance matrix for a Gaussian proposal distribution in a new chain. This procedure can be iterated until the sampling seems to be working efficiently (see Section 3.3). Samples from the preliminary runs are discarded. Alternatively, when new data become available, there are often old data that can be used to generate quite a good initial proposal covariance. In addition, there is considerable freedom in how the covariance is used; for example, using a Gaussian proposal may not be a good idea if there is a danger that the covariance is inaccurate. Also proposing changes in all directions at once (e.g., using an n-D Gaussian) may not be as good as making a series of proposals in lower dimensions. For further discussion of these issues see, e.g., Lewis and Bridle (unpublished).

As mentioned above, the method works poorly if the proposal density does not allow the region of high likelihood to be explored efficiently. In particular, if two parameters are strongly correlated, proposals which vary a single parameter may almost always be rejected, even though a large change to some combination of the parameters would be allowed by exploration along the degeneracy direction.

Using the covariance matrix for a Gaussian proposal distribution allows linear degeneracies to be explored efficiently. However, if the distribution is banana-shaped (or worse), exploration can still be very slow. For this reason, when tight degeneracies are known in advance, it is a very good idea to change variables so that the parameters are as independent as possible.

In the context of cosmology this is a common situation; for example, we know that the CMB data constrains the angular diameter distance to the last scattering surface very well, but that various non-linear combinations of curvature, dark energy and Hubble parameters can all give the same distance. Posterior constraints on these parameters therefore tend to be tightly correlated. By changing variables to the well-constrained combination (i.e., the angular diameter distance) and other combinations, a much more weakly correlated set of parameters can be obtained, greatly improving the convergence properties of the Markov chain (Kosowsky, Milosavljevic & Jimenez 2002). Note that when changing variables the prior should also be modified by the corresponding Jacobian, though in practice people often simply re-define their priors to be flat in the decorrelated variables.

3.2.5 *Other sampling methods*

One distinct disadvantage of the Metropolis method is that the efficiency depends strongly on the choice of the proposal density. A poor choice can mean that the algorithm is hopelessly slow. If we don't know very much about the target distribution, it might therefore be nice to have a sampling method that is more robust, being able to explore the full width of the distribution efficiently, even if an initial guess is very wrong. One general method is called slice sampling (Neal 2003); this uses a 'stepping-out' and 'stepping in' process to automatically adjust the proposal density width at each step. There is however some cost to doing this, and for this reason the method has so far been little used in cosmological parameter estimation, where often we have a very good idea about the distribution and Metropolis methods can be made to work efficiently.

One particularly simple choice for the Metropolis–Hastings proposal density is to use the conditional distributions, assuming they can easily be sampled from; e.g., propose changes to x by drawing from $P(x|y)$ and propose changes to y by drawing from $P(y|x)$. By substituting into the acceptance probability we see that these proposals are always accepted. This sampling method that alternately samples from the conditional distributions is called Gibbs sampling (Geman & Geman 1984). It can be very efficient if x and y are not too correlated and work even in very high dimensions. However, if the parameters are strongly correlated it can take many Gibbs steps to random walk in a diagonal direction by taking many steps parallel to the axes.

One of the main problems of the Metropolis method is that the exploration of highly correlated directions can be quadratically slow due to the random walk behaviour. If possible, it would be much better to use a partly directed walk, so that the chain could traverse the distribution much more quickly. One such method is called Hamiltonian sampling, or the hybrid Monte Carlo algorithm. This method is useful when derivatives of the likelihood can be calculated easily, and works in analogy to classical Newtonian mechanics by giving the chain position a 'momentum' vector, so that the current state has some 'memory' of the direction that it came from encoded in this momentum. For a detailed discussion of the method and scaling with dimension see, e.g., Neal (1993). The method can be applied both when the derivatives can be calculated analytically, and when the derivatives can be calculated approximately; see, e.g., Hajian (2007), and Taylor, Ashdown and Hobson (2007) for potential applications in the cosmological context.

3.2.6 Thermodynamic and flat-histogram methods

As discussed in this book's opening chapter, many sampling methods are primarily solutions to the question of how to calculate the probability marginalized over all parameters within a model (evidence). This is particularly true in solid state physics where the equivalent quantity is the free energy of the system, a quantity of physical importance. There is a substantial literature giving sophisticated sampling methods to calculate the free energy in a variety of different problems. Chapter 1 briefly described nested sampling and thermodynamic integration. Another class of methods are the 'flat-histogram' methods (e.g., multi-canonical (Berg 2000; Gubernatis & Hatano 2000) and Wang–Landau (Wang & Landau 2001a & 2001b) sampling), which aim to generate approximately equal numbers of samples for each log-range in likelihood (in the solid state context, equal numbers of samples at each energy). Like nested sampling and thermodynamic integration, these methods generate samples far into the tails of the distribution and probe a much wider range of likelihood values than direct Metropolis sampling (which generates samples only where the likelihood is high). They are therefore potentially of use both for parameter estimation, and answering questions about the tails of the distribution that may not be easy to answer accurately from a set of samples directly from the posterior, such as high-significance confidence limits.

If the target distribution is multi-modal, sampling methods that probe more of the distribution are much more likely to find the multiple modes than purely local methods making a random walk in the posterior. When asking questions about the extreme tails of the distribution, or sampling complicated multi-modal or awkwardly shaped distributions, using samples from an evidence-estimation method can therefore be a good idea, even if the evidence value is not actually required.

A detailed discussion is beyond the scope of this chapter, but most of the methods can produce weighted samples from which parameters of interest can be calculated straightforwardly (see the importance sampling section below and Chapter 1).

3.2.7 Baby and toy

As a final example of a clever sampling algorithm, we discuss briefly one method for sampling even when the likelihood cannot be calculated easily at a single point. We assume the likelihood distribution can be factored as $L(\theta|\mathbf{d}) = f(\mathbf{d},\theta)/Z(\theta)$ into a part $f(\mathbf{d},\theta)$ that depends on the data $\mathbf{d}$ and parameters θ which is easy to calculate, and some normalizing factor $Z(\theta)$ that is very hard to compute. In addition, it must be possible to simulate fake data $\mathbf{y}$ from L; this has a probability $L(\theta|\mathbf{y})$. There is also a prior $P_\mathrm{r}(\theta)$ that must be applied to calculate the posterior probability from which we wish to sample. For example, the likelihood may be calculable to within a normalizing constant from a predicted mean vector $\mathbf{d}_m(\theta)$ and a covariance matrix C as $f(\mathbf{d},\theta) = \exp(-(\mathbf{d}-\mathbf{d}_m(\theta))^\mathrm{T} C^{-1}(\mathbf{d}-\mathbf{d}_m(\theta))/2)$ and $Z(\theta)$ is a normalization that depends on hard-to-compute determinants.

The baby and toy algorithm (Murray, Ghahramani & MacKay 2006) may be executed as follows. Generate some typical fake data $\mathbf{y}$ from the model at θ_{i+1} and calculate $f(\theta_i,\mathbf{y})/f(\theta_{i+1},\mathbf{y})$, which is a one-shot importance sampling estimator for the hard-to-compute $Z(\theta_i)/Z(\theta_{i+1})$ ratio (see Eq. (3.16) below). Then ask: Does the real data prefer θ_{i+1} relative to θ_i more than the typical fake data prefers θ_{i+1} relative to θ_i? If it does, i.e., $f(\mathbf{d},\theta_{i+1})/f(\mathbf{d},\theta_i) > f(\mathbf{y},\theta_{i+1})/f(\mathbf{y},\theta_i)$, then θ_{i+1} is likely to be preferred and the move is looking promising, pending taking into account priors and transition ratios. We therefore use a Metropolis–Hastings method, where the acceptance ratio is

$$\alpha(\mathbf{y},\theta_i,\theta_{i+1}) = \min\left\{1, \frac{f(\mathbf{d},\theta_{i+1})P_\mathrm{r}(\theta_{i+1})q(\theta_{i+1},\theta_i)}{f(\mathbf{d},\theta_i)P_\mathrm{r}(\theta_i)q(\theta_i,\theta_{i+1})} \frac{f(\mathbf{y},\theta_i)}{f(\mathbf{y},\theta_{i+1})}\right\}. \quad (3.12)$$

The total transition probability is then

$$T(\theta_i,\theta_{i+1}) = \int \mathrm{d}\mathbf{y}\, \frac{f(\mathbf{y},\theta_{i+1})}{Z(\theta_{i+1})}\alpha(\mathbf{y},\theta_i,\theta_{i+1})q(\theta_i,\theta_{i+1}). \quad (3.13)$$

We can easily check that detailed balance is satisfied for the target distribution $P_\mathrm{r}(\theta)f(\mathbf{d},\theta)/Z(\theta)$, so the sampling method is valid. The sampling method is sometimes called 'baby and toy' or the 'single variable exchange algorithm'. Generalizations of the algorithm could use a more refined estimator of the $Z(\theta_i)/Z(\theta_{i+1})$ ratio.

3.3 Have I taken enough samples yet?

A general problem with Markov chain based sampling methods is that the points in the chain are correlated. Only asymptotically is one position of the chain guaranteed to be an independent sample from the target distribution. For example, near the beginning of a chain there will generally be a steady progression of the chain likelihoods towards higher values as the chain walks towards the high-likelihood region. How do we know when it has got there and, once there, how long does it take to generate another independent sample?

Unfortunately there is no general sufficient test for whether a set of samples are independent. In particular, if the distribution is multi-modal it is quite possible for the chain to walk directly towards one mode and then appear to be producing independent samples, when in fact the other modes are not being explored at all (i.e., we have *no* independent samples). Having said that, it is quite easy to write down a string of tests that are (on average over realizations) necessary for the samples to be independent. We just outline a few here, and you should remember that they are only necessary but not sufficient; for this reason it is often a good idea to look at more than one test.

Often the easiest way to tell if a chain is producing samples from the correct distribution is to produce multiple chains and compare results. Any difference between chains gives a measure of the statistical error in the quantity being estimated. If we are interested in some function $f(\theta)$, we can estimate f from each chain and calculate its variance. For the estimates of f from a single chain to be reliable we want the rms between the chains to be much smaller than the posterior uncertainty on f.

The 'Gelman and Rubin' ratio statistic $R = $ (variance between chains)/(mean variance within the chains) is often quoted as a measure of convergence, where we want $R - 1 \sim 0$ for the Monte Carlo estimators potentially to be reliable (Gelman & Rubin 1992; Gelman, Roberts & Gilks 1996). R is most often quoted for the means of the parameters, but it may also be used for any quantity of interest (for example, higher confidence limits). If there are many variables, the between-chain covariance can be diagonalized and R quoted for the worst eigenvector, allowing the statistic to show up poor convergence in correlated directions. Since chains will take a while to 'burn in' as they converge on the high-likelihood region, the Gelman–Rubin statistic is usually quoted for the last halves of the chains. If $R - 1 \sim 0$, quantities estimated from the samples might then be reliable if the first half of each chain is discarded.

Convergence can also be assessed from a single chain. One way is simply to split the chain up and then apply a Gelman and Rubin test to each chunk. We can also calculate the correlations between chain parameters as a function of distance along the chain. This allows an estimate of the correlation length, the distance r_c at which the correlation falls to $1/e$. A high value for the correlation length for one parameter indicates that the chain is not exploring that direction efficiently. The correlation length can also be used to give an upper limit to the number of independent samples we might have, n_{step}/r_c. Clearly, for reliable results we need $n_{step} \gg r_c$ for all the parameters.

A similar idea is used by the Raftery and Lewis (1992) convergence test for percentile limits. This calculates roughly how much the chain must be thinned in order for occurrences of parameter values above a given percentile to have a Markovian or independent distribution. The method gives an estimate of the distance between independent samples, similar to the correlation length.

3.4 What do I do with the samples?

We now discuss how to calculate some quantities of interest from your samples, and how to combine your original dataset with other information without having to resample. We continue with the example of parameter estimation from the WMAP CMB data. You can download sets of samples for various cosmological models yourself.[2]

3.4.1 Parameter constraints

The samples give you a full multidimensional map of your probability distribution, since the number density of samples is proportional to the probability. However, it is difficult to display this in more than three dimensions. The expectation values of parameters can be calculated easily, as discussed in Section 3.1. Equation (3.2) shows that the desired quantity can be calculated for each sample, and the average taken. In the simplest case, we are interested in the expectation value of one of our parameters in θ. In this case we just take the mean of the sample positions for that parameter, and can ignore all the other parameters.

Traditional probability analyses often start by finding the best-fit point in parameter space. However such high-dimensional statistical properties cannot be calculated accurately from the samples.[3] If the probability value has been stored with each sample position, then the sample with the highest probability can be identified. There may be very few samples close to the best-fit point, and these will be scattered along the directions of degeneracy. Therefore, due to the random

---

[2] http://lambda.gsfc.nasa.gov/product/map/current/parameters.cfm.
[3] The maximum of an n-dimensional unit Gaussian is at the origin, but the region $r < \epsilon \ll 1$ only contains a fraction $\propto \epsilon^n$ of the probability – in high dimensions there will almost certainly be no samples within $r < 1$ of the origin, where r is the radius from the origin.

nature of sampling, the sample with the highest probability will move around from sampling realization to realization.

The original probability density can be estimated from the samples to within a normalization constant because the number density of samples is proportional to the probability. For example, $f(\theta)$ could be one in some area of interest around θ_p and zero otherwise; $\hat{E}_f$ then has expectation proportional to $P(\theta_p)$. Note that if using just the number density of samples, the normalization has been lost, so we can only easily estimate *ratios* of probability densities at different points. Also, the estimate will only be accurate to within $\mathcal{O}(1/\sqrt{n_s})$ of the posterior range of P. If $P(\theta_p)$ is small relative to the peak, then a very large number of samples would be required to estimate it to within a small fractional error. Nonetheless, the probability density can be estimated easily where the probability is large, and density plots of samples are a useful way to show the shape of the distribution in the most likely region.

It is also difficult to calculate conditional probability distributions from samples. For example, we may wish to calculate the probability of the Hubble constant from WMAP, conditional on all other cosmological parameters being fixed, e.g., with $\Omega_m = 0.3$. We would therefore need to keep only the samples for which $\Omega_m = 0.3$. However, there will most likely be no samples with $\Omega_m = 0.300000$ precisely. Therefore we could instead decide to use a narrow prior $0.29 < \Omega_m < 0.31$ to achieve a similar effect. Again there may be few samples in this range, so the answer will be correspondingly noisy.

Fortunately it is extremely easy to calculate marginalized probability distributions. For example, a histogram of sample values for a single parameter is proportional to the probability distribution marginalized over all other parameters that were sampled. Similarly, a plot of sample positions for two parameters shows the probability density marginalized over all other parameters. The points in the left-hand panel of Figure 3.2 show this for the WMAP ΛCDM samples using the natural parameters for the CMB, in which the contours are relatively uncorrelated.

We can estimate the underlying probability distribution from the sample positions, and thus draw lines enclosing 68 and 95 per cent of the probability. The simplest method is to bin the samples on a grid (2-D histogram). This will be noisy due to the finite number of samples, therefore, if the probability distribution is believed to be slowly varying, then the histogram could be smoothed to produce a more realistic set of contours. This is shown by the solid lines in Figure 3.2.

Probability distributions of new parameters can often be calculated easily. The right-hand panel of Figure 3.2 shows the values of the more conventional cosmological parameters H_0 and Ω_m. For each sample, the values of H_0 and Ω_m have been calculated from the natural CMB parameters. Since these parameters are more

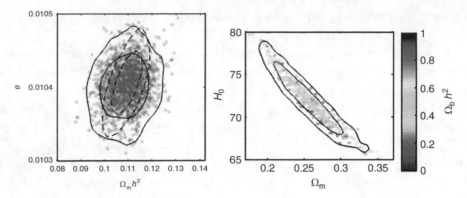

Fig. 3.2. Samples generated from the WMAP 5-year data. Samples are plotted corresponding to two of the parameters, and the solid lines enclose approximately 68% and 95% of the samples (and hence probability). The dashed lines show the probability after importance sampling using a prior of $H_0 = 72 \pm 1\,\mathrm{km\,s^{-1}\,Mpc^{-1}}$. The left-hand plot shows the natural CMB parameters in which the samples were taken. The right-hand plot shows the transformation to more conventional parameters, which are more correlated and therefore not good parameters to use when sampling.

correlated it would be more difficult to sample along the degeneracy directions in parameter space using these. It is also possible to illustrate a third dimension, for example, by colour coding the samples according to the value of a third parameter. Note that if a flat prior was assumed on the natural CMB parameters when calculating the samples, then this is still present. It would be necessary to apply a correction to replace this with, e.g., a flat prior on H_0 and Ω_m, as discussed below.

3.4.2 Importance sampling

We have already discussed how generating a set of samples from a distribution $P(\theta)$ can be an efficient way to represent the information in the distribution. From these samples expectations of functions of the parameters can be calculated easily. But what happens if we want to test how the results change when we change $P(\theta)$? For example, we might want to change our choice of prior slightly, update the distribution due to the arrival of more data, or perhaps test robustness by removing parts of the data.

If the new distribution $P'(\theta)$ is not too dissimilar to $P(\theta)$ we can 'importance sample' the original samples from $P(\theta)$ without having to produce a new set of samples from $P'(\theta)$. The idea is essentially the same as rejection sampling, except that instead of actually rejecting samples we weight them in proportion to the probability ratios. This effectively gives a collection of non-integer weighted samples for computing Monte Carlo estimates. For example, the expected value of a function $f(\theta)$ under $P'(\theta)$ is given by

$$\langle f(\theta) \rangle_{P'} = \int \mathrm{d}\theta\, P'(\theta) f(\theta) = \int \mathrm{d}\theta \frac{P'(\theta)}{P(\theta)} P(\theta) f(\theta)$$

$$= \left\langle \frac{P'(\theta)}{P(\theta)} f(\theta) \right\rangle_P . \tag{3.14}$$

Given a set $\{\theta_i\}$ of n_{s} samples from $P(\theta)$ a Monte Carlo estimate is therefore

$$\langle f(\theta) \rangle_{P'} \approx \frac{1}{n_{\mathrm{s}}} \sum_{i=1}^{n_{\mathrm{s}}} \frac{P'(\theta_i)}{P(\theta_i)} f(\theta_i) = \frac{1}{n_{\mathrm{s}}} \sum_{i=1}^{n_{\mathrm{s}}} w_i f(\theta_i). \tag{3.15}$$

For this to work it is essential that $P(\theta) > 0$, where $P'(\theta) \neq 0$, and in practice the weights $w_i \equiv P'(\theta)/P(\theta)$ should not be very large. The method works best when $\{w_i\} \sim$ const as we then have the same number of importance samples as we started with. If $P'(\theta)$ is zero over some fraction F of $P(\theta)$, then the number of samples will be reduced by at least a fraction F. If there is a significant mismatch between the distributions it is therefore necessary to start with many more samples from $P(\theta)$ than would be required if sampling directly from $P'(\theta)$.

If the distributions are not normalized, so that $\int \mathrm{d}\theta\, P(\theta) = Z$, the ratio of the normalizing constants can be estimated using

$$\frac{Z'}{Z} = \left\langle \frac{P(\theta)'}{P(\theta)} \right\rangle_P \approx \frac{1}{n_{\mathrm{s}}} \sum_{i=1}^{n_{\mathrm{s}}} \frac{P'(\theta_i)}{P(\theta_i)} = \frac{1}{n_{\mathrm{s}}} \sum_{i=1}^{n_{\mathrm{s}}} w_i, \tag{3.16}$$

and hence

$$\langle f(\theta) \rangle_{P'} \approx \frac{\sum_{i=1}^{n_{\mathrm{s}}} w_i f(\theta_i)}{\sum_{i=1}^{n_{\mathrm{s}}} w_i}. \tag{3.17}$$

This is the obvious generalization of the normal Monte Carlo estimator to weighted samples. The estimator for the normalization in Eq. (3.16) is precisely what you would calculate when performing Monte Carlo integration of $P'(\theta)$ (and $P(\theta)$ is chosen to be analytically integrable so that you know Z).

Importance sampling is illustrated by the dashed lines in Figure 3.2. We have taken the samples shown by the circles, and applied importance sample weights using a Gaussian centred on $H_0 = 72\,\mathrm{km\,s^{-1}\,Mpc^{-1}}$ of unit variance. Ideally, this kind of information would come from another dataset (although in practice current data yield much larger uncertainties on the Hubble constant).

Given a new likelihood function we only need to do n_{s} likelihood calculations to re-weight our samples for the new posterior distribution; importance sampling is very fast. However, like rejection sampling, it faces serious problems in high dimensions. It is only possible to importance sample between two high-dimensional distributions if the distributions are very similar in most dimensions so that the effective number of samples $\sum_i w_i / \max(\{w_i\}) \gg 1$.

Combining consistent cosmological datasets is one instance where importance sampling can be useful. Often, any one dataset constrains only one or two combinations of parameters at all well. Combining one dataset with another therefore usually only shrinks the distribution in one or two dimensions. If this is the case, importance sampling can work reliably. It can also be useful for rapidly checking how small changes in the likelihood propagate into the parameters, for instance, small changes to the observational noise model or different methods for correcting for systematics. Note, however, that if there is any inconsistency in the datasets, importance sampling can be disastrous: if $P(\theta) \sim 0$ but $P'(\theta)$ is significant, there are going to be almost no samples to importance weight and the effective number of weighted samples will be tiny.

Note that unlike direct sampling, where the likelihood values are not actually used to construct the estimators of the expectation values, to use importance sampling the likelihood values are used. So before importance sampling the compression is into a set of parameter values and the (potentially un-normalized) posterior probability at each point $\{\theta_i, P(\theta_i)\}$. In most sampling methods, the probability values are calculated in order to do the sampling, but in some cases calculating the probability values may take much more work than just sampling.

3.4.3 *Inference from simulation*

A common situation in observations is that we have a huge stream of data $\mathbf{d}$ with approximately Gaussian noise, and we are interested in measuring some linear function of $\mathbf{d}$ that encapsulates the relevant physics, say, $\mathbf{a} = M\mathbf{d}$. For example, $\mathbf{a}$ may be just the large-scale modes of the data or the principal components. Typically M is a rectangular matrix because $\mathbf{a}$ is much smaller than $\mathbf{d}$. Since the process is linear, Gaussian noise on $\mathbf{d}$ gives Gaussian noise on $\mathbf{a}$ with some co-variance. If M is a large complicated matrix, computing the noise covariance of $\mathbf{a}$ directly may be very hard, since calculating $N = \langle \mathbf{a}\mathbf{a}^T \rangle = M \langle \mathbf{d}\mathbf{d}^T \rangle M^T$ requires calculating the full $\mathbf{d}$-covariance and then performing large matrix multiplications. Instead, it is often easy to simulate multiple pure-noise data streams $\mathbf{d}_i$. Each simulation gives one sample of $\mathbf{a}$ in its noise distribution. From a set of Monte Carlo simulations of the noise, we can generate an estimator for the noise covariance,

$$\hat{N} = \frac{1}{n_s} \sum_{i=1}^{n_s} \mathbf{a}_i \mathbf{a}_i^T. \tag{3.18}$$

This is an instance of applying parameter estimation to the parameters of cosmological observations rather than cosmological parameters themselves. Of course, having estimated the noise we might then want to use its likelihood to calculate a posterior for other parameters of more direct interest. This estimator is a sufficient statistic for the likelihood of the true noise covariance N, since for Gaussian noise,

$$\mathcal{L}(N|\hat{N}) \propto \frac{\exp(-\sum_i \mathbf{a}_i^{\mathrm{T}} N^{-1} \mathbf{a}_i / 2)}{|2\pi N|^{n_s/2}} \propto \frac{e^{-n_s \mathrm{Tr}(\hat{N}N^{-1})/2}}{|2\pi N|^{n_s/2}}. \qquad (3.19)$$

This is called an inverted Wishart distribution (for details see, e.g., Gupta and Nagar (1999)). At least $n_s = n$ samples are required for $\hat{N}$ to even be invertible (because we need at least one sample per eigenvalue). For the likelihood to be normalizable, we need $n_s > 2n$. In general, the expected value (for $n_s > 2n + 2$ and a flat prior on N) is $E(N|\hat{N}) = \hat{N}n_s/(n_s - 2n - 2)$, so the the actual covariance is likely to be somewhat larger than the maximum likelihood estimate $\hat{N}$. The inverse is however unbiased, $E(N^{-1}|\hat{N}) = \hat{N}^{-1}$. The central limit theorem means that $\mathcal{L}(N|\hat{N})$ can be approximated by a multivariate Gaussian distribution for large numbers of samples.

Either the full distribution of N should be used and integrated out, or we must have $n_s \gg 2n$ in order to safely assume that $N \sim \hat{N}$. If we are calculating the noise N in order to estimate a signal covariance component,

$$S \sim \frac{1}{n_{\mathrm{obs}}} \sum_{i=1}^{n_{\mathrm{obs}}} \mathbf{a}_i \mathbf{a}_i^{\mathrm{T}} - N$$

accuracy can be critical. The fractional uncertainty on N from simulations is $\sim\sqrt{2/n_s}$, so we need $n_s \gg (N/S)^2$ simulations to distinguish S from noise sampling uncertainty. If the signal to noise is low, a great many simulations may be required if the uncertainty in N is to be a small contribution to the posterior uncertainty in S. For further discussion and possible ways to improve convergence in a cosmological context, see, e.g., Hartlap, Simon and Schneider (2007) and Pope and Szapudi (2007).

3.4.4 Model selection as parameter estimation

If we have a set of cosmological models $\{M_i\}$, each potentially with different sets of parameters $S_m = \{\theta_{m,i}\}$, there are a variety of different questions that we can ask. For example, the straight parameter estimation question asks: 'Given the model M, what are the parameters?' But we could also ask, 'Which model is correct?', or more quantitatively, 'What are the posterior odds on each model being correct given the data?'. All of these questions can be recast in terms of some larger parameter estimation framework. For instance, we could introduce a discrete model index parameter h, so $h = 1$ corresponds to model M_1. Adding this 'hyperparameter' to all of the parameters of all the models gives an enlarged set of parameters that completely describes all the possibilities, $\{h, \{S_m\}\}$. To formulate this correctly as a posterior probability problem it is necessary to include the relative priors on each of the models. Doing parameter estimation on this enlarged set would give us the posterior distribution of h and thus tell us about our posterior belief in the different models. We could also compute conditional distributions; for example, $P(S_1|h = 1)$ would tell us the distribution of the parameters of model 1 given model 1 is correct. Our posterior odds of model 1 compared with model 2 would just be $P(h = 1)/P(h = 2)$, where the distributions have all the other parameters integrated out. If all models are considered equally likely a priori (same prior weight to each model) then this corresponds to the ratio of 'evidence', discussed further in later chapters.

Casting model selection problems as parameter estimation can be particularly useful when there is a large overlap between models, for example, when many parameters are common between the models. This is a frequent situation in cosmology; for instance, M_2 may just be a special case of M_1 where one of the parameters is fixed at a particular value. As a concrete example, $M_2(\theta)$ may correspond to a flat universe model where the curvature parameter $\Omega_K = 0$, and M_1 has a free curvature parameter $M_1(\Omega_K, \theta)$, so that $M_2(\theta) = M_1(0, \theta)$. In this case, knowing $P(\theta, \Omega_K|M_1)$ is sufficient to also know $P(\theta|M_2)$. But can we work out the posterior odds $P(M_2)/P(M_1)$? The general model selection approach would be to work out $P(M_1)$ and $P(M_2)$ separately by integrating out all the other parameters, which may be a very time consuming process, and $P(M_1)$ will of course depend on the prior chosen over Ω_K. Instead, we could view M_2 as a special case of M_1, so that for a separable prior[4]

$$\frac{P(M_2|\mathbf{d})}{P(M_1|\mathbf{d})} = \frac{P(\mathbf{d}|M_2)}{P(\mathbf{d}|M_1)} \frac{P_r(M_2)}{P_r(M_1)} \tag{3.20}$$

$$= \frac{P(\mathbf{d}|M_1, \Omega_K = 0)}{\int d\Omega_K P(\mathbf{d}|M_1, \Omega_K) P_r(\Omega_K|M_1)} \frac{P_r(M_2)}{P_r(M_1)}. \tag{3.21}$$

This result depends only on the un-normalized marginalized density $P(\mathbf{d}|\Omega_K)$ and the priors. So from a parameter estimation run in M_1, calculating the marginalized likelihood $P(\mathbf{d}|\Omega_K)$ (e.g., estimated from the sample density as a function of Ω_K, where sampling is done with flat prior on Ω_K), we can easily work out the posterior model odds given a particular choice of prior $P_r(\Omega_K|M_1)$. In fact, in this case, simply plotting $P(\mathbf{d}|\Omega_K)$ is generally much more informative that computing the odds for a particular choice of prior. In particular, if $P(\mathbf{d}|\Omega_K)$ peaks at $\Omega_K = 0$, it is clear that for any choice of prior $P(M_2|d)/P(M_1|d) > 0$, so the restricted model will be favoured.

---

[4] This is essentially the same as the Savage–Dickey density ratio, see, e.g., Trotta (2007).

In the above example we can of course only estimate $P(\Omega_K = 0|\mathbf{d})$ by assuming $P(\Omega_K|\mathbf{d})$ is a smooth function, and the estimate might be quite noisy. If this is an issue, as it may be in other cases, we could run our parameter estimation process in the enlarged parameter space with the model index parameter $h \in (1, 2)$. In practice, this would mean that the sampling method would have to arrange to give some samples on the fixed hypersurface $\Omega_K = 0$, but this is often not difficult.

Even in more general cases where the differences between models are more radical, it may still be helpful to treat model selection as a parameter estimation problem. For instance, sampling to compute the relevant posterior model odds ratio may be much easier than computing each evidence value separately. In some cases, the model index parameter h can also usefully be extended into a continuous parameter. This can mean that its posterior distribution may give useful additional information, and it may also improve sampling efficiency in some instances. For example, M_1 might use a likelihood with a simple noise model, and M_2 might be a likelihood using an alternative model that includes an extra contribution from some noise source that was neglected in M_1. By allowing the amplitude of this extra noise to be a free parameter A, we can pose the model selection problem of M_2 versus M_1 as a question about the ratio $P(A)/P(0)$. In this case, if we originally thought that $A = 1$ in M_2, but by plotting $P(A)$ we see that $P(A)$ peaks at $A = 2$, this is giving us useful information that we should reconsider our noise model.

3.5 Conclusions

We described the computational difficulties of mapping out probability distributions in multidimensional space and showed how sampling is often a good way to obtain parameter estimates and error bars. We outlined a range of sampling techniques, and emphasized the various potential problems and the importance of convergence tests when using Markov chain results. When the underlying probability distribution changes, importance sampling techniques can be used to generate a new set of weighted samples very quickly.

Samples from a distribution are a useful resource for answering many questions that could only be calculated directly from the likelihood function at much greater computational cost. Monte Carlo simulation samples are also often a useful way to learn about the low-dimensional statistical properties of high-dimensional distributions. In simple cases, model selection problems can also usefully be viewed as simple parameter estimation problems.

References

Berg, B. A. (2000). *Fields Inst. Commun.*, **26**, 1.

Dunkley, J., Bucher, M., Ferreira, P. G., Moodley, K. and Skordis, C. (2005). *Mon. Not. Roy. Astron. Soc.*, **356**, 925.

Gamerman, D. and Lopes, H. F. (2006). *Markov Chain Monte Carlo: Stochastic Simulation for Bayesian Inference*. London: Chapman and Hall/CRC.

Gelman, A. and Rubin, D. B. (1992). *Stat. Sci.*, **7**, 457.

Gelman, A., Carlin, J. B., Stern, H. S. and Rubin, D. B. (2003). *Bayesian Data Analysis*. London: Chapman and Hall/CRC.

Gelman, A., Roberts, G. O. and Gilks W. R. (1996). In J. M. Bernado *et al.*, eds., *Bayesian Statistics 5*. Oxford: Oxford University Press, p. 599.

Geman, S. and Geman, D. (1984). *IEEE Trans. Pattern Anal. Mach. Intell.*, **6**, 721.

Gilks, W. R., Richardson, S. and Spiegelhalter, D. J. (1996). *Markov Chain Monte Carlo in Practice*. London: Chapman and Hall/CRC.

Gubernatis, J. E. (2000). *Comput. Sci. Eng.*, **2**, 95.

Gupta, A. K. and Nagar, D. K. (1999). *Matrix Variate Distributions*. London: Chapman & Hall.

Hajian, A. (2007). *Phys. Rev. D*, **75**, 083525.

Hartlap, J., Simon, P. and Schneider P. (2007). *Astron. Astrophys.*, **464**, 399.

Hastings, W. K. (1970). *Biometrika*, **57**, 97.

Kosowsky, A., Milosavljevic, M. and Jimenez, R. (2002). *Phys. Rev. D*, **66**, 063007.

Lewis, A. and Bridle S., CosmoMC Notes, unpublished, available at http://cosmologist. info/notes/cosmomc.ps.gz.

Liu, J. S. (2001). *Monte Carlo Strategies in Scientific Computing*. New York: Springer.

MacKay, D. J. C. (2003). *Information Theory, Inference and Learning Algorithms*. Cambridge: Cambridge University Press; http://www.inference.phy.cam.ac.uk/mackay/ itila/book.html.

Metropolis, N., Rosenbluth, A. W., Rosenbluth, M. N., Teller, A. H. and Teller, E. (1953). *J. Chem. Phys.*, **21**, 1087.

Murray, I., Ghahramani, Z. and MacKay, D. J. C. (2006). *Proceedings of the 22nd Annual Conference on Uncertainty in Artificial Intelligence (UAI-06)*. Arlington, VA: AUAI Press.

Neal, R. M. (1993). Probabilistic inference using Markov chain Monte Carlo methods, available at http://cosmologist.info/Neal93.

Neal, R. M. (2003). *Ann. Stat.*, **31**, 705.

Pope, A. C. and Szapudi, I. (2007). arXiv:0711.2509 [astro-ph].

Raftery, A. E. and Lewis, S. M. (1992). In J. M. Bernado, ed., *Bayesian Statistics*. Oxford: Oxford University Press, p. 765.

Roberts, G. O., Gelman, A. and Gilks, W. R. (1997). *Ann. Appl. Probab.*, **7**, 110.

Taylor, J. F., Ashdown, M. A. J. and Hobson, M. P. (2007). arXiv:0708.2989 [astro-ph].

Thomas, D. B., Luk, W., Leong, P. H. W. and Villasenor, J. D. (2007). *ACM Comput. Surv.*, **39**, 4, 11.1.

Trotta, R. (2007). *Mon. Not. Roy. Astron. Soc.*, **378**, 72.

Wang, F. and Landau, D. P. (2001a). *Phys. Rev. Lett.*, **86**, 2050.

Wang, F. and Landau, D. P. (2001b). *Phys. Rev. E*, **64**, 5, 056101.

4

Model selection and multi-model inference

Andrew R. Liddle, Pia Mukherjee and David Parkinson

4.1 Introduction

One of the principal aims of cosmology is to identify the correct cosmological model, able to explain the available high-quality data. Determining the best model is a two-stage process. First, we must identify the set of parameters that we will allow to vary in seeking to fit the observations. As part of this process we need also to fix the allowable (prior) ranges that these parameters might take, most generally by providing a probability density function in the N-dimensional parameter space. This combination of parameter set and prior distribution is what we will call a *model*, and it should make calculable predictions for the quantities we are going to measure. Having chosen the model, the second stage is to determine, from the observations, the ranges of values of the parameters which are compatible with the data. This second step, parameter estimation, is described in the cosmological context by Lewis and Bridle in Chapter 3 of this volume. In this article, we shall concentrate on the choice of model.

Typically, there is not a single model that we wish to fit to the data. Rather, the aim of obtaining the data is to choose between competing models, where different physical processes may be responsible for the observed outcome. This is the statistical problem of model comparison, or model selection. This is readily carried out by extending the Bayesian parameter estimation framework so that we assign probabilities to models, as well as to parameter values within those models. The model probabilities can be consistently updated, as usual, by applying Bayes' theorem.

It is necessary to tackle the issue of cosmological model selection head-on, because of the large number of parameters that one could conceive of including within a cosmological model. It is currently well-accepted that the best cosmological data – a compilation of cosmic microwave background (CMB) anisotropy,

79

Table 4.1. *Parameter constraints of the Standard Cosmological Model, reproduced from Spergel et al. (2007, WMAP collaboration) with some additional rounding. The values quoted are mean values and 68 per cent confidence intervals. All columns assume the ΛCDM cosmology with a power-law initial spectrum, no tensors, spatial flatness, and a cosmological constant as dark energy. Three different data combinations are shown to highlight the extent to which this choice matters. The parameters are $\Omega_m h^2$ (physical matter density), $\Omega_b h^2$ (physical baryon density), h (Hubble parameter), n (density perturbation spectral index), τ (optical depth to last-scattering surface), and σ_8 (density perturbation amplitude).*

	WMAP alone	WMAP + 2dF	WMAP + all
$\Omega_m h^2$	0.128 ± 0.008	0.126 ± 0.005	0.132 ± 0.004
$\Omega_b h^2$	0.0223 ± 0.0007	0.0222 ± 0.0007	0.0219 ± 0.0007
h	0.73 ± 0.03	0.73 ± 0.02	$0.704^{+0.015}_{-0.016}$
n	0.958 ± 0.016	0.948 ± 0.015	0.947 ± 0.015
τ	0.089 ± 0.030	0.083 ± 0.028	$0.073^{+0.027}_{-0.028}$
σ_8	0.76 ± 0.05	0.74 ± 0.04	0.78 ± 0.03

galaxy clustering, and supernova luminosity distance data – can well fit with a rather small number of parameters, indeed just 6 or perhaps even 5, as listed in Table 4.1.

However, the list of candidate parameters which might be required by future data is very large indeed. Table 4.2 lists an (incomplete) set of parameters that have already been considered, and there are upwards of twenty of these. One cannot simply include all possible parameters in fits to data, as each one will bring new parameter degeneracies that weaken the constraints on the basic parameters above, until we find that we haven't actually constrained those parameters at all, which surely contradicts common sense. We require a consistent scheme to discard parameters whose inclusion is not motivated by the data to hand. We will argue here for a fully Bayesian methodology, rather than an ad hoc method such as a chi-squared per degree of freedom approach.

4.2 Levels of Bayesian inference

The most familiar application of Bayesian methods is to estimate the allowed parameter values of a model. Having decided what the correct model is, we want to know what values of its parameters are consistent with observational data. This can be done, for instance, by sampling the posterior probability distributions using Markov chain Monte Carlo methods, as described in Chapter 3. This approach is widespread in cosmology.

Table 4.2. *Candidate parameters: those that might be relevant for cosmological observations, but for which there is presently no convincing evidence requiring them. They are listed so as to take the value zero in the base cosmological model. Those above the line are parameters of the background homogeneous cosmology, and those below describe the perturbations. Of the latter set, the first five refer to adiabatic perturbations, the next three to tensor perturbations, and the remainder to isocurvature perturbations. This table is taken from Liddle (2004).*

Ω_k	spatial curvature
$N_\nu - 3.04$	effective number of neutrino species (CMBFAST definition)
m_{ν_i}	neutrino mass for species 'i'
	[or more complex neutrino properties]
$m_{\rm dm}$	(warm) dark matter mass
$w + 1$	dark energy equation of state
dw/dz	redshift dependence of w
	[or more complex parameterization of dark energy evolution]
$c_{\rm S}^2 - 1$	effects of dark energy sound speed
$1/r_{\rm top}$	topological identification scale
	[or more complex parameterization of non-trivial topology]
$d\alpha/dz$	redshift dependence of the fine structure constant
dG/dz	redshift dependence of the gravitational constant
$dn/d\ln k$	running of the scalar spectral index
$k_{\rm cut}$	large-scale cut-off in the spectrum
$A_{\rm feature}$	amplitude of spectral feature (peak, dip or step) ...
$k_{\rm feature}$	... and its scale
	[or adiabatic power spectrum amplitude parameterized in N bins]
$f_{\rm NL}$	quadratic contribution to primordial non-Gaussianity
	[or more complex parameterization of non-Gaussianity]
r	tensor-to-scalar ratio
$r + 8n_{\rm T}$	violation of the inflationary consistency equation
$dn_{\rm T}/d\ln k$	running of the tensor spectral index
$\mathcal{P}_S$	CDM isocurvature perturbation S ...
n_S	... and its spectral index ...
$\mathcal{P}_{S\mathcal{R}}$	... and its correlation with adiabatic perturbations ...
$n_{S\mathcal{R}} - n_S$	... and the spectral index of that correlation
	[or more complicated multi-component isocurvature perturbation]
$G\mu$	cosmic string component of perturbations

However, usually we do not in fact know what the correct model is. Instead, a common data analysis goal is to decide which parameters need to be included in order to explain the data. That is, we want to know the physical effects to which our data are sensitive, in the hope of relating those effects to fundamental physics. So, we have to acknowledge the possibility of more than one model. This brings

us to the second level of Bayesian inference, model selection/comparison. Given a
dataset, we want to know what the best model is. This is achieved by computing
the posterior model probability, which is obtained from the model likelihood, better
known in cosmology circles as the *Bayesian evidence E*. Having figured out which
the best model is, we can then do parameter estimation within that model in order
to find the acceptable parameter ranges.

Even this, however, may not be the last word, because we might well find that
there isn't a single best model (indeed, the authors' current experience is that there
never is). Instead, two or more models may all have non-negligible posterior probabilities. Yet we still might want to know how probable the parameter values are.
This can be achieved using *multi-model inference*, which is the process of consistently tracking and combining probabilities at both model and parameter level.
Parameter probability distributions can be obtained using *Bayesian model averaging* (Hoeting *et al.* 1999), where the posteriors in each model are added together
weighted by the model probabilities.[1]

4.3 The Bayesian framework

The full multi-model inference framework can be built up directly from Bayes
theorem:

$$P(B|A) = \frac{P(A|B)P(B)}{P(A)} . \tag{4.1}$$

Here A and B could be anything at all, but we will take A to be the set of data
D, and B to be the parameter values θ (where θ is the N-dimensional vector of
parameters being varied in the model under consideration), hence writing

$$P(\theta|D) = \frac{P(D|\theta)P(\theta)}{P(D)} . \tag{4.2}$$

As rehearsed in earlier articles in this volume, one of our objectives is to use
this equation to obtain the posterior probability of the parameters given the data,
$P(\theta|D)$. This is achieved by computing the *likelihood $P(D|\theta)$*. The result depends
on our prior belief on the probability of θ, $P(\theta)$, before the data came along, which
is often taken to be flat.

In parameter estimation the normalizing factor $P(D)$ is irrelevant and commonly
ignored, but it turns out to be crucial for model selection. Let us now explicitly

---

[1] This is somewhat analogous to quantum mechanics, with the parameter following a superposition of probability distributions (possibly including delta-functions if the parameter has a fixed value in some models). Future
observations may collapse this probability into fewer viable models, perhaps only one.

acknowledge that our probabilities are conditional not just on the data but on our assumed model M, writing

$$P(\theta|D, M) = \frac{P(D|\theta, M)P(\theta|M)}{P(D|M)}.$$ (4.3)

The denominator, the probability of the data given the model, is by definition the model likelihood, also known as the Bayesian evidence. It is useful because it appears in yet another rewriting of Bayes theorem, this time as

$$P(M|D) = \frac{P(D|M)P(M)}{P(D)}.$$ (4.4)

The left-hand side is the posterior model probability (i.e., the probability of the model given the data), which is just what we want for model selection. To determine it, we need to compute the Bayesian evidence, and we need to specify the prior model probability. It is a common convention to take the prior model probabilities to be equal (the model equivalent of a flat parameter prior), but this is by no means essential.

To obtain an expression for the evidence, consider Eq. (4.3) integrated over all θ. Presuming we have been careful to keep our probabilities normalized, the left-hand side integrates to unity, while the evidence on the denominator is independent of θ and comes out of the integral. Hence

$$P(D|M) = \int P(D|\theta, M)P(\theta|M)\mathrm{d}\theta$$ (4.5)

or, more colloquially,

$$\text{Evidence} = \int (\text{Likelihood} \times \text{Prior}) \, \mathrm{d}\theta$$ (4.6)

In words, the evidence is the average likelihood of the parameters averaged over the parameter prior. For the distribution of parameter values you thought reasonable before the data came along, what was the average value of the likelihood?

The Bayesian evidence rewards model predictiveness. For a model to be predictive, observational quantities derived from it should not depend very strongly on the model parameters. That being the case, if it fits the actual data well for a particular choice of parameters, it can be expected to fit fairly well across a significant fraction of its prior parameter range, leading to a high average likelihood. An unpredictive model, by contrast, might fit the actual data well in some part of its parameter space, but because other regions of parameter space make very different predictions it will fit poorly there, pulling the average down. Finally, a model, predictive or otherwise, that cannot fit the data well anywhere in its parameter space will necessarily get a poor evidence.

Often predictiveness is closely related to model simplicity; typically the fewer parameters a model has, the more limited the variety of predictions it can make. Consequently, model selection is often portrayed as tensioning goodness of fit against the number of model parameters, the latter being thought of as an implementation of Ockham's razor. However the connection between predictiveness and simplicity is not always a tight one. Consider for example a situation where the predictions turn out to have negligible dependence on one of the parameters (or a degenerate combination of parameters). This is telling us that our observations lack the sensitivity to tell us anything about that parameter (or parameter combination). The likelihood will be flat in that parameter direction and it will factorize out of the evidence integral, leaving it unchanged. Hence the evidence will not penalize the extra parameter in this case, because it does not change the model predictiveness.

The ratio of the evidences of two models M_0 and M_1 is known as the Bayes factor (Kass & Raftery 1995):

$$B_{01} \equiv \frac{E_0}{E_1} , \tag{4.7}$$

which updates the prior model probability ratio to the posterior one. Some calculational methods determine the Bayes factor of two models directly. Usual convention is to specify the logarithms of the evidence and Bayes factor.

4.3.1 Priors

Bayesian inference requires that the prior probabilities be specified, giving the state of knowledge before the data was acquired to test the hypothesis. These priors are to be chosen, and different people may well not agree on them – priors are where physical intuition comes into play. In our view, one shouldn't hope to identify a single 'right' prior. Rather, one should test how robust the conclusions are under reasonable variation of the priors. Eventually, sufficiently good data will overturn incorrect choices of prior.[2]

Parameter priors are relatively uncontroversial, though different researchers may certainly have different opinions as to which choice is the most appropriate. This is because they have to be stated explicitly in order for any useful calculation to be possible, for instance, that of the posterior parameter probability distribution. Model prior probabilities are a little more subtle, because they are not required for computation of the evidence, and results are commonly expressed under the (often implicit) assumption that the prior model probabilities are equal. However, in many cases there may be very good reasons to take the prior model probabilities to be

---

[2] 'Incorrect' here means that one's original belief concerning a parameter's value turns out not to match the actual value.

different, for instance, if one model has a strong physical underpinning and is well supported by other data not being considered, while the other is not. Further, within the common Bayesian philosophy of constant updating of probabilities in response to emerging data, one may have inherited unequal prior model probabilities from an earlier comparison with independent data (though admittedly in cosmology usual practice is to refit from scratch with a large combined dataset each time a new piece of data emerges).

In any event, one always ought to consider how robust the conclusions being drawn are to variations in the model prior probabilities. This can usefully be thought of as a decomposition into prior theoretical prejudice, modulated by the evidence which tells us how the new data has changed our previous view. People may disagree on the original position, but should all agree on the extent to which the data has modified it. If our original prejudice were weighted in favour of one model, we might well still end up favouring that model overall, even if the evidence from the data was leaning in the opposite direction.

4.3.2 *Information and complexity*

In this section we have already spoken about how simplicity and predictiveness are not trivially related. Often, parameters can be added to a model that the data can make no predictions about at all, leaving the evidence unchanged. For example, if we added the price of fish at Billingsgate market as a parameter to our cosmological model, it would change none of the predictions for the cosmological datasets, and so its likelihood would be completely flat. If it was included in a model selection analysis, it would have the evidence unchanged, no matter how large a prior was placed on it.

This example may seem slightly absurd, but consider the case of including the galaxy bias as a free parameter in an analysis that only used CMB data. There are also the cases of parameters that we cannot measure well now, but may be measurable in the future, such as the tensor to scalar ratio r, or $w_a = \mathrm{d}w/\mathrm{d}a$, the CPL parameter for the evolution of the equation of state of the dark energy with redshift.

In this case we ask a slightly different question than the standard model selection question. Instead of 'Do we need these parameters?', we can ask, 'Have we measured these parameters?' In this case, the information of a model is defined as a distance (defined by the Kullback–Leibler divergence) between the source measure (the prior) and the destination measure (the posterior), given by the equation (taken from Chapter 1 by Skilling)

$$ H = \int P(\theta) \log[P(\theta|D)/P(\theta)]\mathrm{d}\theta \,. \tag{4.8} $$

Here the information H is (minus) the logarithm of the compression ratio, i.e., the fraction of prior space that contains the bulk of the posterior volume.[3] In this way it represents how much we learn about the model by analysing the data. If we introduce a new parameter, the prior space will expand, but if this new parameter is well measured, then the ratio of prior to posterior will be larger and we gain more information. If, on the other hand, the new parameter is useless at making predictions, then the prior to posterior ratio will be unity, and the log of this ratio will be zero, meaning no new information is learnt. Usually $H \simeq N \log(\text{signal}/\text{noise})$, where N is the number of parameters.

A similar quantity is the complexity, introduced by Spiegelhalter *et al.* (2002) and applied to cosmology by Kunz, Trotta and Parkinson (2006). It is defined as

$$C = -2 \left(D_{\mathrm{KL}}(p, \pi) - \widehat{D_{KL}} \right), \tag{4.9}$$

where D_{KL} is once again the Kullback–Leibler divergence. The first term in the brackets is the gain in information under the model, i.e., H as defined above. The second term $\widehat{D_{KL}}$ is an estimate for the maximum information gain we can expect under the model. By taking the difference between these quantities (with some suitable multiplying factor), we estimate the number of parameters being measured by the data under the model.

We can use Eq. (4.8) and Bayes' theorem to rewrite Eq. (4.9) as

$$C = -2 \int P(\theta|d) \log P(D|\theta) + 2 \log P(D|\hat{\theta}). \tag{4.10}$$

Defining an effective χ^2 through the likelihood as $P(D|\theta) \propto \exp(-\chi^2/2)$ (any constant factors drop out of the difference of the logarithms in Eq. 4.10) we can write the *effective number of parameters as*

$$C = \overline{\chi^2(\theta)} - \chi^2(\hat{\theta}), \tag{4.11}$$

where the mean is taken over the posterior PDF. This quantity can be computed fairly easily from a Markov chain Monte Carlo (MCMC) run, which is nowadays widely used to perform the parameter inference step of the analysis.

When using the complexity, one must be aware that it is not really a Bayesian quantity. The point estimator for the maximum information gain, $\widehat{D_{KL}}$, suffers from the fact that it is only well defined in the best case, where the posterior is well described by a multidimensional Gaussian. The information, on the other hand, is both well defined and well behaved.

Both quantities, the information H and the complexity C, encode information about how well the analysis has done at measuring the different parameters of the

---

[3] Since it involves an averaging over the prior space, it can be computed simultaneously with the evidence.

model, on scales set by the prior. As such they are measures of the predictiveness of the model, independent of how well the model fits the data. They can be used in combination with the evidence to confirm that, in the case where many models have very similar evidence values, the best model really is the simplest.

4.4 Computing the Bayesian evidence

As we have seen, the problem of computing the evidence is that of carrying out a multidimensional integral, which is of course a standard numerical problem. In this case, however, the problem is made difficult for two reasons. One is that the integral is commonly extremely sharply peaked, with the location of the peak not known in advance. The second is that the evaluations of the likelihood, if requiring inclusion of cosmic microwave anisotropy or galaxy clustering data, are time-consuming; the typical requirement of several seconds per evaluation limits the number of points at which the integrand can be sampled to perhaps 10^5 or 10^6, even with multiprocessor capability.

There are a range of methods available, which fall into three broad classes:

- General Monte Carlo methods (which would become exact in the limit of arbitrarily many likelihood evaluation).
- Monte Carlo methods holding in restricted circumstances (but still potentially arbitrarily accurate in those circumstances).
- Approximate methods.

We consider these in turn.

4.4.1 General Monte Carlo methods

Monte Carlo methods invoke a strategy for random sampling of the likelihood in a way that allows an approximation to the integral to be made. To date, three such methods have been applied in cosmology:

Nested sampling: This algorithm was devised by Skilling (2006), and is described in more detail in Chapter 1 in this volume. The method populates the prior with a number of randomly chosen points, and then uses replacement to 'walk' the points to the high-likelihood region. We have implemented this algorithm for cosmology in the publicly-available code COSMONEST (Parkinson *et al.* 2006, www.cosmonest.org) and it is the method used to obtain the results described in this chapter. A more sophisticated sampling strategy, known as clustered nested sampling, has been described by Shaw, Bridges and Hobson (2007), and Feroz and Hobson (2008) .

Thermodynamic integration: A standard technique in the statistical community, which is also described in Skilling's chapter, and sometimes called simulated annealing, this method probes the distribution by 'heating' a Monte Carlo Markov chain in order to probe the entire prior volume. It is somewhat less versatile than nested sampling, and so far has proven less fast in cosmological applications, though this may be due to lack of proper optimization.

VEGAS: This is a multidimensional integration package popular with particle physicists, and deployed in cosmology by Serra *et al.* (2007). It has been less widely tested than other methods but is promising both in terms of speed and accuracy.

In addition to these three methods, there are other techniques in the statistics literature that have not yet been used in cosmology. An example is parallel tempering, described in Gregory (2005), where Markov chains at different temperatures are run in parallel and are permitted to swap chain elements under certain conditions.

4.4.2 Restricted Monte Carlo methods

The most important method here is the Savage–Dickey density ratio, introduced into cosmological usage by Trotta (2007). It can be used where one model is nested within another (i.e., it is obtained from the larger model by fixing one or more parameters to fixed values), provided that the prior distribution of those parameters obeys the factorization condition,[4]

$$\pi_1(\theta, \psi | M_1) = \pi_0(\theta | M_1)\tilde{\pi}(\psi | M_0), \tag{4.12}$$

where π indicates the priors, θ is the vector of parameters varied in both models, and ψ the parameters varied in model M_1 but held fixed in M_0. This factorization condition holds in many cosmological cases, though certainly not all (and may be broken by prior boundaries even with apparently uniform priors).

It can then be shown that the evidence ratio between the two models is given by evaluating the marginalized posterior parameter distribution at the special parameter value(s) $\psi = \psi_*$:

$$B_{01} = \frac{P(\psi | D, M_1)}{\pi_1(\psi | M_1)} \bigg|_{\psi = \psi_*}. \tag{4.13}$$

This then allows the evidence ratio to be estimated from the posterior parameter distribution of ψ, which typically has few enough dimensions that the correct normalization can be obtained, for instance, in a Markov chain. The general

---

[4] A slightly weaker condition is actually required, but usually priors that satisfy it will be in this form.

problem of normalizing the full posterior is the same as computing the evidence; the advantage here is that only the normalization of the marginalized posterior is needed.

The main problem with this method, however, is that if the nested model is disfavoured by the data, there may be few Markov chain samples in the vicinity of the nested model, meaning that the estimate becomes noisy. The method is thus better suited to cases where the simpler model will be supported by the data than to those where it is not. It may be possible to redeem this situation using Markov chains run at higher temperatures, but this possibility has not been formulated.

4.4.3 Approximate methods

Approximate methods can be used to estimate the evidence. For instance, one could fit a multivariate Gaussian to the likelihood in the vicinity of its maximum (the so-called Laplace approximation), and then use this fit to evaluate the integral. Understanding the accuracy of such an approximation may be challenging, however.

Related to this method is an approximate model selection statistic known as the Bayesian Information Criterion (BIC), defined by (Schwarz 1978; Liddle 2004)

$$\text{BIC} = -2 \ln \mathcal{L}_{\text{max}} + k \ln N, \tag{4.14}$$

where $\mathcal{L}_{\text{max}}$ is the maximum likelihood achievable within the model, k the number of parameters, and N the number of data points. The difference in BIC between models estimates the log-evidence difference, but only under highly restricted circumstances (Gaussian likelihood well contained within the prior, identically and independently distributed data points) that are unlikely to hold in practice. The BIC should therefore be avoided unless it really is impossible properly to evaluate the evidence itself.

Outside the Bayesian framework there are a number of other possible model selection tools, most notably the Akaike Information Criterion (Akaike 1974) and the principles of Minimum Message/Description Length (Wallace 2005), but as this book is about Bayesian methods we do not need to discuss these further here.

4.5 Interpretational scales

The standard scale used in interpreting the evidence is the Jeffreys scale, which appeared in Appendix B of his famous 1961 textbook (we have omitted some of his finer levels of distinction).

	Jeffreys	This volume
$\Delta \ln E < 1$	Not worth more than a bare mention.	
$1 < \Delta \ln E < 2.5$	Significant.	Weak.
$2.5 < \Delta \ln E < 5$	Strong to very strong.	Significant.
$5 < \Delta \ln E$	Decisive.	Strong.

While the verbal descriptions he gives are useful, we have found that they may be a little too strong if there is significant uncertainty over the suitable choice of priors, as is typical in cosmological problems. For a result to prove robust under a range of reasonable priors the ratio must be stronger. Accordingly, we propose the alternative verbal descriptions in the rightmost column.

In practice, we find the divisions at 2.5 (corresponding to posterior odds of about 13:1) and 5 (corresponding to posterior odds of about 150:1) the most useful. Generally speaking, a difference of less than one is not very interesting, while differences above one are worth paying attention to.

The Jeffreys scale is commonly described, as above, as applying to the evidence ratios, but should more properly be applied to the ratio of posterior model probabilities. Common convention sets the prior model probabilities equal, so that the evidence and posterior model probability ratios coincide, but one should always consider the possible effect of varying that assumption. In many cosmological contexts, there are good physical reasons to favour one model over another before the data are considered, though this may be hard to quantify. Indeed, in extreme cases this can allow theories to be considered as favoured even in the absence of any observational evidence, as has long been the case with superstring theory.

4.6 Applications

Cosmological applications of Bayesian model selection and multi-model inference fall into two broad categories:

- Application to observational data to obtain model and parameter probabilities.
- Forecasting of the ability of future experiments to answer model selection questions. This second category also includes survey optimization using model selection methods.

4.6.1 Applications to real data

A key goal of cosmology is to establish which models, or sets of parameters, to investigate. At the time of writing the standard cosmological model consists of a Universe around 14 billion years old, composed of a mixture of baryonic matter and cold dark matter, whose rate of expansion is being accelerated by a mysterious dark energy, and whose structure was seeded at early time by a near scale-invariant spectrum of Gaussian, adiabatic, superhorizon perturbations. This model is described by the parameters given in Table 4.1, and supported by a range of observations led by cosmic microwave anisotropies from the WMAP satellite.

Some of the most interesting questions concern the properties of those primordial perturbations, which one hopes to link to physical processes taking place in the early Universe, such as cosmological inflation. Of especial interest is the spectral index of the primordial power spectrum, whose deviation from the scale-invariant value would indicate dynamical processes taking place. There are also other, more radical models for the primordial perturbations, such as isocurvature perturbations, and causal seed models such as cosmic strings. Another major sector of model uncertainty is in the modelling of dark energy, where a straightforward cosmological constant is to be set against various models for dynamical dark energy and/or modified gravity.

The spectral tilt

The favoured mechanism for generating the primordial perturbations is inflation, which can accommodate a range of values for the spectral tilt parameter ($n - 1 \equiv d \ln \mathcal{P}_{\mathcal{R}} / d \ln k$ where $\mathcal{P}_{\mathcal{R}}$ is the spectrum of curvature perturbations) under various conditions. However, the Harrison–Zel'dovich model makes the simpler assumption of a scale-invariant spectrum ($n = 1$), and this model predates inflation. Using the WMAP1 data, the HZ model achieved the largest evidence by virtue of having one less parameter than a model with n (Mukherjee, Parkinson & Liddle 2006). The datasets used were the CMB TT and TE anisotropy power spectrum data from the WMAP experiment (Bennett *et al.* 2003; Spergel *et al.* 2003; Kogut *et al.* 2003), together with higher l CMB temperature power spectrum data from VSA (Dickinson *et al.* 2004), CBI (Pearson *et al.* 2003) and ACBAR (Kuo *et al.* 2004), matter power spectrum data from SDSS (Tegmark *et al.* 2004) and 2dFGRS (Percival *et al.* 2001), and supernovae apparent magnitude–redshift data from the High-z Supernovae Search Team (Riess *et al.* 2004). The priors used are stated in Table 4.3. The Bayes factor between the HZ and n models was found to be -0.58, slightly favouring the simpler HZ model, but with no significance.

The three-year WMAP3 data (Spergel *et al.* 2007; Hinshaw *et al.* 2007; Page *et al.* 2007), however, for the first time suggested a preference for a tilted spectrum, i.e., a scale-dependent primordial power spectrum with lesser power on small scales. Parameter constraints from WMAP3 alone indicated $n = 0.958 \pm 0.016$, nearly three-sigma away from unity. A model selection analysis by Parkinson, Mukherjee and Liddle (2006), and by Trotta (2007) indicated that WMAP3 data on its own showed little evidence for n deviating from one, but with external data

Table 4.3. *Priors for the different parameters, as discussed in the text.*

Parameter	Type	Min.	Max.
$\Omega_b h^2$	Uniform	0.018	0.032
$\Omega_{cdm} h^2$	Uniform	0.04	0.16
Θ	Uniform	0.98	1.1
τ	Uniform	0	0.5
$\ln(A_s \times 10^{10})$	Uniform	2.6	4.2
n	Uniform	0.8	1.2
r	Uniform	0.0	1.0
r	Logarithmic	e^{-80}	1

from smaller scale CMB experiments and LSS experiments (the same as the other datasets for the WMAP1 analysis, with the addition of the Boomerang dataset by Jones *et al.* 2006) the log-evidence for the n model went up to almost 2, which is a substantial increase. The priors are the same as in the WMAP1 case. These results are shown in Table 4.4.

However, slow-roll inflation models predict not only a spectral tilt but also a non-zero tensor-to-scalar ratio r, and a proper model selection motivated by inflationary predictions should contrast its predictions regarding n and r to a simpler ($n = 1$, $r = 0$ Harrison–Zel'dovich) model. r can take values between 0 and 1 (case 1), or, because its order is entirely unknown, linked to the energy scale of inflation, one can use a logarithmic prior for r which allows inflation to occur at the very small electroweak scale, giving rise to an r of about 10^{-34} (case 2). Thus, under the logarithmic prior, the allowed range for $\ln r$ is from -80 to 0. Most of this range, though, corresponds to an observationally negligible r, and hence this prior generates an evidence that is close to the n-only model, whereas in case 1, the n–r model is disfavoured over the simplest HZ model, as seen from the table.

A last point to note is that measurements of n are especially prone to a number of systematic uncertainties as of now. Assumptions made during foreground (in particular, point source) subtraction, corrections for other small-scale effects such as the SZ effect, and reionization, all affect the estimate of n, urging caution in arriving at a verdict about n just yet.

Dark energy

Another very interesting application of model selection is in making deductions regarding the nature of dark energy. This was studied in detail by Liddle *et al.* (2006). Here the simplest model to consider is the cosmological constant, corresponding to the energy of the vacuum, with a constant equation of state w of -1.

Table 4.4. *Evidence ratios for the different models and different datasets, as discussed in the text.*

Datasets	Model	$\ln B_{10}$
WMAP1 + all	HZ	0.0
	n	-0.58 ± 0.12
WMAP3 only	HZ	0.0
	n	0.34 ± 0.26
WMAP3 + all	HZ	0.0
	n	1.99 ± 0.26
	$n + r$ (uniform prior)	-1.45 ± 0.45
	$n + r$ (log prior)	1.90 ± 0.24

Table 4.5. *Evidence ratios for the different dark energy models.*

Model	$\ln B_{10}$	Prob
I: ΛCDM	0.0	63%
II: $-1 \le w \le -0.33$	-1.3 ± 0.1	17%
III: $-2 \le w \le -0.33$	-1.8 ± 0.1	10%
IV: $-2 \le w_0 \le -0.33, -1.33 \le w_a \le 1.33$	-2.0 ± 0.1	9%
V: $-1 \le w(a) \le 1$ for $0 \le z \le 2$	-4.1 ± 0.1	1%

A model with a different constant value of w is possible, though not well motivated theoretically. Two-parameter models of the equation of state dark energy are better motivated; these, for instance, describe the evolution of w with epoch or scale factor a as $w(a) = w_0 + w_a(1 - a)$. Then there is the issue of whether to allow w to cross the so-called phantom divide of -1 in either of these models/parameterizations. Results for various cases are shown in Table 4.5.

Overall, the simplest model continues to be a good fit to the data and is rewarded by the evidence for its predictiveness, and thus remains the favoured model. Other models have smaller evidence, though none of them are decisively ruled out. But the two-parameter models, considered more natural, are quite substantially disfavoured. Leaving all the models on the table, and converting the Bayes factor into probabilities, results in the probabilities shown in the final column of Table 4.5. Probabilities of 77%, 18% and 5% are obtained if we consider each parameterization as referring to a model and, within each parameterization, marginalize over the different prior choices we consider viable. This would avoid penalizing the ΛCDM

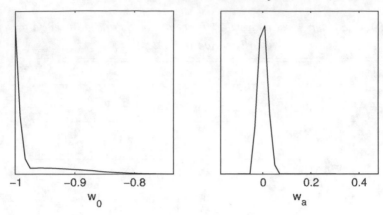

Fig. 4.1. The parameter likelihood distributions of w_0 and w_a after Bayesian model averaging over models I, II and V (reproduced from Liddle *et al.* 2006).

model for having only one prior choice. Or we could disallow the crossing of the phantom divide and then models I, II, and V results in probabilities of 78%, 21% and 1%, respectively. This is to say that given model uncertainties the indicated probabilities can change somewhat, but ΛCDM remains preferred by current data.

Future data will be quite significantly more constraining of dark energy parameters. This is often taken to mean that they have a high chance of detecting dark energy evolution, because in evaluating future surveys one assumes that a w_0, w_a model is the true model, ignoring other models, and evaluates a parameter estimation Figure of Merit (FoM) based on how much the allowed parameter space in these parameters would shrink given a future survey. However, a model selection analysis of current, relatively poor quality, data already indicates that the ΛCDM is the favoured model, and if it is indeed the true model there is no dark energy evolution to detect. In that event, the merit of future experiments will lie in favouring in the ΛCDM model decisively against its alternatives. A model selection based Figure of Merit (as discussed in Mukherjee *et al.* 2006) must then be used to rank future surveys based on their ability to do this, as parameter estimation based FoMs cannot be used to represent positive support for a simpler model.

One final point to note in this context is that when multiple models are on the table, Bayesian model averaging, consisting of superimposing the parameter constraints within each model, weighted by the model's probability, should be used to arrive at parameter constraints that thus include model uncertainty. For dark energy, this results in distributions for w_0 and w_a that are already pretty tightly constrained about -1 and 0, respectively, their values in a ΛCDM model, as seen in Figure 4.1.

Other applications

One question of interest is the nature of primordial perturbations, which would reflect in turn the nature of the mechanisms that could have produced such perturbations. Adiabatic (or density) perturbations are most naturally produced, but there are, for instance, multi-field scenarios of inflation which would lead to isocurvature (or entropy) perturbations. The observable signature of such a component is a shift in the peaks in the CMB spectrum. The issue of constraining models that contain a variety of isocurvatures modes has been considered in Beltràn *et al.* (2005) and Trotta (2007), who considered various priors on the modes. These models are currently disfavoured. There is only a small chance that CMB polarization data, which will improve dramatically with Planck, can affect this result.

Similarly, one can consider models that contain a subdominant contribution from cosmic strings. Cosmic strings are theoretically motivated by supersymmetric and grand-unified theories, and actively source additional anisotropies in the CMB affecting the resultant temperature and polarization spectra. After performing detailed simulations of the string to arrive at firm predictions, Bevis *et al.* (2008) do not find significant evidence for strings at present.

Another question of interest is whether our Universe has a non-trivial topology. A variety of non-trivial topologies (here it is hard to exhaust all possibilities) were considered by Niarchou and Jaffe (2007), and Niarchou *et al.* (2004). The observable signatures in this case are multipole space correlations in the CMB. No evidence was found for a non-trivial topology.

The Bayesian evidence has also been used to, for example: argue for the presence of residual foreground contamination in the CMB by Eriksen *et al.* (2007); determine the optimal level of regularization in component separation of multi-frequency CMB observations (Hobson *et al.* 1998; Stolyarov *et al.* 2002); determine the optimal relative weighting of cosmological datasets in joint analyses (Hobson, Bridle & Lahav 2002); determine the level of multual consistency of cosmological datasets (Marshall, Rajguru & Slosar 2006); determine the probability that localized features in CMB maps correspond to Sunyaev–Zel'dovich clusters (Hobson & McLachlan 2003) and extragalactic point sources (Carvalho, Rocha & Hobson 2009); determine whether a cold spot seen in the CMB could be due to a cosmic texture (Cruz *et al.* 2007); investigate whether the WMAP CMB data is well described by a Bianchi model (Bridges *et al.* 2007) and investigate the form of the primordial perturbation spectrum (Bridges, Lasenby & Hobson 2006).

Hence model selection tools have been used to address a variety of questions that we were unable to address robustly before. Future better-quality data will bring further interesting opportunities for applying such tools.

4.6.2 Forecasting and survey design

In addition to making inferences from actual observational data, model selection tools are well adapted to understanding the capabilities of proposed future experiments, where they complement and extend parameter estimation techniques, such as Fisher matrix forecasting of expected parameter errors.

Parameter estimation offers a fairly limited view of experimental prospects, based as it is on the assumption that we already know the correct model and are interested only in determining its parameters. It also encourages the incorrect (or at least non-Bayesian) view that fixing the parameters of a larger model to specific values is the same as considering a more restricted model (from a Bayesian viewpoint, the two are distinguished by the different model predictiveness). Moreover, parameter estimation methods are at best restricted to the case where one model is embedded as a special case of the other, and offers no guidance on how an experiment might distinguish between two competing alternatives. For instance, if we imagine a future experiment that manages to rule out the pure cosmological constant as the model of dark energy, one might wish to know whether it would then be able to say whether quintessence or modified gravity is the preferred explanation.

Once we allow multiple models, the questions we ask of future experiments become much broader. We can ask how well two models can be distinguished, under the assumption that either one is the true model, and also as a function of the true values of model parameters. For instance, we could ask how far the values of dark energy parameters have to be from the cosmological constant limit before a model comparison would be able to rule out the latter model.

Model selection forecasting for cosmology was first discussed by Trotta (2007) in an implementation predicting the probability of future experiments yielding different model selection outcomes (based on present knowledge), and by Mukherjee *et al.* (2006) in an implementation studying the dependence of model selection outcome on true model parameters. The two approaches were combined in Pahud *et al.* (2006) and Liddle *et al.* (2006), the latter also employing Bayesian model averaging.

Model selection forecasting, and its further extension to Bayesian experimental design, are described in the article by Trotta *et al.* in Chapter 5 of this volume.

4.7 Conclusions

There is a growing body of literature in the use of Bayesian model selection to make inferences about cosmological models. Model selection allows one to pose and answer questions which are not available within the parameter estimation framework alone, for instance, in comparing two models where one is not embedded within the other. As in Bayesian parameter estimation, these developments are driven by the recent availability of powerful calculational algorithms and sufficient computing power to deploy them. In our view, developing a robust understanding of cosmological models requires the use of model-level Bayesian inference, both in analyzing the viability of models in light of existing data, and in designing forthcoming experiments to maximize their impact.

References

Akaike, H. (1974). *IEEE Trans. Automat. Contr.*, **19**, 716.

Beltrán, M., Garcia-Bellido, J., Lesgourgues, J., Liddle, A. R. and Slosar, A. (2005). *Phys. Rev. D*, **71**, 063532.

Bevis, N., Hindmarsh, M., Kunz, M. and Urrestilla, J. (2008). *Phys. Rev. Lett.*, **100**, 021301.

Bridges, M., Lasenby, A. N. and Hobson, M. P. (2006). *Mon. Not. Roy. Astron. Soc.*, **369**, 1123.

Bridges, M., McEwen, J. D., Lasenby, A. N. and Hobson, M. P. (2007). *Mon. Not. Roy. Astron. Soc.*, **377**, 1473.

Carvalho, P., Rocha, G. and Hobson, M. P. (2009). *Mon. Not. Roy. Astron. Soc.*, **393**, 681.

Cruz, M., Turok, N., Vielva, P., Martinez-Gonzalez, M. and Hobson, M. (2007). *Science*, **318**, 1612.

Dickinson, C. *et al.* (2004). *Mon. Not. Roy. Astron. Soc.*, **353**, 732.

Eriksen, H. K., Banday, A. J., Gorski, K. M., Hansen, F. K. and Lilje, P. B. (2007). *Astrophys. J.*, **660**, L81.

Feroz, F. and Hobson, M. P. (2008). *Mon. Not. Roy. Astron. Soc.*, **384**, 449.

Gregory, P. (2005). *Bayesian Logical Data Analysis for the Physical Sciences*. Cambridge: Cambridge University Press.

Hinshaw, G. *et al.* (2007). *Astrophys. J. Supp.*, **170**, 288.

Hobson, M. P., Bridle, S. L. and Lahav, O. (2002). *Mon. Not. Roy. Astron. Soc.*, **335**, 377.

Hobson, M. P., Jones, A. W., Lasenby, A. N. and Bouchet, F. R. (1998). *Mon. Not. Roy. Astron. Soc.*, **300**, 1.

Hobson, M. P. and McLachlan, C. (2003). *Mon. Not. Roy. Astron. Soc.*, **338**, 765.

Hoeting, J. A., Madigan, D., Raftery, A. E. and Volinsky, C. T. (1999). *Stat. Sci.*, **14**, 382.

Jeffreys, H. (1961). *Theory of Probability*, 3rd edn. Oxford: Oxford University Press.

Jones, W. C. (2006). *Astrophys. J.*, **647**, 823.

Kass, R. E. and Raftery, A. E. (1995). *J. Am. Stat. Assoc.*, **90**, 773.

Kogut, A. *et al.* (2003). *Astrophys. J. Supp.*, **148**, 161.

Kunz, M., Trotta, R. and Parkinson, D. (2006). *Phys. Rev. D*, **74**, 023503.

Kuo, C. J. *et al.* (2004). *Astrophys. J.*, **600**, 32.

Liddle, A. R., 2004, *Mon. Not. Roy. Astron. Soc.*, 351, L49

Liddle, A. R., Mukherjee, P., Parkinson, D. and Wang, Y. (2006). *Phys. Rev. D*, **74**, 123506.

Marshall, P., Rajguru, N. and Slosar, A. (2006). *Phys. Rev. D*, **73**, 7302.

Mukherjee, P., Parkinson, D., Corasaniti, P. S., Liddle, A. R. and Kunz, M. (2006). *Mon. Not. Roy. Astron. Soc.*, **369**, 1725.

Mukherjee, P., Parkinson, D. and Liddle, A. R. (2006). *Astrophys. J. Lett.*, **638**, L51.

Niarchou, A. and Jaffe A. (2007). *Phys. Rev. Lett.*, **99**, 081302.

Niarchou, A., Jaffe, A. and Pogosian, L. (2004). *Phys. Rev. D*, **69**, 063515.

Page, L. *et al.* (2007). *Astrophys. J. Supp.*, **170**, 355.

Pahud, C., Liddle, A. R., Mukherjee, P. and Parkinson, D. (2006). *Phys. Rev. D*, **73**, 123524.

Parkinson, D., Mukherjee, P. and Liddle, A. R. (2006). *Phys. Rev. D*, **73**, 123523.

Pearson, T. J. *et al.* (2003). *Astrophys. J.*, **591**, 556.

Percival, W. J. *et al.* (2001). *Mon. Not. Roy. Astron. Soc.*, **327**, 1297.

Riess, A. G. *et al.* (2004). *Astrophys. J.*, **607**, 665.

Schwarz, G. (1978). *Ann. Stat.*, **5**, 461.

Serra, P., Heavens, A. and Melchiorri, A. (2007). *Mon. Not. Roy. Astron. Soc.*, **379**, 1169.

Shaw, J. R., Bridges, M. and Hobson, M. P. (2007). *Mon. Not. Roy. Astron. Soc.*, **378**, 1365.

Skilling, J. (2006). *Bayesian Analysis*, **1**, 833.

Spergel, D. N. *et al.* (2003). *Astrophys. J. Supp.*, **148**, 175.

Spergel, D. N. *et al.* (2007). *Astrophys. J. Supp.*, **170**, 377.

Spiegelhalter, D. J. *et al.* (2002). *J. R. Stat. Soc. B*, **64**, 583.

Stolyarov, V., Hobson, M. P., Ashdown, M. A. J. and Lasenby, A. N. (2002). *Mon. Not. Roy. Astron. Soc.*, **336**, 97.

Tegmark, M. *et al.* (2004). *Astrophys. J.*, **606**, 702.

Trotta, R. (2007). *Mon. Not. Roy. Astron. Soc.*, **378**, 72.

Wallace, C. S. (2005). *Statistical and Inductive Inference by Minimum Message Length.* Berlin: Springer.

5

Bayesian experimental design and model selection forecasting

Roberto Trotta, Martin Kunz, Pia Mukherjee and David Parkinson

5.1 Introduction

Common applications of Bayesian methods in cosmology involve the computation of model probabilities and of posterior probability distributions for the parameters of those models. However, Bayesian statistics is not limited to applications based on existing data, but can equally well handle questions about expectations for the future performance of planned experiments, based on our current knowledge.

This is an important topic, especially with a number of future cosmology experiments and surveys currently being planned. To give a taste, they include: large-scale optical surveys such as Pan-STARRS (Panoramic Survey Telescope and Rapid Response System), DES (the Dark Energy Survey) and LSST (Large Synoptic Survey Telescope), massive spectroscopic surveys such as WFMOS (Wide-Field Fibre-fed Multi-Object Spectrograph), satellite missions such as JDEM (the Joint Dark Energy Explorer) and EUCLID, continental-sized radio telescopes such as SKA (the Square Kilometer Array) and future cosmic microwave background experiments such as B-Pol searching for primordial gravitational waves. As the amount of available resources is limited, the question of how to optimize them in order to obtain the greatest possible science return, given present knowledge, will be of increasing importance.

In this chapter we address the issue of experimental forecasting and optimization, starting with the general aspects and a simple example. We then discuss the so-called Fisher Matrix approach, which allows one to compute forecasts rapidly, before looking at a real-world application. Finally, we cover forecasts of model comparison outcomes and model selection Figures of Merit. Throughout, the theory is illustrated with examples of cosmological applications.

99

5.2 Predicting the effectiveness of future experiments

The optimization of experiments can be usefully considered from the point of view of decision theory. First, one needs to define a quantity of interest, generally called the Figure of Merit (FoM), associated with the proposed experiment and its science return. The choice of this figure of merit is a priori free, and it clearly depends on the question that is being asked. In some cases it may be obvious – if we try to forecast the outcome of financial strategies, we are probably trying to maximize the expected profit. For cosmology the choice is less obvious, and we will discuss a few possibilities further below. We call the FoM a *utility* U (if we want to maximize it), or *loss* L (in which case minimization would seem preferable). The task is to determine which experiment maximizes the utility, subject to external constraints and – crucially – to the uncertainty about the true nature of the Universe.

5.2.1 Utility, expected utility and optimization

Let us assume that we have a choice of experiments e that we can build, and that within a given model $\mathcal{M}$, the physical system under consideration is described by a set of parameters θ. We assume that other experiments o have been performed, so that our present knowledge of the system before we build the new experiment is described by the posterior[1] $P(\theta|o)$. The quantity of interest to us is the utility $U(\theta, e, o)$, which in general depends on what we have observed so far (the data o), on the assumed values for the parameters θ, and on the characteristics of the future experiment, e. From the utility we can build the expected utility,

$$\mathcal{E}[U|e, o] = \int \mathrm{d}\hat{\theta}\, U(\hat{\theta}, e, o) P(\hat{\theta}|o). \tag{5.1}$$

This expression should be interpreted as follows: If the Universe is correctly described by the set of parameters $\hat{\theta}$ (that we call *fiducial parameters*) and we build the experiment e, then we can compute the utility for that experiment, $U(\hat{\theta}, e, o)$. However, our knowledge of the Universe is limited, and it is described by the current posterior distribution $P(\hat{\theta}|o)$. Averaging the utility over the posterior accounts for present uncertainty as to the value of the parameters. If, as is usually the case in cosmology, there is also uncertainty as to the underlying true model, then one has to average over possible models, as well:

$$\mathcal{E}[U|e, o] = \sum_i P(\mathcal{M}^{(i)}|o) \int \mathrm{d}\hat{\theta}^{(i)} U(\hat{\theta}^{(i)}, e, o) P(\hat{\theta}^{(i)}|o, \mathcal{M}^{(i)}), \tag{5.2}$$

where $\hat{\theta}^{(i)}$ represents the set of parameters under model $\mathcal{M}^{(i)}$. The aim is to select the experiment e that maximizes the expected utility in Eq. (5.2). Notice that

---

[1] If e is the first experiment of its kind then $P(\theta|o)$ becomes the prior distribution for the parameters.

$\mathcal{E}[U|e, o]$ takes into full account current model and parameter uncertainty, and therefore maximizing it is a way of 'hedging' over current lack of information as to the true model for the Universe (see Loredo (2004) and Bassett (2005) for other examples).

5.2.2 Choosing the best experiment – an example

The above ideas can be illustrated in the following toy example. Suppose we want to measure a single parameter θ, which is the slope of a linear function of $x > 0$. In cosmology, we might think of x as being the redshift and θ the local Hubble constant. This simple linear model can be written as

$$y = \theta x + \epsilon, \tag{5.3}$$

where ϵ is a noise term. The experiment o has been performed by measuring the quantity y at two locations, x_0, x_1, resulting in the data y_0, y_1. We assume that the two measurements are uncorrelated, and that the noise from experiment o is described by a Gaussian of 0 mean and variance σ^2. Then one can show that the likelihood as a function of the parameter θ is given by

$$\mathcal{L}(o|\theta) = \mathcal{L}_{\max}\exp\left[-\frac{1}{2}(\theta - \theta_0)^2 L\right], \tag{5.4}$$

where $\theta_0 = (x_0 y_0 + x_1 y_1)/(x_0^2 + x_1^2)$ is the maximum likelihood value for the parameter and the inverse variance L is given by $L = (x_0^2 + x_1^2)/\sigma^2$. The posterior from o follows by applying Bayes' theorem. If we take a prior centred around 0 for θ and of unit variance, then the posterior $P(\theta|o)$ is again a Gaussian, with mean given by $\bar{\theta} = (x_0 y_0 + x_1 y_1)/(\sigma^2 + x_0^2 + x_1^2)$ and inverse variance (denoted by F) given by the sum of the prior and likelihood inverse variances, i.e., $F = 1 + L$.

We now have a choice as to how to design a new experiment e in the light of what we have learned from o. The noise of the future measurement, τ, might depend in a known way on the location on the x-axis, for example, because measurements further out along x come at a greater experimental cost and hence their noise increases as some function of x. For a known function $\tau(x_f)$, the optimization problem for e is then reduced to the choice of location x_f that will yield the best science return. This is described in terms of the utility function we choose to adopt. Since we are trying to determine θ as best we can, it is sensible to use as utility function the inverse of the posterior variance of θ, denoted by $\mathcal{F}$. Recalling that the error on θ is given by $1/\sqrt{\mathcal{F}}$, maximizing the utility function means minimizing the posterior error on the parameter of interest.

Let us now compute the utility function from e, $U(\hat{\theta}, e, o) = \mathcal{F}$. By repeating the reasoning above, we obtain that the posterior for θ from the future experiment

e is again a Gaussian distribution. Its inverse variance (our utility function) is given by

$$\mathcal{F}(x_f) = 1 + L + \frac{x_f^2}{\tau^2(x_f)}. \tag{5.5}$$

Because this is a linear model, the posterior inverse variance does not depend on the assumed fiducial value for the parameter, $\hat{\theta}$. Then the expected utility is

$$\mathcal{E}[U|o, e] = \int d\hat{\theta} P(\hat{\theta}|o)\mathcal{F}(x_f) = \mathcal{F}(x_f), \tag{5.6}$$

where the second equality derives from the facts that our utility function does not depend on $\hat{\theta}$ for this particular model and that the posterior from current data, $P(\hat{\theta}|o)$, integrates to unity. Therefore, in this case, maximization of the expected utility is equivalent to maximization of the utility itself.

If the experiment e has a noise that is approximately constant as a function of the x-axis location, i.e., for $\tau^2(x_f) = \tau^2$, then $\mathcal{E}[U|o, e] = 1 + L + x_f^2/\tau^2$, which is maximized if we take x_f as large as we can, compatibly with our experimental set-up. This is simply expressing the fact that we should use the longest possible lever arm when trying to measure a slope, an eminently sensible result. If the future measurement is correlated with the previous ones, the expected utility becomes $\mathcal{E}[U|o, e] = 1 + X^t C^{-1} X$, where $X = (x_0, x_1, x_f)$ and C is the data correlation matrix (which itself might depend on x_f). Then the maximization of the expected utility reduces to the choice of x_f that maximizes $X^t C^{-1} X$.

Consider now the case where we could build a more accurate experiment, a, which could measure y with noise $\tau_\star \ll \tau$, but only if the signal falls around a certain value, $y_\star$. In other words, $y_\star$ is a 'sweet spot' where experiment a can deliver a very precise measurement. We can model this situation by writing for the noise of a,

$$\tau_a^2 = \tau_\star^2 \exp\frac{(y - y_\star)^2}{2\Delta^2}. \tag{5.7}$$

We consider the parameters $y_\star, \tau_\star, \Delta$ as fixed quantities, but in principle one could imagine optimizing over them as well. Here, we are interested in finding the optimal location x_f for a new measurement and in comparing the performance of a with the experiment considered above, e.

In building a, the problem is that we do not know precisely to which value x_f on the x-axis the sweet spot $y_\star$ corresponds, because of our finite precision on θ. If we wanted simply to minimize the noise τ_a, we would pick x_f in such a way that $y = \bar{\theta} x_f = y_\star$, hence $x_f = y_\star/\bar{\theta}$, recalling that $\bar{\theta}$ is the posterior mean for the slope from the current data. However, this is ignoring the lesson drawn from consideration of the utility, Eq. (5.5), which told us that the quantity one has to maximize is actually $x_f^2/\tau_a(x_f)^2$, in order to maximize the lever arm of the measurement. Now the utility depends on what we assume for θ, since $y = \theta x_f$. If we take the current posterior mean as the fiducial value for the slope (i.e., using $y = \bar{\theta} x_f$) and we maximize the utility Eq. (5.5) with the noise given by (5.7), we get for the optimal location,

$$x_f = \frac{y_\star}{\bar{\theta}} \frac{1 + \sqrt{1 + 8\Delta^2/y_\star^2}}{2}.$$ (5.8)

This is shifted to a larger value than if we had simply minimized the noise, exploiting the fact that the lever arm of the new datum is increased if x_f is pushed to a larger value, while not degrading the performance of the experiment too much by staying within a width $\sim\Delta$ of the sweet spot.

Finally, we can compute the expected utility, which accounts for the fact that there is uncertainty around the relation $y = \bar{\theta} x_f$, since $\bar{\theta}$ is known only with accuracy σ from present data. From the first equality in Eq. (5.6) we get

$$\mathcal{E}[U|o, a] = 1 + L + \frac{x_f^2}{\tau_\star^2} \exp\left[-\frac{(y_\star - \bar{\theta} x_f)^2}{2(\Delta^2 + \Delta y^2)} \right] \frac{\Delta}{\sqrt{\Delta^2 + \Delta y^2}},$$ (5.9)

where we have defined $\Delta y \equiv \Sigma x_f$, where $\Sigma \equiv 1/\sqrt{F}$ is the posterior error on θ from the current data. Hence Δy represents the uncertainty in y at the location $x = x_f$ stemming from our present uncertainty on the value of the slope. If $\Delta^2 \gg \Delta y^2$, then our uncertainty on the slope is much less than the width of the sweet spot, Δ, and the expected utility (5.9) reduces to the utility assuming $\bar{\theta}$ as the fiducial value, leading again to Eq. (5.8). However, the opposite case $\Delta^2 \ll \Delta y^2$ describes a sweet spot that is much narrower than our current uncertainty on its location. Then the expected utility is maximized by the choice

$$x_f = \frac{y_\star}{\bar{\theta}} \frac{1 + \sqrt{1 - 4\Sigma^2/\bar{\theta}^2}}{2\Sigma^2/\bar{\theta}^2}.$$ (5.10)

Notice that x_f is real only for $|\bar{\theta}/\Sigma| > 2$, i.e., if the slope from current data is measured to better accuracy than two sigma. If this is not the case, inspection of Eq. (5.9) reveals that $\mathcal{E}[U|o, a]$ is a monotonically increasing function of x_f (see the example in the lower panel of Figure 5.1), which is therefore maximized by taking x_f as large as possible, even though this means carrying out an experiment with very poor accuracy, as $\tau_a^2 \to \infty$ for $x_f \to \infty$. The interpretation of this result is that designing an experiment that exploits a sweet spot is only feasible if our current uncertainty on the parameter is small enough compared with the width of the sweet spot window of opportunity.

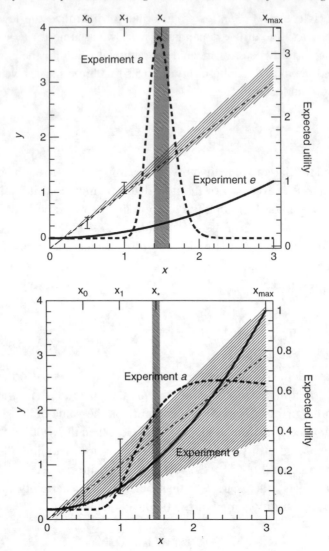

Fig. 5.1. Optimization of a future measurement using the expected utility. The choice is between a more accurate experiment a, with a 'sweet spot' given by the vertical dashed band, and a less accurate experiment e that has constant noise across the whole x range. The diagonal/dashed line is the true value of the slope while the diagonal/triangle shaped dashed region is the 68.3% region as inferred from the two measurements (data points at x_0, x_1). The dashed/thick curve is the expected utility of a while the solid/thick curve is for experiment e (arbitrary units on the right vertical axis). The ratio of the maxima of the expected utilities determines which experiment is better. For the case in the top panel, experiment a is to be preferred, while in the bottom panel, experiment e is the better choice.

It is instructive to examine the numerical values of a concrete case. Let us consider the situation where data have been obtained at $x_0 = 0.5$ and $x_1 = 1.0$ with noise $\sigma = 0.1$, when the true underlying slope is $\theta = 1.0$. For the data realization shown in the upper panel of Figure 5.1, the posterior from experiment o has mean $\bar{\theta} = 1.04$ and standard deviation $\Sigma = 0.09$. The x-range of future measurements is limited by $x_f < x_{\max} = 3.0$. Experiment e consists of collecting a further datum at larger x with the same noise as o, hence $\tau = \sigma = 0.1$. From the considerations above, we know that the utility for e is maximized by choosing the largest possible lever arm, hence by carrying out the measurement at $x_f = x_{\max}$. For experiment a, we assume that $\tau_\star = \sigma/5$, $y_\star = 1.5$ and $\Delta = 0.1$ in Eq. (5.7), hence the width of the sweet spot is comparable to the present uncertainty in its location (for $\Delta y \approx \Sigma y_\star/\bar{\theta} = 0.128$). If we knew the exact location of the sweet spot, by targeting it with experiment a (i.e., by choosing $x_f = x_\star = 1.5$), the measurement of θ would improve by a factor

$$\mathcal{R}_U \equiv \frac{\max(U(a))}{\max(U(e))} = \frac{1 + L + x_\star^2/\tau_\star^2}{1 + L + x_{\max}^2/\tau^2} = 5.2 \qquad (5.11)$$

with respect to what could be obtained with experiment e. However, since there is uncertainty on the location of the sweet spot, the relevant ratio to decide which experiment is best is the one between the maxima of the *expected* utilities, i.e.,

$$\mathcal{R}_{\mathcal{E}U} \equiv \frac{\max(\mathcal{E}[U|o, a])}{\max(\mathcal{E}[U|o, e])} = 3.3. \qquad (5.12)$$

The expected utility for a (Eq. (5.9)) is plotted in the upper panel of Figure 5.1, and it is maximized by the choice $x_f = 1.46$. For comparison, the expected utility for e peaks at $x_{\max}$. It is clear that in this case experiment a is the better choice, since $\mathcal{R}_{\mathcal{E}U} > 1$. For the case shown in the lower panel of Figure 5.1, however, present-day uncertainty is much larger than the width of the sweet spot (here we have chosen $\Delta = 0.05$, $\sigma = 0.5$, $\tau = 0.1$ and all other quantities as before). Experiment a is still much more powerful than e, as at the sweet spot it would deliver an improvement by a factor $\mathcal{R}_U = 6.2$ over what e can obtain at $x_f = x_{\max}$. However, the expected utility for a (plotted in the lower panel of Figure 5.1) also peaks at the largest possible lever arm, as the location of the sweet spot cannot be reliably determined given the uncertainty in the slope. Indeed, the estimate of $x_\star$ from the current posterior mean for the slope is $x_\star = 1.35$, where the noise of a is degraded by a factor of ~ 90 with respect to its best performance, which occurs at $x_\star = 1.5$. In this case, the ratio of the expected utilities at their maximum is $\mathcal{R}_{\mathcal{E}U} = 0.65$, indicating that experiment e ought to be preferred.

Many other elements can be incorporated in the utility function as appropriate. For example, in the above toy problem we might want to add a dimension of 'risk' in the utility, describing the probability that the experiment will not deliver an improvement over present-day knowledge. In this case, strategy e has a risk which is close to 0, as its accuracy is constant as a function of x_f and the resulting datum will improve current constraints. Experiment a might instead have a non-negligible risk factor, stemming from the fact that it might miss its sweet spot altogether and hence deliver a very noisy measurement that will not add to current data. Furthermore, every real measurement costs time and money, and this sets up a tension between the desired outcome and the resources available to achieve that outcome. Therefore, a real-world experiment optimization has to cope with complicated boundary conditions in high-dimensional parameter spaces, describing for example the accuracy of a future measurement given a fixed observational time for different observation strategies. This is a highly non-trivial computational problem, and in the next section we introduce the Fisher matrix approach which allows a much faster evaluation of the quantities involved. Then we look in more detail at a specific example, the optimization of the Wide-field Fibre-fed Multi-Object Spectroscopic (WFMOS) survey.

5.3 Experiment optimization for error reduction

The conceptually straightforward approach to evaluating the expected utility is to construct virtual experiments, by simulating data D from a proposed experiment e, assuming a fiducial value for the parameters of interest, $\hat{\theta}$. The procedure is then repeated for all possible values of the fiducial parameters and each outcome is weighted by the corresponding current posterior probability (both within and across models, as in Eq. (5.2)). The utility is computed in each case and the experimental choice that maximizes it is selected. In most cases, however, this procedure is computationally prohibitive. A massive shortcut in computational time is provided by the Fisher matrix method, which gives a lower bound (the Cramér-Rao bound) to the accuracy with which the parameters can be measured using a quadratic estimator. Here we employ it as a simple approximation to the likelihood for a future experiment, allowing one to assess the performance of a future measurement without time-consuming generation of virtual realizations of data.

5.3.1 Fisher matrix error forecast

Consider the likelihood function for a future experiment with experimental parameters e, $\mathcal{L}(\theta|e) \equiv P(D_{\hat{\theta}}|\theta, e)$, where $D_{\hat{\theta}}$ are simulated data from the future experiment assuming that the true parameter values are $\hat{\theta}$ (within a given model). We can Taylor expand the log-likelihood around its maximum-likelihood value. By definition, at an extremum the first derivatives vanish, and the shape of the log-likelihood in parameter space is approximated by the Hessian H,

$$\ln \mathcal{L}(\theta|e) \approx \ln \mathcal{L}(\theta^{\mathrm{ML}}) + \frac{1}{2} \sum_{ij} (\theta_i - \theta_i^{\mathrm{ML}})^t H_{ij} (\theta_j - \theta_j^{\mathrm{ML}}), \tag{5.13}$$

where

$$H_{ij} \equiv \frac{\partial^2 \ln \mathcal{L}}{\partial \theta_i \partial \theta_j} \bigg|_{\theta^{\mathrm{ML}}}, \tag{5.14}$$

and the derivatives are evaluated at the maximum-likelihood point. By taking the expectation of Eq. (5.13) with respect to many data realizations, we can replace the maximum-likelihood value with the fiducial value, $\hat{\theta}$, as the maximum-likelihood estimate is unbiased, i.e., $\langle \theta^{\mathrm{ML}} \rangle = \hat{\theta}$. We then define the Fisher information matrix as the expectation value of the Hessian,

$$F_{ij} \equiv \langle H_{ij} \rangle. \tag{5.15}$$

The inverse of the Fisher matrix, F^{-1}, is an estimate of the covariance matrix for the parameters, and it describes how fast the log-likelihood falls (on average) around the maximum likelihood value. (Kendall & Stuart 1977; Tegmark *et al.* 1997). In general, the derivatives depend on where in parameter space we take them (except for the simple case of linear models), hence it is clear that F is a function of the fiducial parameters.

Once we have the Fisher matrix, we can give estimates for the accuracy on the parameters from a future measurement. Interpreting the likelihood in a Bayesian sense, and taking the prior as a Gaussian with inverse covariance matrix (i.e., Fisher matrix) Π, then the above approximation for the likelihood means that the posterior will also be a Gaussian with Fisher matrix given by $\mathcal{F} = F + \Pi$. Furthermore, if the likelihood is much more informative than the prior, the posterior Fisher matrix is approximately equal to the likelihood Fisher matrix, $\mathcal{F} \approx F$. If we are only interested in a subset of the parameters, then we can marginalize easily over the others: computing the Gaussian integral over the unwanted parameters is the same as inverting the Fisher matrix, dropping the rows and columns corresponding to those parameters (keeping only the rows and columns containing the parameters of interest) and inverting the smaller matrix back. The result is the marginalized Fisher matrix $\bar{F}$. For example, the one-sigma error for parameter i, marginalized over all other parameters, is simply given by $\sigma_i = \sqrt{(\mathcal{F}^{-1})_{ii}}$. It is also straightforward to combine constraints from different independent experiments: multiplication of their likelihoods just leads to the sum of the corresponding Fisher matrices.

It remains to compute the Fisher matrix for the future experiment. This can be done analytically for the case where the likelihood function is approximately Gaussian in the data, which is a good approximation for many applications of interest. We can write for the log-likelihood

$$-2\ln\mathcal{L} = \ln|C| + (D-\mu)^t C^{-1}(D-\mu), \tag{5.16}$$

where D are the (simulated) data that would be observed by the experiment and, in general, both the mean μ and covariance matrix C may depend on the parameters θ we are trying to estimate. The expectation value of the data corresponds to the true mean, $\langle D \rangle = \mu$, and similarly the expectation value of the data matrix $\Delta \equiv (D-\mu)^t(D-\mu)$ is equal to the true covariance, $\langle \Delta \rangle = C$. Then it can be shown (see, e.g., Tegmark *et al.* 1997) that the Fisher matrix is given by

$$F_{ij} = \frac{1}{2}\mathrm{tr}\left[A_i A_j + C^{-1}\langle\Delta_{,ij}\rangle\right], \tag{5.17}$$

where $A_i \equiv C^{-1}C_{,i}$ and the comma denotes a derivative with respect to the parameters, for example $C_{,i} \equiv \partial C/\partial\theta_i$. The fact that this expression depends only on *expectation values*, and not on the particular data realization, means that the Fisher matrix can be computed from knowledge of the noise properties of the experiment without having to go through the step of actually generating any simulated data. The specific form of the Fisher matrix then becomes a function of the type of observable being considered and of the experimental parameters e.

Explicit expressions for the Fisher matrix for cosmological observables can be found in Tegmark *et al.* (1997) for cosmic microwave background data, in Tegmark (1997) for the matter power spectrum from galaxy redshift surveys (applied to baryonic acoustic oscillations in Seo & Eisenstein 2003) and in Hu and Jain (2004) for weak lensing. A useful summary of Fisher matrix technology is given in the Dark Energy Task Force report (Albrecht *et al.* 2006).

5.3.2 *Utility functions for error minimization*

Using the Fisher matrix technology sketched above, we can compute very quickly what the expected posterior error on the parameters will be (or at least a lower bound) for a given fiducial choice of cosmological parameters. From the Fisher matrix $\mathcal{F}$ (its dependence on the fiducial parameters is understood), we can define several interesting utility functions associated with the experiment. Some popular choices are:

- The determinant of the Fisher matrix, $|\mathcal{F}|$ (often called *D-optimality*), which is inversely proportional to the square of the parameter volume enclosed by the posterior. This is close to the Dark Energy Task Force (Albrecht *et al.* 2006) Figure of Merit, $[\sigma(w_a) \times \sigma(w_p)]^{-1}$, which is in fact $\sqrt{|\bar{F}_{DE}|}$ where $\bar{F}_{DE}$ is the marginalized 2×2 Fisher matrix for the dark energy parameters w_0 and w_a (defined in Eq. (5.20) below).

- The logarithm of the determinant, $\ln |\mathcal{F}|$. For a fixed fiducial model it does not matter if we maximize the determinant or its logarithm, but it may be important when averaging over fiducial models when computing the expected utility.
- The trace of the Fisher matrix, $\mathrm{tr}\mathcal{F}$, or its logarithm: this is proportional to the sum of the variances, and is often called *A-optimality*.
- The information gain H from performing the experiment (also often called Kullback–Leibler divergence),

$$H(\hat{\theta}, e, o) = \int P(\theta|\hat{\theta}, e, o) \ln \frac{P(\theta|\hat{\theta}, e, o)}{P(\theta|o)} \mathrm{d}\theta \qquad (5.18)$$

$$= \frac{1}{2}\left(\ln |\mathcal{F}| - \ln |\Pi| - \mathrm{tr}[1 - \Pi\mathcal{F}^{-1}]\right), \qquad (5.19)$$

where the prior $P(\theta|o)$ has Fisher matrix Π and $P(\theta|\hat{\theta}, e, o)$ is the posterior from future data assuming the fiducial parameters $\hat{\theta}$, with the Fisher matrix given by $\mathcal{F}$.

In many cases, D-optimality is a good indicator of the overall size of the error across all parameter space. But it does not tell us whether we are dealing with a roughly spherical error region, or with a very narrow and elongated one with the same volume, which is the hallmark of strongly degenerate directions in parameter space. A-optimality, on the other hand, will prefer a spherical error region, but may not necessarily select the one with the smallest volume. The information gain represents a compromise between these two possibilities and also has a direct information-theoretical interpretation. Additionally, we may be able to tolerate larger errors in some parameters than in others, depending on the science goals of the experiment. All these considerations have to enter the choice of the utility function. Below we discuss a somewhat different case, where we do not try to minimize error bars but are instead interested in model comparison questions, leading to a new class of utility functions, discussed in Section 5.4.

We now have a framework at our disposal for computing the utility at a given fiducial point in parameter space, $\hat{\theta}$. We now turn to discussing an application of this procedure to a specific real-world example, which tries to assess quantitatively the expected performance of future experiments in terms of their ability to reduce the error on certain cosmological parameters as a function of experimental parameters e.

5.3.3 Application to cosmology: optimization of the WFMOS survey

We consider an experiment for measuring the parameters of the mysterious dark energy based on the WFMOS design (Glazebrook *et al.* 2005; Bassett, Nichol & Eisenstein 2005), following Parkinson *et al.* (2007). WFMOS measures the

Table 5.1. *Survey parameters in each regime (high- and low-redshift observations) for the WFMOS optimization problem.*

Survey parameter	Symbol
Survey time	τ
Area covered	A
Number of redshift bins	n_{bin}
Midpoint of ith redshift bin	z_i
Half-width of ith redshift bin	dz_i

expansion history of the Universe through geometrical measurements of standard rulers, in this case Baryon Acoustic Oscillations (BAO) frozen into the large-scale structure of the Universe at early times. To measure the BAO, it is necessary to conduct a survey of the positions and redshifts of millions of galaxies over large volumes of the Universe.

The questions of experimental design are now much more complicated. Instead of just two different experiments (as in the example above) we have a vast ensemble of possible experimental configurations to consider. We can ask questions such as:

- What are the optimal redshifts for BAO observations?
- What is the best combination of areal coverage and target density?
- What type of galaxies are the best targets?

To answer these questions, each different survey geometry s is defined in terms of *survey parameters* that completely specify the survey. The survey is split into observational regimes (high- and low-redshift observations), and then for each regime we define τ (survey time), A (survey area), n_{bin} (number of redshift bins), z_i (median redshift of the ith redshift bin), and dz_i (half-width of the ith redshift bin). All these parameters are listed in Table 5.1.

In selecting a survey, we are also constrained by the technical characteristics of the proposed instrumentation. These form *constraint parameters* that limit the possible range of values of the survey parameters. Here we only consider a WFMOS-like experiment, whose important properties are summarized in Table 5.2. These include the field of view of the telescope (FoV), which gives the amount of area that can be observed per telescope pointing, and the number of fibres (n_{fibres}), which limits the number of objects that can be observed simultaneously. The total observing time serves as a good example of a constraint parameter. Since we treat these two redshift regimes separately, they have separate areas and exposure times, so the sums of the two observing times must equal this total.

The baryon acoustic oscillations offer two different measurements of the cosmological parameters, depending on which direction they are measured in. The

Table 5.2. *Constraint parameters for the WFMOS survey.*

Constraint parameter	Value
n_{fibres}	3000
FoV	1.5° diameter
Aperture	8 m
Fibre diameter	1″
Overhead time	10 minutes
Minimum exposure time	15 minutes
Maximum exposure time	10 hours
Total observing time	1500 hours

oscillations tangential to our line of sight allow us to measure the angular-diameter distance to that redshift, $d_A(z)$, while measurement of the radial modes allows us to measure the Hubble expansion rate $H(z)$ directly. The angular-diameter distance in a flat universe containing only matter and dark energy is given by

$$d_A(z) = \frac{c}{(1+z)} \int_0^z \frac{dz'}{H(z')},$$

where $H(z')$ is the Hubble function given by

$$H(z') = H_0 \left[\Omega_{\text{m}}(1+z')^3 + \Omega_{\text{DE}} f(z') \right]^{1/2},$$

where we have defined

$$f(z') \equiv \exp \int_0^{z'} \frac{3(1+w(x))}{1+x} dx.$$

Thus both the angular-diameter distance and the Hubble rate depend on the amount of matter (Ω_{m}) and dark energy (Ω_{DE}) in the Universe, and on how $w(z)$, the equation of state of dark energy, evolves with redshift. In a flat universe, $\Omega_{\text{DE}} = 1 - \Omega_{\text{m}}$, which leaves only H_0, Ω_{m} and $w(z)$ to be measured. For our optimization, we employ the following parameterization for the equation of state (Chevalier & Polarski 2001; Linder 2003):

$$w(a) = w_0 + w_a(1-a), \tag{5.20}$$

where a is the scale factor of the Universe (normalized to unity today), $a = 1/(1+z)$. We therefore wish to optimize the WFMOS survey so as to obtain the best constraints on w_0 and w_a, while marginalizing over H_0 and Ω_{m}. For these last two parameters, we recast them as $\Omega_{\text{m}} h^2$ and Ω_{m} (where $h = H_0/100 \, \text{km s}^{-1} \, \text{Mpc}^{-1}$), and we further assume a flat universe.

For the utility (or FoM) appearing in Eq. (5.1) we use

$$U(e, \hat{\theta}) = \ln |\mathcal{F}|, \tag{5.21}$$

(see, e.g., Bassett 2005; Bassett, Parkinson & Nichol 2005), the logarithm of the determinant of the posterior Fisher matrix for the parameters (see Section 5.3), which we estimate using the method outlined in Glazebrook and Blake (2005). Since $|\mathcal{F}|$ is inversely proportional to the square of the volume of the posterior ellipsoid, a larger value for our utility function corresponds to smaller errors on the parameters of interest.

The computational problem of searching through the survey parameter space (S) is that the available volume of possible surveys is very large. Therefore, we use a Monte Carlo Markov chain (MCMC) procedure to find the maximal value of the FoM, and the survey configuration associated with it. As there may be a number of degenerate minima in this parameter space (i.e., surveys with very different configurations but similar FoMs), we use the simulated annealing algorithm (Kirkpatrick *et al.* 1983; Cerny 1985; also described in Chapter 1) to cool and heat the chains generated by the MCMC process.

The optimization proceeds as follows for a fixed choice of fiducial parameters $\hat{\theta}$, which in our case correspond to the cosmological constant model, i.e., $\hat{w}_0 = -1, \hat{w}_a = 0$:

  (i) Select a test survey configuration (s) of survey parameters (area coverage, exposure time, redshifts etc.) from the experimental parameter space (S).
 (ii) Estimate the number density of galaxies that will be observed by this test survey using luminosity functions.
(iii) Estimate the error on $d_A(z)$ and $H(z)$ using scaling relations given in Blake *et al.* (2006).
 (iv) Calculate the Fisher matrix of parameters, using distance data plus other external constraints that will be available from future experiments.
  (v) Use the Fisher matrix to calculate the Figure of Merit for survey s, as given by Eq. (5.21).
 (vi) Repeat steps 1 through 5, conducting an MCMC search over the survey configuration parameter space S, attempting to maximize the utility.

We first focus on finding the optimal trade-off between exposure time and areal coverage for different galaxy populations. At low redshift, galaxies are observed either by line emission (active, star-forming 'blue' galaxies) or by continuum emission (passive, 'red' galaxies). At high redshift, the LBGs (Lyman Break Galaxies) can be observed using either method, but we assume line emission only in this case. We fix the two redshift bins at $0.5 < z < 1.3$ for 'low redshift' (or $z_{\text{low}} = 0.9$ and $dz_{\text{low}} = 0.4$) and $2.5 < z < 3.5$ for 'high redshift' ($z_{\text{high}} = 3.0$ and $dz_{\text{high}} = 0.5$) and the total time spent observing in each regime, $\tau_{\text{low}} = 800\,\text{hr}$ and $\tau_{\text{high}} = 700\,\text{hr}$. This leaves us with only two free survey parameters in each

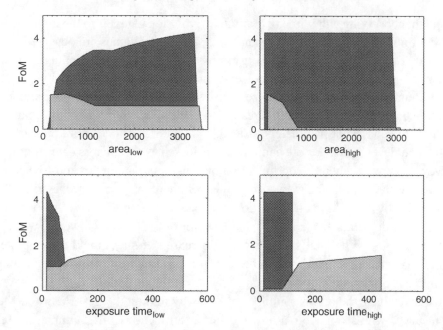

Fig. 5.2. The distribution of survey parameters by FoM for the area (in square degrees, top panels) and exposure time (in minutes, bottom panels) in the low ($z \sim 1$) and high ($z \sim 3$) redshift bins. The shaded regions delimit surveys compatible with experimental constraints. The darker shade is for surveys observing line-emission galaxies at low redshift, and the lighter shade for continuum emission. Both cases observe line-emission galaxies at high redshift. Line emission galaxies appear to be a superior target in terms of the FoM used here.

redshift bin, namely the area and exposure time for each pointing. Note that since the total observing time is fixed, the exposure time and area are linked through the number of repeat pointings to each field of view. This parameter is selected by an algorithm that makes best use of the available number of fibres.

The results shown in Figure 5.2 indicate that at low redshift, the 'blue' galaxy population is a significantly better choice than the 'red' population, as the best FoM for surveys targeting 'blue' galaxies is a factor of $3.2 \simeq \ln(25)$ higher. This corresponds to a factor of $\sqrt{25} \simeq 5$ improvement in the area of the error ellipse in the w_0–w_a parameter space. This advantage is due to the speed at which redshifts can be obtained for these 'blue' galaxies (from the O[II] emission line), which allows the survey to cover more area per unit time compared with targeting 'red' galaxies (even though they are more biased). On the other hand, the flatness of the FoM for the high-redshift bin (both as a function of area and exposure time) indicates that these parameters do not have a large impact on the FoM. It is also found that observing either line or continuum emission galaxies in the high-redshift bin does not improve the FoM, no matter what type of galaxies are observed at low redshift. This is due to the small energy density of the dark energy at high redshift, hence to the smaller constraining power of the observations in this regime.

Although it is possible to optimize simultaneously for all the survey parameters in Table 5.1, it is perhaps more interesting to consider the possibility of optimizing some of the constraint parameters instead. Constraint parameters are built in at an instrument level, and therefore cannot be considered as parameters determining the survey strategy – rather, the optimization of constraint parameters must be carried out before the instrument is even designed. Some constraint parameters will be unbounded by the FoM in one direction. For example, a survey that runs for 5 years will always do better than a survey that lasts for only 3 years. However, while the FoM increases as the total survey time is increased, it may quickly asymptote to some maximum as a function of other constraint parameters, such as for example the number of fibres in the instrument.

We investigate the optimal number of fibres by finding the best Figure of Merit for a survey with a single bin at low redshift as a function of the number of fibres, while all other survey parameters are allowed to vary. We also test the dependence on our models of the galaxy number counts, by including pessimistic (and optimistic) models which translate into roughly a 50% deficit (surplus) of galaxies. Here we assume that the fibres are able to access any part of the field of view.

The results can be seen in Figure 5.3. Looking at the left panel, the optimal value of the number of fibres for a single low-redshift bin is found to be equal to or greater than around 10 000 when sampling 'blue' galaxies and ≥ 750 for 'red' galaxies. At this point the galaxies are well sampled (including the effects of bias) so that shot noise is negligible. Neither strategy is affected much by the change in number counts because the amount of excess/deficit in number density is small.

This optimal number of 10 000 fibres for line-emission galaxies may seem rather large, especially as the instrument being considered so far in the design proposal only has 3000 fibres. Furthermore, a detailed analysis shows that the optimal survey configuration for 10 000 fibres is only actually observing around 6000 galaxies, with the remaining targets possessing emission-line fluxes that fall below the minimum S/N level and produce failed redshifts. Hence such a configuration wastes roughly 40% of the instrument capacity. We might try to find the most efficient use of fibres by plotting the FoM per fibre as a function of fibre number, as shown on the right-hand plot of Figure 5.3. For line-emission galaxies (blue band), the most efficient number is about 2000, but this has a lower FoM (5.0, compared with 5.7 for 10 000 fibres). A happy medium would be around 3000–4000, where the FoM is higher, but the efficiency is still reasonable.

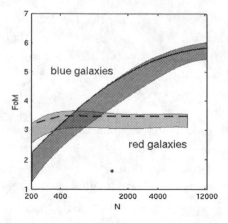
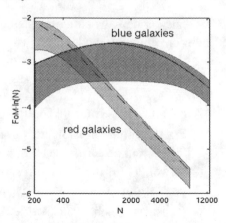

Fig. 5.3. The behaviour of the FoM of the best survey as the total number of spectroscopic fibres (N) changes for line emission (solid curve) and continuum (dashed) galaxies for a single bin at low redshift. We also include the effect of changing the number count models of the galaxies ('blue galaxies' band for line-emission galaxies and 'red galaxies' band for continuum-emission galaxies). Changes in the number count models leave the optimal value largely unchanged. On the right we plot the FoM$-\ln(N)$ against the number of fibres to find the most efficient use of the instrument. For line-emission (blue) galaxies, the most efficient number of fibres is around 2000.

5.4 Experiment optimization for model selection

In cosmology, often the question is less about the precise value of a given param-eter, and more about the probability of models. For example, we are more inter-ested in the question whether space is flat or not, and whether the dark energy is a cosmological constant or not. These are questions of model selection rather than parameter constraints. In many cases we may want to optimize our experiment to-wards deciding that issue rather than to deliver minimal error bars. This section is dedicated to the issue of experimental optimization in view of answering model selection questions.

5.4.1 Quantifying experimental capabilities using Bayes factors

Let us recall that the prime tool for Bayesian model comparison is the Bayes factor B, giving the change in relative plausibility for two models in the light of the data (as discussed in detail in Chapter 4). A Bayes factor plot chalks out the region in parameter space over which a simulated future survey can convincingly rule out one model in favour of the other (Mukherjee *et al.* 2006). The aim is to design the experiment so as to maximize the region where the models in question can be clearly distinguished. In a later section, we consider instead the problem of obtaining probabilities for the different verdicts, i.e., the problem of Bayes factor forecast.

We consider here the case of nested models, i.e., we assume that the parameters of the more complex model ($\mathcal{M}_1$) can be written as $\theta = (\phi, \psi)$ with the simpler model ($\mathcal{M}_0$) obtained by setting the extra parameter(s) $\psi = 0$. Then the procedure for producing a Bayes factor plot is as follows:

(i) Select an experimental configuration, e.

(ii) Fix the common parameters in both models to a fiducial value, $\hat{\phi}$, usually the current posterior mean. In principle, one should marginalize over them, as well, but if the future error on the parameters of interest ψ does not depend strongly on the assumed value for ϕ, this approximation is sufficient.

(iii) Consider possible values for ψ that cover the entire theoretically motivated range for them under $\mathcal{M}_1$.

(iv) For each fiducial choice, $\psi = \hat{\psi}$ generate simulated data with the properties expected of experiment e. When employing a Fisher matrix approach, simulated data are further averaged over realizations.

(v) From the simulated data D, compute the evidences for models $\mathcal{M}_0$ and $\mathcal{M}_1$. Build the Bayes factor B_{01} as the ratio of the evidences, $B_{01} \equiv P(D|\mathcal{M}_0)/P(D|\mathcal{M}_1)$.

(vi) Plot $\ln B_{01}$ as a function of $\hat{\psi}$.

(vii) Define a suitable model comparison utility function; for example, the volume in ψ parameter space where $\ln B_{01}$ exceeds a certain threshold, corresponding to the region where e will be able to discriminate strongly between the two models (see next section for an example).

For the nested models considered here, a computationally economic method to compute Bayes factors is the Savage–Dickey density ratio (SDDR; see Trotta (2007a) and references therein). Under mild conditions for the priors, the Bayes factor between $\mathcal{M}_0$ and $\mathcal{M}_1$ is given by

$$B_{01} = \left. \frac{P(\psi|D, \mathcal{M}_1)}{P(\psi|\mathcal{M}_1)} \right|_{\psi=0}, \tag{5.22}$$

i.e., the evaluation of the Bayes factor of two nested models only requires the properly normalized value of the marginal posterior for the extended model at $\psi = 0$ (the value predicted under the simpler model). We can then estimate the Bayes factor from a future observations from the predicted Fisher matrix alone (see, e.g., Kunz *et al.* 2006; Heavens *et al.* 2007), as we can approximate the posterior as a Gaussian with mean given by the fiducial parameters $\hat{\psi}$ and Fisher matrix $\mathcal{F}$, obtained as explained in Section 5.3. Then using Eq. (5.22), the Bayes factor in favour of the simpler model that one would obtain by performing the experiment is

$$\ln \mathcal{B}_{01}(\hat{\psi}, e, o) = \frac{1}{2} \ln \frac{|\bar{\mathcal{F}}|}{|\bar{\Pi}|} - \frac{1}{2} \hat{\psi}^t \bar{\mathcal{F}} \hat{\psi}, \qquad (5.23)$$

where $\bar{\Pi}$ and $\bar{\mathcal{F}}$ are the marginalized prior and posterior Fisher matrices for the parameters ψ, respectively. Recalling that $\ln \mathcal{B}_{01} > (<) 0$ favours the simpler (more complex) model, the first term on the right-hand side represents the 'Occam's razor' effect, which penalizes models with a large volume of wasted parameter space, i.e., those for which the parameter space volume $|\bar{\mathcal{F}}|^{-1/2}$ that survives after arrival of the data is much smaller than the initially available parameter space under the model prior, $|\bar{\Pi}|^{-1/2}$. The second term, on the contrary, disfavours the simpler model if $\hat{\psi}^t \bar{\mathcal{F}} \hat{\psi} \gg 0$, i.e., if the fiducial value of the parameters $\hat{\psi}$ is far away from the predicted value $\psi = 0$ under the simpler model.

The use of the Bayes factor plots to quantify experimental capabilities is quite distinct, both philosophically and operationally, from the use of parameter error forecasts we considered above. In fact, many experiments (e.g., dark energy experiments) are motivated principally by model selection questions (e.g., does the dark energy density evolve with time?) and so their performance should be quantified by their ability to answer such questions. Bayesian model selection accords special status to the $\psi = 0$ model as being a well-motivated lower-dimensional model, which in Bayesian terms is rewarded for its predictiveness in having a smaller prior volume. Parameter estimation analyses do not recognize a special status for such models. Furthermore, model selection criteria provide a more conservative threshold than significance tests for the inclusion of new parameters in a model, as discussed in Trotta (2008) and Gordon and Trotta (2007). Another feature is that model selection analyses can also accrue positive support for the simpler model, whereas significance tests can only fail to reject the simpler model. Finally, we notice that in Bayes factor plots, the data are simulated at each point of the parameter space of the more complex model and then compared with the simpler model, whereas parameter error forecasts are plotted around only selected fiducial reference values for the parameters (often just one). In particular, in the latter case the data are usually simulated for a model that people hope to exclude (i.e., the simpler model) rather than the true model which would allow that exclusion.

Set against these advantages, the only disadvantages of the Bayes factor method are that it is computationally more demanding, and that its conceptual framework has yet to become as familiar as that of parameter estimation.

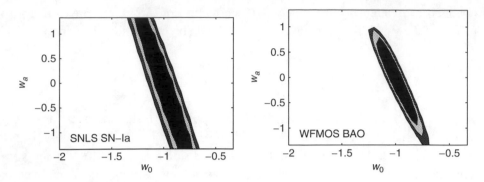

Fig. 5.4. Bayes factor forecasts for the SNLS supernovae survey and the WFMOS BAO survey. Contours are shown for $\ln \mathcal{B}_{01} = 0$, -2.5 and -5 (from the inside out). They delineate regions where the experiment will gather positive evidence in favour of a cosmological constant ($\ln \mathcal{B}_{01} > 0$), or on the contrary moderate ($-5.0 < \ln \mathcal{B}_{01} < -2.5$) or strong ($\ln \mathcal{B}_{01} < -5.0$) evidence in favour of evolving dark energy. An independent measurement of Ω_{m} to ± 0.01 accuracy is assumed.

5.4.2 Application: dark energy vs. a cosmological constant

A relevant example from cosmology is again the dark energy equation of state. Parameterizing the equation of state as in Eq. (5.20), one would like to determine which future survey will be able to distinguish more strongly between a model with $(w_0 = -1, w_a = 0)$ (i.e., dark energy in the form of a cosmological constant, or the ΛCDM model), and one in which both these parameters need to be varied in order to fit the data better (dynamical dark energy, Liddle *et al.* 2006). In this section we give examples of Bayes factor plots in the w_0-w_a space and construct model selection based utility functions with respect to which future surveys can be optimized.

As an example, consider the Bayes factor plots shown in Figure 5.4, based on Mukherjee *et al.* (2006). The innermost contoured region is where the evidence of the ΛCDM model is greater than that of the evolving dark energy model ($\ln \mathcal{B}_{01} > 0$). The outer contours show $\ln \mathcal{B}_{01} = -2.5$ and -5, levels that suggest moderate and strong evidences in favour of the evolving dark energy model, respectively. As with parameter estimation contours, the smaller the contours the more powerful the experiment is. Since the logarithms of the Bayes factors are additive, if more than one of these surveys is realized, or if there are two independent parts to a survey, their Bayes factor plots can be added together to give a net Bayes factor plot. Flat priors in the ranges $-2 < w_0 < -0.333$ and $-1.333 < w_a < 1.333$ have been used for the model comparison, and we comment on the prior dependence further below.

Table 5.3. *Two experimental Figures of Merit based on model selection capabilities: the inverse of the area in the $\hat{w}_0$–$\hat{w}_a$ plane where $\ln \mathcal{B}_{01}$ exceeds -2.5, and the value of $\ln \mathcal{B}_{01}$ at $\hat{w}_0 = -1$ and $\hat{w}_a = 0$ (cosmological constant model). A larger value for the former means that the experiment has a better model discrimination capability, i.e., it will not confuse an evolving dark energy model for ΛCDM. The second FoM is called 'MaxEv' and it measures the maximum evidence with which the experiment would support ΛCDM were it the true model. Both FoMs are additive between surveys and for independent probes of dark energy within the same survey. The value in parenthesis is normalized to SNLS.*

Experiment	Inverse area	MaxEv	Type of observation
SNLS	2.0 (1.0)	3.7 (1.0)	Supernovae (ground)
SNAP	2.9 (1.4)	4.5 (1.2)	Supernovae (space)
JEDI (SN)	5.3 (2.6)	5.0 (1.3)	Supernovae (space)
ALPACA	12.5 (6.2)	6.1 (1.6)	Supernovae (ground)
WFMOS	3.8 (1.9)	4.8 (1.3)	BAO (ground)
JEDI (BAO)	25.0 (12.5)	6.0 (1.6)	BAO (space)

In addition to plotting Bayes factor contours, one can further compress the information on how powerful an experiment is by computing the inverse of the area, within a particular contour level, to give a single number summarizing its utility. Table 5.3 summarizes these inverse areas, expressed in coordinate units, for six experiments considered in Mukherjee *et al.* (2006), showing the inverse of the area where $\ln \mathcal{B}_{01}$ exceeds -2.5. Note that this region corresponds to the parameter area in which an experiment *cannot* strongly exclude ΛCDM (i.e., regions where the experiment can confuse the models) and hence a small area corresponds to more discriminative experiments, hence to a larger utility.

We can also consider the ability of the experiments to rule out the evolving dark energy model in favour of ΛCDM, rather than the opposite which we have focused on thus far. Unlike parameter estimation methods, Bayesian model selection can offer positive support in favour of the simpler model. Because the simpler model is nested within the dark energy model, it can never fit the data better, but it can benefit from the Occam's razor effect of its smaller parameter space. All one needs to do is read off the Bayes factor for the case where the fiducial model is ΛCDM. Table 5.3 shows $\ln \mathcal{B}_{01}$ at $\hat{w}_0 = -1$ and $\hat{w}_a = 0$, i.e., when ΛCDM is the true model. We call this FoM 'MaxEv', for 'Maximum Evidence'. We find that this value is above 2.5 for all surveys, and above 5 for several of them. Thus many of the surveys are capable of accumulating strong evidence supporting ΛCDM over evolving dark energy. Thus the MaxEv can be seen as another utility function quantifying the power of experiments. Note that this utility is also additive.

Clearly, the absolute values of the two FoMs above do depend on the chosen prior range. While the choice adopted here is somewhat arbitrary, if the prior space were reduced for instance by a factor of 2, that would decrease $\ln \mathcal{B}_{01}$ by at most $\ln 2 \simeq 0.69$, and this would not significantly affect our contours or conclusions, which are based on differences in $\ln \mathcal{B}_{01}$ of 2.5 and 5. As a consequence, the inverse area FoM would not be greatly affected. The MaxEv FoM is more sensitive to the prior ranges chosen for the dark energy parameters, which control the Occam's razor effect. However, the relative strength of surveys, as given by the ratio of their FoMs, is largely unaffected by prior changes. It is important to remember that these results also retain some parameterization dependence – dark energy may turn out to be something entirely different from the parameterization adopted here, a problem common to more traditional uncertainty forecasts.

5.5 Predicting the outcome of model selection

The Bayes factor plots show the predicted outcomes of a future experiment as a function of the fiducial parameters. A further step is to predict a probability distribution for the future value of the Bayes factor itself – in other words, to derive a predictive distribution over the outcome of a future model comparison, accounting for both the current parameters and model uncertainty.

5.5.1 Predictive distributions

Assume for simplicity that we are considering only two models (the generalization to an arbitrary number of nested models is simple), with model 0 nested within model 1. Then the *predictive distribution* for the value of the posterior mean for the parameters $\bar{\theta}$ from a future experiment e given the current data o is

$$P(\bar{\theta}|o, e) = \sum_i P(\bar{\theta}|o, e, \mathcal{M}^{(i)}) P(\mathcal{M}^{(i)}|o) \tag{5.24}$$

$$= \sum_i P(\mathcal{M}^{(i)}|o) \int P(\bar{\theta}|\hat{\theta}^{(i)}, e, \mathcal{M}^{(i)}) P(\hat{\theta}^{(i)}|o, \mathcal{M}^{(i)}) \mathrm{d}\hat{\theta}^{(i)}.$$

Here $P(\bar{\theta}|\hat{\theta}^{(i)}, e, \mathcal{M}^{(i)})$ is the probability distribution for the posterior mean of θ assuming $(\hat{\theta}^{(i)}, \mathcal{M}^{(i)})$ are the true parameters and true model (the superscripts (i) designate different model choices and the corresponding parameter sets) and for a given choice of experimental parameters e for the future observation. $P(\hat{\theta}^{(i)}|o, \mathcal{M}^{(i)})$ is the current posterior under model $\mathcal{M}^{(i)}$, while different models are averaged over with relative weight given by the outcome of the current model comparison result, $P(\mathcal{M}^{(i)}|o)$. On the right-hand side, the posterior for the future experiment can be estimated by using the Fisher matrix forecast technique explained in Section 5.3. In words, Eq. (5.24) is a probability distribution for a future observation e to measure a value $\bar{\theta}$ averaging over present-day uncertainty.

Let us consider a simple example, where the predictive distribution can be written down analytically under some simplifying assumptions, but which is still relevant for many real-world cases. As before, we consider a more complex model $\mathcal{M}_1$ with free parameters $\theta = (\phi, \psi)$ and a simpler, nested model $\mathcal{M}_0$ which is obtained by setting the extra parameter $\psi = 0$. We further assume that the likelihood Fisher matrix F does not depend on ϕ – in other words, that future errors do not depend on the location in the subspace of parameters common to both models. Then one can marginalize over the common parameters ϕ and therefore we can assume without loss of generality that $\mathcal{M}_1$ has only one free parameter (ψ) compared with model $\mathcal{M}_0$ with no free parameters. We take a Gaussian prior on the extra parameter, ψ, centred around 0 and of unity standard deviation. The present-day likelihood is a Gaussian of mean μ and variance σ^2 (both are understood to be expressed in units of the prior width and are thus dimensionless). Finally, the marginal posterior for ψ from the future experiment can be computed using Fisher matrix techniques. It follows that the distribution of the posterior mean, $\bar{\psi}$, is a Gaussian centred around the fiducial parameter value $\hat{\psi}$ and has a variance τ^2 (which will itself depend on the experimental parameters e). Then plugging this into Eq. (5.24) and integrating over the value for the fiducial parameter $\hat{\psi}$ we obtain

$$P(\bar{\psi}|o, e) \propto \frac{1}{\tau\sigma} \exp\left(-\frac{\bar{\psi}^2}{2\tau^2} - \frac{\mu^2}{2\sigma^2}\right)$$
$$+ \frac{1}{(\tau^2 + \sigma^2 + \tau^2\sigma^2)^{1/2}} \exp\left(-\frac{1}{2}\frac{(\bar{\psi} - \mu)^2 + \sigma^2\bar{\psi}^2 + \tau^2\mu^2}{\tau^2 + \sigma^2 + \tau^2\sigma^2}\right),$$

(5.25)

where we have dropped irrelevant constants and assumed that the two models have equal prior probability, $P(\mathcal{M}_0) = P(\mathcal{M}_1) = 1/2$. Equation (5.25) gives the probability of obtaining a value $\bar{\psi}$ from a future measurement of precision τ, conditional on the present accuracy σ around a measured value μ. The first term on the right-hand side stems from the simpler model. If either measurement finds a strong discrepancy with the predictions of the simpler model that $\bar{\psi} = 0$ (i.e., for $\mu^2/\sigma^2 \gg 1$ or $\bar{\psi}^2/\tau^2 \gg 1$), then this term will be negligible in the model averaging, because the exponential $\rightarrow 0$. The second term represents the prediction assuming that the more complex model is true. If the future experiment is considerably better than the current one (i.e., for $\tau \ll \sigma < 1$), then this term becomes small if the two experiments give discordant outcomes (i.e., for $(\bar{\psi} - \mu)/\sigma^2 \gg 1$). Notice that in the scenario $\tau \ll \sigma$, the spread in the predictive distribution is dominated by σ, i.e., the accuracy of our prediction is limited by the extent of our present uncertainty, σ, and not by how well the future experiment will be able to do (τ). This is very different from the result one would get by simply considering the usual Fisher matrix forecast around a single fiducial point in parameter space, which does not take into account the current uncertainty at all.

5.5.2 Predictive posterior odds distribution

We now wish to use the predictive distribution for the parameters to derive a probability distribution for the Bayes factor that a future experiment can obtain. This is done as follows. For every value of the fiducial parameter $\hat{\psi}$, one can derive the posterior distribution that the future experiment would obtain if that was the true value of the parameters, as we did above. To each posterior corresponds a value of the Bayes factor that one would derive from such a measurement, as given in Eq. (5.23). Each of such values of $\ln \mathcal{B}_{01}$ has a weight given by the current posterior probability that $\hat{\psi}$ is the correct value of the parameters, averaged over possible choices of models, as well. If we have a series of MCMC samples drawn from the current posterior for each model, then the distribution of samples is automatically proportional to the probability of $\hat{\psi}$ being correct within that model. We only need to weight the samples from each model by the corresponding model probability obtained from the present-day model comparison. Then producing a histogram of the values of $\ln \mathcal{B}_{01}$ obtained from the samples gives a probability distribution for the Bayes factor that will be obtained by the future experiment e. This procedure has been introduced in Trotta (2007b), where it was called 'Predictive Posterior Odds Distribution', or PPOD for short.

Going back to the one-dimensional example of the previous section, we can write down the Bayes factor corresponding to a choice of fiducial parameters $\hat{\psi}$, cf. Eq. (5.23), as

$$\ln \mathcal{B}_{01}(\hat{\psi}, e, o) = \frac{1}{2} \ln \frac{1 + \tau^2}{\tau^2} - \frac{1}{2} \hat{\psi}^2 \frac{1 + \tau^2}{\tau^2}. \tag{5.26}$$

Notice that setting $\hat{\psi} = 0$ (corresponding to the future observation measuring the predicted value of ψ under $\mathcal{M}_0$) the above expression gives the maximum odds in favour of the simpler model one can hope to gather from a future measurement of accuracy τ. This is the MaxEv utility introduced in Section 5.4.2, giving the strength of evidence that an experiment can gather in favour of the simpler model.

5.5.3 Application: spectral index from the Planck satellite

As an example of the application of the PPOD technique, we consider the predictive probability for the Planck satellite to gather significant evidence against a

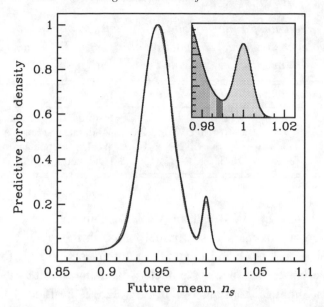

Fig. 5.5. Predictive distribution for the Planck satellite nominal mission, conditional on current knowledge, showing the probability distribution of obtaining a certain value of the spectral tilt, n_S. The bump at $n_S = 1$ corresponds to the probability associated with the HZ model, obtained by model averaging. The nearly identical curves show the numerical result from the MCMC chains and form the Gaussian approximation, Eq. (5.25). In the inset, the four shaded areas, reading from left to right, delimit regions where the Bayes factor from Planck will deliver strong evidence in favour of $n_S \neq 1$ ($\ln \mathcal{B}_{01} < -5.0$, this region extending to all smaller n_S values), moderate evidence for $n_S \neq 1$ ($-5.0 \leq \ln \mathcal{B}_{01} \leq -2.5$), weak evidence for $n_S \neq 1$ ($-2.5 \leq \ln \mathcal{B}_{01} \leq 0.0$) or favour $n_S = 1$ ($\ln \mathcal{B}_{01} > 0$).

scale-invariant spectral index of perturbations, $n_S \neq 1$ (further details are given in Trotta 2007b). Thus we have a more complex model $\mathcal{M}_1$ with a non-scale invariant index and a simpler, nested model $\mathcal{M}_0$ predicting $n_S - 1 = 0$ (the so-called 'Harrison–Zel'dovich' (HZ) model). For the more complex model, considerations of slow-roll inflation lead us to believe that deviations from scale invariance cannot be too large, otherwise slow roll would break down. This can be approximately quantified by selecting a Gaussian prior for n_S under $\mathcal{M}_1$ centred around $n_S = 1$ and with standard deviation $\Delta n_S = 0.2$. Current data then give a model comparison result that favours $\mathcal{M}_1$ with odds of order 10:1, or equivalently $\ln B_{01} \sim -2.5$ (Trotta 2007a; Pahud *et al.* 2007; Kunz *et al.* 2006). This result seems to indicate a moderate preference for a non-scale invariant spectrum, although it still falls short of the 'strong' evidence threshold. The question is then to predict what model selection result the Planck satellite will obtain from its higher quality measurements of the CMB temperature and polarization power spectra.

Figure 5.5 shows the predictive probability distribution for the Planck satellite to measure a certain value of n_S. The two curves compare the full numerical

Table 5.4. *Prediction for the probability of Planck to obtain different levels of evidence in a model comparison of $n_S = 1$ (HZ) against an inflationary-motivated model with $n_S \neq 1$.*

Integrated probability	Bayes factor range	Interpretation of Bayes factor
0.928	$\ln \mathcal{B}_{01} < -5$	Strong evidence against HZ
0.005	$-5 < \ln \mathcal{B}_{01} < -2.5$	Moderate evidence against HZ
0.006	$-2.5 < \ln \mathcal{B}_{01} < 0$	Weak evidence against HZ
0.061	$\ln \mathcal{B}_{01} > 0$	Evidence for HZ

computation using Eq. (5.24) and an MCMC chain and the approximation in Eq. (5.25). The approximation is extremely good in this case, because the current posterior is close to Gaussian, and the future errors forecast for Planck vary only very mildly over the range of parameter space singled out by the present posterior, hence assuming a constant τ as in Eq. (5.25) is justified. The bump in the predictive distribution around $n_S = 1$ corresponds to the probability associated with the HZ model. We can then obtain the PPOD numerically and integrate it to get the probability of the model comparison result from future data. This is given in Table 5.4 as the probability for Planck to obtain different levels of evidence. We can conclude that there is a very high probability (about 93%) that Planck will not only confirm but even strengthen the current model selection outcome favouring $n_S \neq 1$, by obtaining evidence above the 'strong' threshold. On the contrary, there is only about 6% of probability that future data will overturn the current verdict, by favouring the HZ model instead ($\ln \mathcal{B}_{01} > 0$). An analysis of the dependence of PPOD on the prior width for n_S shows that these results are robust for physically plausible choices of prior (Trotta 2007b).

5.6 Summary

We have seen that Bayesian methods offer powerful tools for the prediction of the outcome of future experiments. The science return of a future measurement can be encapsulated in utility functions that can be targeted at both error reduction and model comparison questions. By exploring the dependence of the utility function on experimental parameters and maximizing it, we can design measurements that are optimally suited to the science goal being pursued.

The technique of model averaging can be employed to 'hedge' our predictions against current parameters and model uncertainties that are thereby fully accounted for in the optimization procedure. We can also produce probability distributions for the outcome of a future observation and the ensuing model comparison, thereby allowing one to assess proposed experiments in terms of their ability to answer model selection questions.

The exploration of those techniques is only just beginning in cosmology and many exciting developments will take place, given the necessity of optimizing our science return from a limited amount of resources.

References

Albrecht, A. *et al.* (2006). Report of the Dark Energy Task Force, available as astro-ph/0609591.

Bassett, B. A., 2005, *Phys. Rev. D*, **71**, 083517.

Bassett, B. A., Nichol, R. C. and Eisenstein, D. J. (2005). *Astron. Geophys.*, **46** (5), 5.26.

Bassett, B. A., Parkinson, D. and Nichol, R. C. (2005). *Astrophys. J.* **626**, L1.

Blake, C., Parkinson, D., Bassett, B., Glazebrook, K., Kunz, M. and Nichol, R. C. (2006). *Mon. Not. Roy. Astron. Soc.*, **365**, 255.

Cerny, V. (1985). *J. Opt. Theory Appl.*, **45**:1, 41.

Chevalier, M. and Polarski, D. (2001). *Int. J. Mod. Phys. D*, **10**, 213.

Glazebrook, K. and Blake, C. (2005). *Astrophys. J.*, **631**, 1.

Glazebrook, K., Eisenstein, D., Dey, A. and Nichol, B. (2005). White paper to the Dark Energy Task Force, available as astro-ph/0507457.

Gordon, C. and Trotta, R. (2007). *Mon. Not. Roy. Astron. Soc.*, **382**, 1859.

Heavens, A. F., Kitching, T. D. and Verde, L. (2007). *Mon. Not. Roy. Astron. Soc.*, **380**, 1029.

Hu, W. and Jain, B. (2004). *Phys. Rev. D*, **70**, 043009.

Kendall, M. G. and Stuart, A. (1977). *The Advanced Theory of Statistics*, 4th edn. London and High Wycombe: Griffin & Co.

Kirkpatrick, S., Gelatt Jr., C. D. and Vecchi, M. P. (1983). *Science*, **220**, 671.

Kunz, M., Trotta, R. and Parkinson, D. (2006). *Phys. Rev. D*, **74**, 023503.

Liddle, A. R., Mukherjee, P., Parkinson, D. and Wang, Y. (2006). *Phys. Rev. D*, **74**, 123506.

Linder, E. V. (2003). *Phys. Rev. Lett.*, **90**, 091301.

Loredo, T. J. (2003). *AIP Conf. Proc.*, **707**, 330, available as astro-ph/0409386.

Mukherjee, P., Parkinson, D., Corasaniti, P. S., Liddle, A. R. and Kunz, M. (2006). *Mon. Not. Roy. Astron. Soc.*, **369**, 1725.

Pahud, C., Liddle, A. R., Mukherjee, P. and Parkinson, D. (2007). *Mon. Not. Roy. Astron. Soc.*, **381**, 489.

Parkinson, D., Blake, C., Kunz, M., Bassett, B. A., Nichol, R. C. and Glazebrook, K. (2007). *Mon. Not. Roy. Astron. Soc.*, **377**, 185.

Seo, H.-J. and Eisenstein, D. (2003). *Astrophys. J.*, **598**, 720.

Tegmark, M. (1997). *Phys. Rev. Lett.*, **79**, 3806.

Tegmark, M., Taylor, A. and Heavens, A. (1997). *Astrophys. J.*, **480**, 22.

Trotta, R. (2007a). *Mon. Not. Roy. Astron. Soc.*, **378**, 72.

Trotta, R. (2007b). *Mon. Not. Roy. Astron. Soc.*, **378**, 819.

Trotta, R. (2008). *Contemp. Phys.*, **49**, 2, 71.

6

Signal separation in cosmology

M. P. Hobson, M. A. J. Ashdown and V. Stolyarov

Signal separation is a common task in cosmological data analysis. The basic problem is simple to state: a number of signals are mixed together in some manner, either known or unknown, to produce some observed data. The object of signal separation is to infer the underlying signals given the observations.

A large number of techniques have been developed to attack this problem. The approaches adopted depend most crucially on the assumptions made regarding the nature of the signals and how they are mixed. Often methods are split into two broad classes: so-called blind and non-blind methods. Non-blind methods can be applied in cases where we know how the signals were mixed. Conversely, blind methods assume no knowledge of how the signals were mixed, and rely on assumptions about the statistical properties of the signals to make the separation. There are some techniques that straddle the two classes, which we shall refer to as 'semi-blind' methods. They assume partial knowledge of how the signals are mixed, or that the mixing properties of some signals are known and those of others are not.

There is a large literature in the field of signal processing about signal separation, using Bayesian techniques or otherwise. For any cosmological signal separation problem, it is almost always the case that someone has already attempted to solve an analogous problem in the signal processing literature. Readers who encounter a problem of this type, which is not already addressed in the cosmological literature, are encouraged to look further afield for existing solutions.

In this chapter, we shall describe the application of Bayesian methods to one particular cosmological signal separation problem, the separation of multi-wavelength observations into their cosmological or astrophysical components. The techniques described here are quite general, and can be applied to other signal separation problems as they stand or with only minor modifications. Throughout this chapter, we will assume that the data are measurements only of the total intensity of the sky.

126

It is straightforward, in principle, to extend the methods presented to the case in which one also measures polarization, but we will not discuss this generalization.

6.1 Model of the data

The first step towards performing a Bayesian signal separation is to define a model for the data. Focusing on our astrophysical application, we assume that (part of) the sky is observed in a number of frequency channels, N_ν, and that the sky signal at each frequency is a linear combination of N_c components of emission. We also assume that the data are convolved by a telescope beam (point spread function), which may be different in each channel, and that the instrumental noise is additive. Thus, the general model for the data in channel ν observed in the direction $\mathbf{x}$ is

$$
d_\nu(\mathbf{x}) = W(\mathbf{x}) \left[\int B_\nu(|\mathbf{x} - \mathbf{x}'|) \sum_{c=1}^{N_c} F_{\nu c}(\mathbf{x}') \, s_c(\mathbf{x}') \, \mathrm{d}\mathbf{x}' + n_\nu(\mathbf{x}) \right], \qquad (6.1)
$$

where $s_c(\mathbf{x})$ is the spatial emission from component c (at some reference frequency ν_0) and n_ν is the noise in channel ν. The coefficients $F_{\nu c}(\mathbf{x})$ determine how the components are mixed to make the data; they form the components of the *mixing matrix*. They may depend on other parameters and be a function of position (as indicated), although they are often assumed to be spatially invariant. The beam $B_\nu(r)$ in channel ν is assumed not to change with position on the sky. Furthermore, we assume that it is circularly symmetric about its peak.[1] The window function $W(\mathbf{x})$ represents the region of the data maps to be analyzed: it may equal unity everywhere; it may be zero in masked regions (or regions missing data) and unity elsewhere; or a more general (apodized) form for the window function may be used. We assume here that the same window function is applied to all frequency channels. Finally, it is worth noting that if the noise term $n_\nu(\mathbf{x})$ in Eq. (6.1) is set to zero, the resulting expression is sometimes termed the predicted data, noiseless data or data model, and is often denoted by $m_\nu(\mathbf{x})$; we shall use this notation in Section 6.5.1.

We note that, in attempting to infer the components $s_c(\mathbf{x})$ from the data defined in Eq. (6.1), one is not only performing a signal separation, but also a simultaneous deconvolution of the instrumental beams. Some methods do attempt this joint task, but many others do not perform the deconvolution. For the latter class of methods, it is usual to convolve all the channel maps to a common resolution (usually set

---

[1] It is possible to extend the methods we discuss below to asymmetric or spatially varying beams, but they become significantly more complicated and so we shall not discuss them here.

by the largest of the observing beams). Thus, assuming that any window function used is applied after the convolution, the data then become

$$\bar{d}_\nu(\mathbf{x}) = W(\mathbf{x}) \left[\int B(|\mathbf{x} - \mathbf{x}'|) \left(\sum_{c=1}^{N_c} F_{\nu c}(\mathbf{x}') \, s_c(\mathbf{x}') + n_\nu(\mathbf{x}) \right) d\mathbf{x}' \right]$$

$$\equiv W(\mathbf{x}) \left[\bar{F}_{\nu c}(\mathbf{x}) \bar{s}_c(\mathbf{x}) + \bar{n}_\nu(\mathbf{x}) \right], \tag{6.2}$$

where $B(r)$ is the same for all channels ν, and a bar over a variable indicates the spatial 'averaging' resulting from convolution to a common resolution. In the final expression, it should be noted that there is some freedom in defining $\bar{F}_{\nu c}(\mathbf{x})$ and $\bar{s}_c(\mathbf{x})$, but this is not usually a problem. The signal separation task is then to infer the convolved components $\bar{s}_c(\mathbf{x})$ from the data. Once this has been completed, an attempt may be made to deconvolve (partially) the resulting component maps, although this is rarely performed. As above, setting the noise term in Eq. (6.2) to zero results in the predicted data, which we will denote by $\bar{m}_\nu(\mathbf{x})$.

In performing a signal separation, different quantities in Eqs. (6.1) or (6.2) may be (assumed to be) known. The window function $W(\mathbf{x})$ is typically always known, and we will generally assume that the beam $B_\nu(r)$ at each frequency and the statistical properties of the noise $n_\nu(\mathbf{x})$, and hence those of $\bar{n}_\nu(\mathbf{x})$, are also known. Typically, the instrumental noise is well described by a Gaussian process with a known covariance matrix (although we will consider the case in which the noise covariance matrix is not known in Section 6.7.2). These assumptions are usually valid, but can be relaxed at the cost of introducing additional complications that we will not discuss here. The key differences between signal separation methods therefore result from the assumptions made regarding the mixing matrix $F_{\nu c}(\mathbf{x})$ and the statistical properties of the components $s_c(\mathbf{x})$ in Eq. (6.1), or their convolved counterparts in Eq. (6.2). The nature of these assumptions typically depends on the parameterization chosen to represent the mixing matrix and the component fields.

6.2 The hidden, visible and data spaces

The next steps in performing a Bayesian signal separation – steps that are often taken for granted – are (i) to choose an appropriate parameterization of the problem, which may differ from the 'natural' parameterization used in the definition of the model; and (ii) to decide in what domain to analyze the data. Adapting slightly the nomenclature of Gull and Skilling (1990), this means defining three spaces: the 'hidden' space $\mathcal{H}$, the 'visible' space $\mathcal{V}$ and the 'data' space $\mathcal{D}$. For our purposes, the visible space $\mathcal{V}$ may be taken as consisting of the parameters appearing explicitly in the data model Eq. (6.1), namely the mixing matrix elements $F_{\nu c}$ and component amplitudes s_c in each pixel $\mathbf{x}_p$ (of the reconstructed maps), or their convolved counterparts in Eq. (6.2). The parameters defining the hidden space $\mathcal{H}$ are those in terms of which the Bayesian inference problem is actually formulated, and may differ from the 'visible' parameters in the model. It is the values of the hidden parameters over which the corresponding posterior distribution is maximized or explored (usually) via sampling. Any priors adopted are also usually imposed directly on the hidden-space parameters. Finally, the choice of $\mathcal{D}$ determines in what space the model predictions are compared with the data, and hence the form of the likelihood function. The data space is usually chosen as the 'natural' space in which the data are taken, and hence the instrumental noise properties are simplest. Nonetheless, other choices are possible and can lead to useful algorithmic simplifications. The possible choices of the hidden and data spaces are discussed below.

6.3 Parameterization of the hidden space

We now turn to the specification of the hidden space $\mathcal{H}$ for our astrophysical signal separation problem, which requires us to define the parameterizations of the mixing matrix and the component fields.

6.3.1 Mixing matrix

Let us first consider the parameterization of the mixing matrix, which divides into three main areas: (i) position (in)dependence; (ii) modelling of spectral behaviour; and (iii) level of prior knowledge. These are discussed in turn below.

It is clear that the mixing matrix must depend on position to some extent, particularly for diffuse Galactic emission, such as synchrotron, free–free and dust emission, and several methods allow for some (large-scale) spatial variation. Nonetheless, it is not uncommon to assume that the mixing matrix is independent of position, and hence that the spectral behaviour of the components does not vary across the region of sky being analyzed. This assumption cannot be strictly true, but may be a reasonable approximation when analyzing small sky patches separately.

The parameterization of the spectral behaviour of the components can be defined in a large number of ways. In the most general case, the parameters are taken simply to be the mixing matrix components $F_{\nu c}$ themselves (which can vary with position). The resulting proliferation of parameters can be unhelpful, however, and one often adopts some parameterized physical model for the spectral behaviour of each component. In this case $F_{\nu c}(\mathbf{x}) = f_c(\nu; \boldsymbol{\phi}_c(\mathbf{x}))$, where the function f_c defines the physical model for the spectral behaviour of component c and depends on the parameters ϕ_c (which may be position dependent). A commonly adopted form is

$$f_c(\nu; \phi_c(\mathbf{x})) = g(\nu, \nu_0) \left(\frac{\nu}{\nu_0}\right)^{\beta_c(\mathbf{x})}, \tag{6.3}$$

where the only parameter for each component is the (in general, spatially dependent) spectral index β_c. The quantity ν_0 is the reference frequency at which f_c is unity, and $g(\nu, \nu_0)$ is a given function (often just equal to unity). Whatever the precise form chosen for the spectral behaviour, we will denote the spectral parameters for all the components collectively by $\phi(\mathbf{x})$, so that $F_{\nu c} = F_{\nu c}(\phi(\mathbf{x}))$.

Depending on whether the mixing matrix is parameterized in terms of the matrix elements themselves or in terms of a set of physically motivated spectral parameters, the full parameter set $\mathbf{\Phi}$ associated with the mixing matrix is either $\mathbf{\Phi} = \{F_{\nu c}(\mathbf{x}_p)\}$ or $\mathbf{\Phi} = \{\phi(\mathbf{x}_p)\}$, where p counts pixels (from 1 to N_{pix}) in the reconstructed component maps. However, since one expects the spectral behaviour of components to change more slowly with position than the pixel scale, it is often better instead to divide the sky into a set of larger 'patches' and assume that mixing matrix parameters are identical for all pixels lying within each patch. This can significantly reduce the number of parameters in $\mathbf{\Phi}$.

The level of prior knowledge assumed about the mixing matrix is central to the distinction between non-blind, semi-blind and blind signal separation methods. Non-blind methods assume the mixing matrix is known, or equivalently the values of all the spectral parameters. Semi-blind and blind methods relax this assumption to varying degrees. The usual (somewhat arbitrary) distinction made in the literature between semi-blind and blind methods is that the former parameterize the mixing matrix with some (physical) spectral parameters ϕ, whereas the latter use the mixing matrix elements $F_{\nu c}$ themselves as the parameters. In each case, there is a wide range of possibilities for the constraints applied to either set of parameters. One might, for example, fix the values of a subset of the parameters, link parameters in some way, or restrict the ranges that parameters might take. In all cases, one can quantify the wide range of possibilities by simply defining the prior probability assigned to the parameter set $\mathbf{\Phi}$ (see Section 6.5.1). Finally, it should be noted that, in semi-blind or blind methods, the number of components N_c can either be assumed (which is most common) or allowed to vary and hence be determined from the data; how the latter may be achieved is discussed in Section 6.7.4.

6.3.2 Component fields

We now turn to the parameterization of the components fields $s_c(\mathbf{x})$ themselves. The assumed statistical properties of the components $s_c(\mathbf{x})$ vary considerably

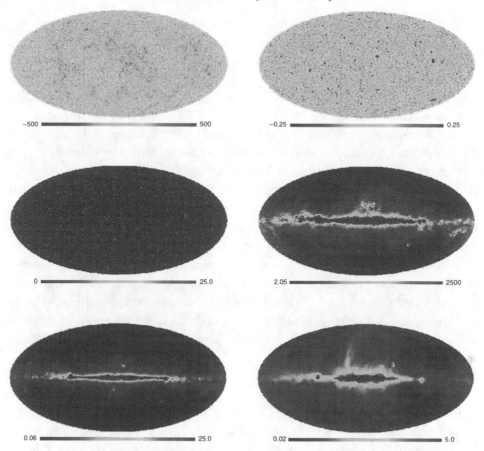

Fig. 6.1. All-sky realizations of the six main physical components contributing to the sky emission at Planck wavelengths. The components are primordial CMB, kinetic and thermal SZ effects from clusters and Galactic dust, free–free and synchrotron emission. Each map is defined in the HEALPix pixelization scheme with $N_{\text{side}} = 2048$, which corresponds to $\sim 50 \times 10^6$ pixels of size 1.7 arcmin. Each map is plotted at 300 GHz in units of μK. (Reproduced from Stolyarov *et al.* 2002.)

between different signal separation methods. In general, each component $s_c(\mathbf{x})$ may be a statistically inhomogeneous, non-Gaussian field with an infinite number of non-vanishing N-point correlation functions; some examples of all-sky component fields are shown in Figure 6.1. It is not possible, however, to model the statistical properties of $s_c(\mathbf{x})$ in such a general way, and so some assumptions or approximations are necessary. The nature of these assumptions typically depends on the parameterization chosen to represent the component fields.

For later convenience, we will denote the parameter set associated with the component field c by the 'signal' vector $\mathbf{s}_c$, and the full signal parameter set by the concatenated (and then possibly re-ordered) vector $\mathbf{s} = \{\mathbf{s}_c\}$. The most natural parameterization of the cth component field is $\mathbf{s}_c = \{s_c(\mathbf{x}_p)\}$, so that the pth element of $\mathbf{s}_c$ is simply the amplitude (at the reference frequency ν_0) of the pth pixel in the component field c; in this case, the hidden space and visible space parameters describing the component fields are identical. An obvious alternative is instead to take $\mathbf{s}_c = \{\widetilde{s}_c(\mathbf{k}_p)\}$, in which case the pth element of $\mathbf{s}_c$ is the amplitude of the pth Fourier mode in the Fast Fourier Transform (FFT) of the (pixelized) component field (when working on the full sky, one instead performs a spherical harmonic expansion, so $\mathbf{s}_c$ contains the discrete set of spherical harmonic coefficients $a_{\ell,m}$). In fact, it is often more convenient to construct $\mathbf{s}_c$ so that it is real, rather than complex. One way to achieve this is simply to fill the first half of $\mathbf{s}_c$ with the real parts of the mode amplitudes $\widetilde{s}_c(\mathbf{k}_p)$ and the second half with their imaginary counterparts. For a real sky, $\widetilde{s}_c^*(\mathbf{k}) = \widetilde{s}_c(-\mathbf{k})$ and so the (maximum required) length of the vector $\mathbf{s}_c$ is again equal to the number of pixels in the real-space component map. One may alternatively choose $\mathbf{s}_c$ to contain the coefficients of the expansion of the component field in some more general basis, such as a wavelet basis (see, e.g., Maisinger, Hobson & Lasenby 2004 and references therein), although we will not pursue this further here. It should be noted that it is not necessary to make the same choice of parameterization for every component c, although we will assume throughout that the vector $\mathbf{s}_c$ is real.

Since the length of the signal parameter vector $\mathbf{s}_c$ is typically rather large, one is usually limited to describing any assumed a-priori statistical properties of $\mathbf{s}_c$ in terms only of its covariance matrix $\mathbf{S}_c = \langle \mathbf{s}_c \mathbf{s}_c^{\mathrm{T}} \rangle$. Moreover, it is usually necessary to take this covariance matrix to be diagonal. These computational restrictions result in natural advantages and disadvantages to each of the parameterizations discussed above. In the pixel-based parameterization, one can include expected pixel-by-pixel rms values and hence it is straightforward to accommodate spatial variations in these quantities. One cannot, however, include any anticipated inter-pixel (i.e., spatial) correlations. In the Fourier-based parameterization, one can include the expected mode-by-mode rms values in the form of a power spectrum, thereby accommodating spatial correlations, but one cannot easily include spatial variations in the statistical properties. Wavelet-based parameterizations lie somewhere between the pixel and Fourier bases, and so allow some inclusion of spatial correlation information, while retaining spatial dependence.

In the construction of the full covariance matrix $\mathbf{S} = \langle \mathbf{s}\mathbf{s}^{\mathrm{T}} \rangle$, one is typically able to include correlations between the corresponding elements of $\mathbf{s}_c$ and $\mathbf{s}_{c'}$, where c' denotes another component (at least when the two components share the same parameterization). Thus, depending on the parameterization chosen, one can include either pixel-by-pixel or mode-by-mode inter-component correlations.

Just as there is considerable freedom in the assumed level of prior knowledge regarding the mixing matrix parameters $\mathbf{\Phi}$, the assumed prior knowledge of the covariance matrix $\mathbf{S}$ can vary significantly. Some methods assume complete knowledge of $\mathbf{S}$ in advance. Others assume knowledge of some subset of the elements of this covariance matrix, make links between the values of various elements, or constrain the range of values they might take. All these cases can be quantified by the imposition of an appropriate prior on the covariance matrix elements. It should be noted, however, that the (unfixed) elements of the covariance matrix then become additional parameters in the hidden space.

6.3.3 Linear and non-linear parameters

Bringing together our discussion of the hidden space, we see that, in general, it is parameterized by the mixing matrix parameters $\boldsymbol{\Phi}$, the component fields signal parameters $\mathbf{s}$ (in some domain) and the elements of the corresponding a-priori covariance matrix $\mathbf{S}$ of the component fields. We shall denote all these hidden-space parameters collectively by $\mathbf{h} = \{\boldsymbol{\Phi}, \mathbf{s}, \mathbf{S}\}$, which we assume always to be real.

It is worth noting that the signal parameters $\mathbf{s}$ are typically related via a linear transformation to the pixel amplitudes of the component fields. Since the model Eq. (6.1) or Eq. (6.2) is linear in the latter quantities, then it is also *linear* in the parameters $\mathbf{s}$. In general, however, there is no such restriction on the spectral parameters $\boldsymbol{\Phi}$. If one is using the parameterization $\boldsymbol{\Phi} = \{F_{\nu c}(\mathbf{x})\}$, then these parameters again enter the model linearly, but if $\boldsymbol{\Phi} = \{\phi(\mathbf{x})\}$ the spectral parameters may enter the model in a *non-linear* manner. Linear and non-linear parameters can be treated differently when we make inferences about the components (see, e.g., Section 6.7.1). Finally, we note that the model Eq. (6.1) or Eq. (6.2) depends only *indirectly* on the parameters $\mathbf{S}$ through the signal vector $\mathbf{s}$.

6.4 Choice of data space

Once the parameterization of the hidden space has been defined, it still remains to specify the data space $\mathcal{D}$, in which the model predictions are compared with the observations. As mentioned above, the data space is usually taken to be the space in which the observations are made, and hence in which the statistical properties of the instrumental noise are simplest, although this is not necessary.

Mirroring our discussion of the parameterization of the component fields in Section 6.3.2, we will denote the parameter set associated with the data channel ν by the 'data' vector $\mathbf{d}_\nu$, and the full data parameter set by the concatenated (and possibly re-ordered) vector $\mathbf{d} = \{\mathbf{d}_\nu\}$. Since most cosmological observations are made with single-dish telescopes, it is most natural to choose the pixel domain as the data space, in which case $\mathbf{d}_\nu = \{d_\nu(\mathbf{x}_p)\}$; examples of all-sky data maps in the lowest eight frequency channels of the Planck satellite are shown in Figure 6.2.

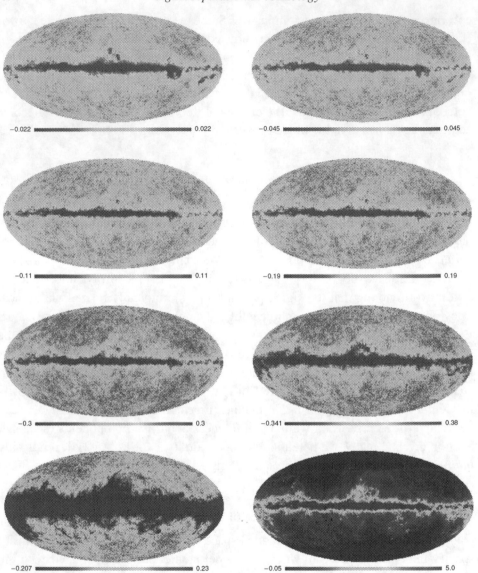

Fig. 6.2. All-sky data maps in the lowest eight frequency channels of the Planck satellite: 30, 44, 70, 100, 143, 217, 353 and 545 GHz. The maps are constructed using the component maps shown in Figure 6.1, with the simple assumption that the spectral dependence of each component is not spatially varying. (Reproduced from Stolyarov *et al.* 2002.)

Once again, however, another possibility is to take $\mathbf{d}_\nu = \{\widetilde{d}_\nu(\mathbf{k}_p)\}$ or, better, a real vector constructed from the complex Fourier mode amplitudes (see Section 6.3.2). Further choices of the elements of $\mathbf{d}_\nu$ include the coefficients of the expansion of the data map in some other, more general basis set; wavelets are again a popular option, but we will not discuss them further. It should be noted that the choice of data space is *independent* of the choice of parameterization for the component fields s. For example, it is not necessary for the data and the component fields both to be defined either in the pixel domain or the Fourier domain, although this can be advantageous in some circumstances, as outlined below.

The advantages and disadvantages of each choice for the data space centre around the resulting statistical properties of the corresponding 'noise' vector $n = \{n_\nu\}$, defined in a similar manner to the data vector d. Throughout this chapter, we will make the reasonably accurate assumption that the (pixel-domain) instrumental noise in each data channel is well described by a multivariate Gaussian distribution with a known covariance matrix. Since the domain chosen as the data space is usually linearly related to the pixel domain, the statistical properties of the corresponding noise vector n are also well described by a multivariate Gaussian distribution with known covariance matrix $N = \langle nn^T \rangle$.

For all analyses except those at very low resolution, the typical length of n means it is necessary to assume that N is diagonal. Hence, if the data space is the pixel domain, one must assume that the noise is not correlated between pixels. If one is working instead in the Fourier or spherical harmonic domain, it is assumed that the noise is not correlated between modes. Both approximations can cause problems, as we discuss below.

6.4.1 Pixel-domain data space

If the data space $\mathcal{D}$ is the pixel domain, the above restriction means that (except for very low resolution analyses) one must assume that the noise is not correlated between pixels. This assumption may not be too severe if the data are analyzed at their original spatial resolution, as in Eq. (6.1). If one performs a convolution as in Eq. (6.2), however, then the approximation can be a poor one. Nonetheless, the advantage of working in the pixel domain is that it is straightforward to accommodate non-stationary noise and data windows (or cuts).

If one additionally chooses the component fields s to be defined in the pixel domain, then it is also straightforward to accommodate a spatially varying mixing matrix, which is very common in practice. Moreover, if one simply ignores any spatial correlations in the components and noise fields (either intrinsic or induced by convolving all the data channels to the same spatial resolution), then one can perform a pixel-by-pixel Bayesian signal separation based on the data model Eq. (6.2), in which the separation is performed entirely separately for each pixel (see, e.g., Section 6.7.1). Assuming for simplicity that any applied window takes only the values zero or unity, then at each pixel $x = x_p$ *separately* in the region with $W(x) = 1$, one obtains the much lower dimensionality inference problem

$$\mathbf{d} = \mathbf{F}(\boldsymbol{\Phi})\mathbf{s} + \mathbf{n}, \tag{6.4}$$

in which the data vector $\mathbf{d}$ now has components d_ν (for $\nu = 1, \ldots, N_\nu$) at that pixel, and similar for $\mathbf{n}$; the signal vector has components s_c (for $c = 1, \ldots, N_c$) at that pixel. Note that $\mathbf{F}(\boldsymbol{\Phi})$ can also be defined on a pixel-by-pixel basis, thus allowing one to model spatially varying spectral behaviour. This pixel-by-pixel approach thus replaces a single Bayesian inference problem with a great many parameters by a large number of individual inference problems, each containing just a few parameters, which is typically far quicker to analyze.

6.4.2 Fourier-domain data space

If one chooses the Fourier domain as the data space, one faces a complementary set of difficulties. Although it is now straightforward to accommodate noise correlations between pixels, it is more difficult to take account of statistically inhomogeneous noise and data cuts. The Fourier transform of the data model Eq. (6.1) reads

$$\widetilde{d}_\nu(\mathbf{k}) = \int d\mathbf{k}' \, \widetilde{W}(\mathbf{k} - \mathbf{k}') \left[\widetilde{B}_\nu(k') \sum_{c=1}^{N_c} \int d\mathbf{k}'' \, \widetilde{F}_{\nu c}(\mathbf{k}' - \mathbf{k}'') \, \widetilde{s}_c(\mathbf{k}'') + \widetilde{n}_\nu(\mathbf{k}') \right],$$
$$\tag{6.5}$$

where $k = |\mathbf{k}|$. From the above expression, we first see the consequences of any data window. Even if the original noise fields and component fields are statistically homogeneous, so that there are no inter-mode correlations, the presence of the window induces correlations between the Fourier modes of the data. Moreover, we see that if the mixing matrix is spatially varying, an additional set of correlations are induced between the modes of the component fields. Computationally, we note that in going from Eq. (6.1) to Eq. (6.5) we have replaced the beam convolutions with a simple multiplication, but also replaced the multiplications by the window and mixing matrix by corresponding convolutions.

It is only when further assumptions are made that choosing the Fourier domain as the data space yields any advantages. If there is no window (or cut) applied to the data and the mixing matrix may be taken as independent of position, the corresponding Fourier transforms in Eq. (6.5) reduce to delta functions and so the expression for the Fourier data reduces to

$$\widetilde{d}_\nu(\mathbf{k}) = \widetilde{B}_\nu(k) \sum_{c=1}^{N_c} F_{\nu c} \widetilde{s}_c(\mathbf{k}) + \widetilde{n}_\nu(\mathbf{k}). \tag{6.6}$$

It is important to note that Eq. (6.6) is satisfied at each Fourier mode **k** *independently*, irrespective of whether the component fields and the noise fields are statistically homogeneous. If both sets of fields are, in fact, statistically homogeneous, or one simply assumes this to be true, then their respective covariance matrices are diagonal in Fourier space. Thus by choosing also to represent the component fields in Fourier space, the entire inference problem can be analyzed mode-by-mode. Hence, at each Fourier mode **k** = **k**$_p$ separately, one has the much smaller inference problem

$$\mathbf{d} = \mathbf{R}(\boldsymbol{\Phi})\mathbf{s} + \mathbf{n}, \tag{6.7}$$

where the data vector **d** now has components $\widetilde{d}_\nu$ (for $\nu = 1, \ldots, N_\nu$) at that Fourier mode, and similar for **n**; the signal vector has components $\widetilde{s}_c$ (for $c = 1, \ldots, N_c$) at that mode. The matrix **R** has components $R_{\nu c}(\boldsymbol{\Phi}) = \widetilde{B}_\nu(\mathbf{k}_p)F_{\nu c}(\boldsymbol{\Phi})$, where $F_{\nu c}(\boldsymbol{\Phi})$ is the same for each mode (since the mixing matrix is assumed to be spatially invariant). In a similar manner to the pixel-by-pixel analysis mentioned above, this replaces a single Bayesian inference problem with many parameters by a large number of inference problems of low dimensionality, which is usually far less computationally demanding.

6.5 Applying Bayes' theorem

Irrespective of the particular forms chosen for the hidden parameters **h** = $\{\boldsymbol{\Phi}, \mathbf{s}, \mathbf{S}\}$ and the data vector **d**, Bayesian inference methods provide a consistent approach to the estimation of the parameters in our model (or hypothesis) H from the data. Bayes' theorem states that

$$\Pr(\mathbf{h}|\mathbf{d}, H) = \frac{\Pr(\mathbf{d}|\mathbf{h}, H)\Pr(\mathbf{h}|H)}{\Pr(\mathbf{d}|H)}, \tag{6.8}$$

where $\Pr(\mathbf{h}|\mathbf{d}, H) \equiv P(\mathbf{h})$ is the posterior probability distribution of the parameters, $\Pr(\mathbf{d}|\mathbf{h}, H) \equiv L(\mathbf{h})$ is the likelihood, $\Pr(\mathbf{h}|H) \equiv \pi(\mathbf{h})$ is the prior, and $\Pr(\mathbf{d}|H) \equiv E$ is the Bayesian evidence.

In parameter estimation, the normalizing evidence factor is usually ignored, since it is independent of the parameters **h**. This (un-normalized) posterior constitutes the complete Bayesian inference of the parameter values. Inferences are usually obtained either by taking samples from the (un-normalized) posterior using Markov chain Monte Carlo (MCMC) methods, or by locating its maximum (or maxima) using numerical optimization routines and approximating the shape around the peak(s) by a multivariate Gaussian.

In contrast to parameter estimation problems, in model selection the evidence takes the central role and is simply the factor required to normalize the posterior over **h**:

$$E = \int L(\mathbf{h})\pi(\mathbf{h})\, d^D\mathbf{h}, \tag{6.9}$$

where D is the dimensionality of the parameter space. As the average of the likelihood over the prior, the evidence is larger for a model if more of its parameter space is likely and smaller for a model with large areas in its parameter space having low likelihood values, even if the likelihood function is very highly peaked. Thus, the evidence automatically implements Occam's razor: a simpler theory with compact parameter space will have a larger evidence than a more complicated one, unless the latter is significantly better at explaining the data. The question of model selection between two models H_0 and H_1 can then be decided by comparing their respective posterior probabilities given the observed dataset **d**, as follows:

$$\frac{\Pr(H_1|\mathbf{d})}{\Pr(H_0|\mathbf{d})} = \frac{\Pr(\mathbf{d}|H_1)\Pr(H_1)}{\Pr(\mathbf{d}|H_0)\Pr(H_0)} = \frac{E_1}{E_2}\frac{\Pr(H_1)}{\Pr(H_0)}, \tag{6.10}$$

where $\Pr(H_1)/\Pr(H_0)$ is the a-priori probability ratio for the two models, which can often be set to unity but occasionally requires further consideration. Unfortunately, evaluation of the multidimensional integral Eq. (6.9) is a challenging numerical task, but sampling-based methods do exist, such as thermodynamic integration (see, e.g., Ó'Ruanaidh & Fitzgerald 1996) or, better, nested sampling (Chapter 1 of this volume; Skilling 2004b; Sivia & Skilling 2006; Feroz & Hobson 2008; Feroz, Hobson & Bridges 2008). Alternatively, a fast approximate method for evidence evaluation is to model the posterior as a multivariate Gaussian centred at its peak(s) (see, e.g., Hobson, Bridle & Lahav 2002).

6.5.1 Defining the posterior distribution

Let us first consider the likelihood $L(\mathbf{h}) \equiv \Pr(\mathbf{d}|\mathbf{h}, H)$ in Eq. (6.8). We will assume that the statistical properties of the instrumental noise vector **n** (in whatever domain is chosen as the data space) are well described by a multivariate Gaussian distribution with known (usually diagonal) covariance matrix $\mathbf{N} = \langle \mathbf{nn}^\mathrm{T}\rangle$. In this case, the likelihood takes the form

$$L(\mathbf{h}) = \frac{\exp\{-\frac{1}{2}[\mathbf{d} - \mathbf{m}(\boldsymbol{\Phi}, \mathbf{s})]^\mathrm{T}\mathbf{N}^{-1}[\mathbf{d} - \mathbf{m}(\boldsymbol{\Phi}, \mathbf{s})]\}}{(2\pi)^{N_{\mathrm{pix}}N_\nu/2}|\mathbf{N}|^{1/2}}, \tag{6.11}$$

where **m** is the data model (also called the predicted or noise-free data) defined by $\mathbf{m} = \lim_{\mathbf{n}\to 0}\mathbf{d}$. Note that the likelihood, in fact, depends only on the sets of parameters $\boldsymbol{\Phi}$ and $\mathbf{s}$. Since we are assuming the noise covariance matrix **N** is known, it will not vary as a function of the hidden space parameters **h**, and so we can neglect the unimportant denominator in Eq. (6.11). Since it is more convenient computationally to work with the log-likelihood, we have

$$\ln L(\mathbf{h}) \propto -\tfrac{1}{2}[\mathbf{d} - \mathbf{m}(\boldsymbol{\Phi},\mathbf{s})]^{\mathrm{T}}\mathbf{N}^{-1}[\mathbf{d} - \mathbf{m}(\boldsymbol{\Phi},\mathbf{s})] \equiv -\tfrac{1}{2}\chi^2(\mathbf{h}), \qquad (6.12)$$

where we have ignored an unimportant arbitrary constant and identified the simple χ^2 misfit function.

Once the likelihood has been defined, all that remains is to specify the prior $\pi(\mathbf{h}) \equiv \Pr(\mathbf{h}|H)$ on the hidden parameters $\mathbf{h} = \{\boldsymbol{\Phi},\mathbf{s},\mathbf{S}\}$ for the model H. The specification of this prior is key to the differences between various signal separation methods used in cosmology. In general, it is reasonable to assume that the mixing matrix parameters $\boldsymbol{\Phi}$ are independent of the signal vector s and its a-priori covariance matrix **S**. Thus, we may write

$$\pi(\mathbf{h}) = \pi(\boldsymbol{\Phi})\pi(\mathbf{s},\mathbf{S}) = \pi(\boldsymbol{\Phi})\pi(\mathbf{s}|\mathbf{S})\pi(\mathbf{S}). \qquad (6.13)$$

A wide range of possibilities are available for the factors $\pi(\boldsymbol{\Phi})$ and $\pi(\mathbf{S})$ above. The priors on the mixing matrix parameters are typically taken to be uniform in some range, or delta functions fixing some of the parameters to pre-defined values. Note that assuming the mixing matrix to be fully known (i.e., in non-blind methods) is equivalent to setting $\pi(\boldsymbol{\Phi}) = \delta(\boldsymbol{\Phi} - \boldsymbol{\Phi}_0)$, where $\boldsymbol{\Phi}_0$ are the pre-defined values of the mixing matrix parameters. This approach is identical to conditioning on the values of $\boldsymbol{\Phi} = \boldsymbol{\Phi}_0$.

The prior $\pi(\mathbf{S})$ must reflect the fact that the individual covariance matrices $\mathbf{S}_c$ for each component are usually assumed to be diagonal, but inter-component correlations at a given pixel or Fourier mode are usually included in the full covariance matrix **S**. The possibilities range from uniform or Jeffreys priors on the elements of **S** to fixing part or all of the covariance matrix. The last option is formally achieved by setting $\pi(\mathbf{S}) = \delta(\mathbf{S} - \mathbf{S}_0)$, which is equivalent to conditioning on $\mathbf{S} = \mathbf{S}_0$.

Once the prior on **S** has been specified, the prior $\pi(\mathbf{s}|\mathbf{S})$ is straightforward to define. The most common choice is to assume a priori that the signal vector s is well described by a multivariate Gaussian with the covariance matrix **S**, in which case,

$$\pi(\mathbf{s}|\mathbf{S}) = \frac{\exp(-\tfrac{1}{2}\mathbf{s}^{\mathrm{T}}\mathbf{S}^{-1}\mathbf{s})}{(2\pi)^{N_{\mathrm{pix}}N_{\mathrm{c}}/2}|\mathbf{S}|^{1/2}}. \qquad (6.14)$$

It is worth noting that it is sometimes useful instead to define a set of a-priori uncorrelated variables $\mathbf{u}$ such that $\mathbf{s} = \mathbf{L}\mathbf{u}$, where $\mathbf{L}$ is the lower triangular matrix in the Cholesky decomposition $\mathbf{S} = \mathbf{L}\mathbf{L}^{\mathrm{T}}$ of the covariance matrix. In this case, since $\langle \mathbf{u}\mathbf{u}^{\mathrm{T}} \rangle$ is the identity matrix a-priori, one has

$$\langle \mathbf{s}\mathbf{s}^{\mathrm{T}} \rangle = \langle \mathbf{L}\mathbf{h}\mathbf{h}^{\mathrm{T}}\mathbf{L}^{\mathrm{T}} \rangle = \mathbf{L}\langle \mathbf{h}\mathbf{h}^{\mathrm{T}} \rangle \mathbf{L}^{\mathrm{T}} = \mathbf{L}\mathbf{L}^{\mathrm{T}} = \mathbf{S}, \tag{6.15}$$

as required. It can occasionally be useful to work in terms of these a-priori uncorrelated variables, especially when using entropic priors (see Section 6.6.2). In such cases, the hidden-space parameters are instead taken as $\mathbf{h} = \{\mathbf{\Phi}, \mathbf{u}, \mathbf{S}\}$, so that $\mathbf{s} \rightarrow \mathbf{u}$ in the expressions (6.11)–(6.13), remembering that $\mathbf{s} = \mathbf{L}\mathbf{u}$. The prior on the uncorrelated variables is then usually written as

$$\pi(\mathbf{u}|\mathbf{S}) \propto \exp\left[\sum_p f(u_p)\right], \tag{6.16}$$

where $f(u_p)$ is some given function and p counts the elements in $\mathbf{u}$. If the component fields are assumed to obey Gaussian statistics, then the prior Eq. (6.14) is recovered by setting $f(u_p) = -u_p^2/2$ to obtain

$$\pi(\mathbf{u}|\mathbf{S}) \propto \exp(-\tfrac{1}{2}\mathbf{u}^{\mathrm{T}}\mathbf{u}). \tag{6.17}$$

6.6 Non-blind signal separation

First, we concentrate on non-blind signal separation where we assume the mixing matrix parameters $\mathbf{\Phi}$ to be completely specified. Moreover, we shall assume throughout that (at least some approximation to) the signal covariance matrix $\mathbf{S}$ is also given in advance. This may seem perverse, as we then require advance knowledge of the full covariance structure of the physical components that we are trying to reconstruct. Nonetheless, it is possible to extract some information concerning the power spectra of the various components, and correlations between them, either from pre-existing observations or by performing an initial approximate separation using, for example, the singular value decomposition (SVD) algorithm (see Bouchet *et al.* 1997; Bouchet & Gispert 1999). This information can then be used to construct an approximate signal covariance matrix. Indeed, the SMICA method discussed in Section 6.7.2 provides a 'pre-processing' step that determines (some of) both the mixing matrix parameters $\mathbf{\Phi}$ and the signal covariance matrix $\mathbf{S}$. The non-blind methods presented below can then be considered merely as the final stage of a larger (semi-)blind approach.

6.6.1 Wiener filtering

The Wiener filter (WF) is the minimum mean squared error (MMSE) optimal stationary linear filter (Kolmogorov 1939; Levinson 1947; Wiener 1943) for convolved images with additive noise. Calculation of the Wiener filter requires the assumption that the signal and noise processes are second-order stationary (in the random process sense). Wiener filtering was extended to multi-frequency and multi-resolution microwave data in Tegmark and Efstathiou (1996); Bouchet *et al.* (1997) and Bouchet and Gispert (1999). Below we will derive the WF in a Bayesian context following Hobson *et al.* (1998).

WF posterior

In applying the WF it is usual to take the data and the component fields to be represented in the Fourier domain (although this is not necessary). Moreover, one assumes (at least formally) that no window (or cut) has been applied to the data, the mixing matrix components are independent of position and fixed, and that each component field and the noise fields in each channel are statistically homogeneous. Thus, the data model is that given in Eq. (6.7), with the further simplification that the mixing matrix parameters are fixed, $\mathbf{\Phi} = \mathbf{\Phi}_0$. Thus, at each Fourier mode $\mathbf{k}$ separately, one has the small inference problem

$$\mathbf{d} = \mathbf{R}\mathbf{s} + \mathbf{n}, \tag{6.18}$$

where the matrix $\mathbf{R}$ has the fixed components $R_{\nu c} = \widetilde{B}_\nu(\mathbf{k}_p) F_{\nu c}(\mathbf{\Phi}_0)$, where $F_{\nu c}$ is the same for each mode (since the mixing matrix is assumed to be spatially invariant). Thus, the χ^2 misfit statistic defined in Eq. (6.12) is a function of $\mathbf{s}$ only and can be written as

$$\chi^2(\mathbf{s}) = (\mathbf{d} - \mathbf{R}\mathbf{s})^{\mathrm{T}} \mathbf{N}^{-1} (\mathbf{d} - \mathbf{R}\mathbf{s}). \tag{6.19}$$

In addition, it is assumed that the signal covariance matrix is also given in advance, so $\mathbf{S} = \mathbf{S}_0$. Thus, the only free parameters in the hidden space are $\mathbf{h} = \mathbf{s}$. Finally, the key assumption in the WF is that the emission from each of the physical components can be modelled by a Gaussian random field. Thus, the prior on $\mathbf{s}$ follows from Eq. (6.14), namely,

$$\pi(\mathbf{s}|\mathbf{S}_0) \propto \exp(-\tfrac{1}{2}\mathbf{s}^{\mathrm{T}}\mathbf{S}_0^{-1}\mathbf{s}). \tag{6.20}$$

Combining Eqs. (6.19) and (6.20) we thus find that the posterior distribution $P(\mathbf{s}) \equiv \Pr(\mathbf{s}|\mathbf{d}, \mathbf{\Phi}_0, \mathbf{S}_0, H)$ on $\mathbf{s}$ is given by

$$P(\mathbf{s}) \propto \exp\left[-\tfrac{1}{2}\chi^2(\mathbf{s}) - \tfrac{1}{2}\mathbf{s}^{\mathrm{T}}\mathbf{S}_0^{-1}\mathbf{s}\right]. \tag{6.21}$$

Completing the square for s in the exponential (see Zaroubi *et al.* 1995), it is straightforward to show that the posterior probability is also a multivariate Gaussian of the form

$$P(\mathbf{s}) \propto \exp\left[-\tfrac{1}{2}(\mathbf{s} - \hat{\mathbf{s}})^{\mathrm{T}}\mathbf{C}^{-1}(\mathbf{s} - \hat{\mathbf{s}})\right], \tag{6.22}$$

which has its maximum value at the estimate $\hat{\mathbf{s}}$ of the signal vector and where $\mathbf{C}$ is the covariance matrix of the reconstruction errors.

Optimal values and error estimates

The estimate $\hat{\mathbf{s}}$ of the signal vector in Eq. (6.22) reads

$$\hat{\mathbf{s}} = \left(\mathbf{S}_0^{-1} + \mathbf{R}^{\mathrm{T}}\mathbf{N}^{-1}\mathbf{R}\right)^{-1}\mathbf{R}^{\mathrm{T}}\mathbf{N}^{-1}\mathbf{d} \equiv \mathbf{W}\mathbf{d}, \tag{6.23}$$

where we have defined $\mathbf{W}$, the Wiener matrix. We thus recover the standard Wiener filter. This optimal linear filter is usually derived by choosing the elements of $\mathbf{W}$ such that they minimize the variances of the resulting reconstruction errors (minimum mean squared error, or MMSE filter). From Eq. (6.23) we see that at a given Fourier mode, we may calculate the estimator $\hat{\mathbf{s}}$ that maximizes the posterior probability simply by multiplying the data vector $\mathbf{d}$ by the Wiener matrix $\mathbf{W}$. The assignment of errors on the Wiener filter reconstruction is also straightforward and the covariance matrix of the reconstruction errors $\mathbf{C}$ in Eq. (6.23) is given by

$$\mathbf{C} \equiv \langle(\mathbf{s} - \hat{\mathbf{s}})(\mathbf{s} - \hat{\mathbf{s}})^{\mathrm{T}}\rangle = \left(\mathbf{S}_0^{-1} + \mathbf{R}^{\mathrm{T}}\mathbf{N}^{-1}\mathbf{R}\right)^{-1}. \tag{6.24}$$

Alternatively, one may adopt an iterative approach and obtain the Wiener filter solution by direct numerical maximization of the posterior Eq. (6.21). Writing $P(\mathbf{s}) \propto \exp[-F_{\mathrm{WF}}(\mathbf{s})]$, this is equivalent to minimizing the function

$$F_{\mathrm{WF}}(\mathbf{s}) = \tfrac{1}{2}\chi^2(\mathbf{s}) + \tfrac{1}{2}\mathbf{s}^{\mathrm{T}}\mathbf{S}_0^{-1}\mathbf{s}, \tag{6.25}$$

where we have retained the factors of one-half for later convenience. Similarly, the covariance matrix $\mathbf{C}$ of the reconstruction residuals can be obtained by evaluating the curvature (or Hessian) matrix $\mathbf{H} = \nabla_{\mathbf{s}}\nabla_{\mathbf{s}} \ln P(\mathbf{s}) = -\nabla_{\mathbf{s}}\nabla_{\mathbf{s}}F_{\mathrm{WF}}(\mathbf{s})$ at the peak $\mathbf{s} = \hat{\mathbf{s}}$ of the posterior, and then setting $\mathbf{C} = (-\mathbf{H})^{-1}$. Since the posterior Eq. (6.21) is in fact a multivariate Gaussian in this case, no approximation is involved.

It should be noted that the linear nature of the Wiener filter and the simple propagation of errors are both direct consequences of assuming that the spatial templates we wish to reconstruct are well described by Gaussian random fields with a known covariance structure.

Finally, the reconstructed maps of the sky emission due to each physical component are obtained by inverse Fourier transformation of the estimated signal vectors $\hat{\mathbf{s}}$ at each Fourier mode. Since this operation is linear, the errors on these maps may therefore be deduced straightforwardly from the above error covariance matrix. Several applications are given in Bouchet *et al.* (1997); an example of component separation of simulated Planck data is shown in Hobson *et al.* (1998).

6.6.2 Harmonic-space maximum-entropy method

The sky emission from most astrophysical components is known to be neither Gaussian nor stationary. This is particularly true for the kinetic and thermal SZ effects, but the Galactic dust and free–free emissions also appear quite non-Gaussian. The maximum-entropy method (MEM) uses non-Gaussian entropic priors that are based on information-theoretic considerations. The concept of entropy in information theory was first introduced by Shannon (1948).

The application of the MEM to cosmological signal separation was first discussed in Hobson *et al.* (1998). In this approach, many of the assumptions made are the same as those outlined above for the WF. In particular, the data and the component fields are represented in the Fourier domain (although this is not necessary). One assumes (at least formally) that no window (or cut) has been applied to the data, the mixing matrix components are independent of position and fixed, and that each component field and the noise fields in each channel are statistically homogeneous. Thus, the data model is precisely that assumed in Eq. (6.18) for the WF, in which the data are analyzed mode-by-mode. In addition, it is again assumed that the signal covariance matrix is also given in advance, so $\mathbf{S} = \mathbf{S}_0$.

Harmonic-space MEM posterior

As discussed in Hobson *et al.* (1998), however, one of the fundamental axioms of the MEM is that it should not itself introduce correlations between individual elements of the image. It is therefore necessary to work in terms of the a-priori uncorrelated variables $\mathbf{u}$ described in Section 6.5.1, which are then the only free parameters in the hidden space, i.e., $\mathbf{h} = \mathbf{u}$. Since $\mathbf{s} = \mathbf{L}\mathbf{u}$, the χ^2 misfit function in Eq. (6.19) becomes

$$\chi^2(\mathbf{u}) = (\mathbf{d} - \mathbf{RLu})^{\mathrm{T}}\mathbf{N}^{-1}(\mathbf{d} - \mathbf{RLu}). \tag{6.26}$$

It may be shown (Skilling 1989; Gull & Skilling 1990; Hobson & Lasenby 1998) that, in the absence of any other information, the prior on the variables $\mathbf{u}$ should take the form of Eq. (6.16) with $f(u_p) = \alpha S(u_p, \mu_p)$, where the dimensional constant α depends on the scaling of the problem and may be considered as a regularizing parameter, and the *cross-entropy* is defined by

$$S(u_p, \mu_p) = \psi_p - 2\mu_p - u_p \ln\left(\frac{\psi_p + u_p}{2\mu_p}\right). \tag{6.27}$$

In this expression $\psi_p = (u_p^2 + 4\mu_p^2)^{1/2}$ and μ_p determines the width of the cross-entropy function as a function of u_p. We note that $S(u_p, \mu_p)$ takes its maximum value of zero at $u_p = 0$, independent of the value of μ_p. Writing $S(\mathbf{u}, \boldsymbol{\mu}) = \sum_p S(u_p, \mu_p)$, the prior thus becomes

$$\pi(\mathbf{u}|\mathbf{S}_0, \alpha) \propto \exp[\alpha S(\mathbf{u}, \boldsymbol{\mu})]. \tag{6.28}$$

One can show (Maisinger *et al.* 2004) that an appropriate choice is to set μ_p proportional to the expected rms of u_p. Since each u_p is constructed to have unit variance, it is convenient (see below) to set $\mu_p = \frac{1}{2}$ for all p.

Optimal values and error estimates

Combining Eqs. (6.26) and (6.28), we learn that the posterior probability $P(\mathbf{u}|\alpha) \equiv \text{Pr}(\mathbf{u}|\mathbf{d}, \boldsymbol{\Phi}_0, \mathbf{S}_0, \alpha, H)$ on $\mathbf{u}$ is given by

$$P(\mathbf{u}|\alpha) \propto \exp\left[-\tfrac{1}{2}\chi^2(\mathbf{u}) + \alpha S(\mathbf{u}, \boldsymbol{\mu})\right]. \tag{6.29}$$

Since the cross-entropy is a non-linear function of $\mathbf{u}$, one must locate the maximum $\mathbf{u} = \hat{\mathbf{u}}$ of the posterior by numerical iterative optimization. Writing $P(\mathbf{u}|\alpha) \propto \exp[-F_{\text{MEM}}(\mathbf{u}; \alpha)]$, this is equivalent to minimizing the function

$$F_{\text{MEM}}(\mathbf{u}; \alpha) = \tfrac{1}{2}\chi^2(\mathbf{u}) - \alpha S(\mathbf{u}, \boldsymbol{\mu}). \tag{6.30}$$

This task can be performed with a variety of numerical optimization methods, such as downhill simplex, conjugate gradient or variable metric methods in multi-dimensions (Press *et al.* 1997). The corresponding estimate of the signal vector, for the given value of α used in Eq. (6.30), is then given by $\hat{\mathbf{s}}(\alpha) = \mathbf{L}\hat{\mathbf{u}}(\alpha)$.

The covariance matrix on the reconstruction residuals can be estimated by first calculating the Hessian matrix $\mathbf{H} = \nabla_{\mathbf{u}}\nabla_{\mathbf{u}} \ln P(\mathbf{u}) = -\nabla_{\mathbf{u}}\nabla_{\mathbf{u}} F_{\text{MEM}}(\mathbf{u}; \alpha)$ at the peak $\mathbf{u} = \hat{\mathbf{u}}$ of the posterior. The required covariance matrix is then given approximately by

$$\mathbf{C} = \langle(\mathbf{s} - \hat{\mathbf{s}})(\mathbf{s} - \hat{\mathbf{s}})^{\text{T}}\rangle \approx -\mathbf{L}\mathbf{H}^{-1}\mathbf{L}^{\text{T}}. \tag{6.31}$$

We note that this procedure is equivalent to making a Gaussian approximation to the posterior Eq. (6.29) about its maximum $\mathbf{u} = \hat{\mathbf{u}}$. As for the WF, the reconstructed map of the sky emission due to each physical component is obtained by inverse Fourier transformation of the estimated signal vectors $\hat{\mathbf{s}}$ at each Fourier mode. Since this operation is linear, the errors on these maps may therefore be deduced straightforwardly from the above error covariance matrix.

Relationship between the MEM and WF

It is worth noting that the WF solution discussed in the previous section can also be expressed in terms of the set of a-priori uncorrelated variables $\mathbf{u}$, and this highlights the relationship between the harmonic-space MEM and WF approaches. In terms of $\mathbf{u}$, the WF prior Eq. (6.20) takes the simple form given in Eq. (6.17). Thus, the corresponding WF posterior then reads $P(\mathbf{u}) \propto \exp[-F_{\mathrm{WF}}(\mathbf{u})]$, where

$$F_{\mathrm{WF}}(\mathbf{u}) = \tfrac{1}{2}\chi^2(\mathbf{u}) + \tfrac{1}{2}\mathbf{u}^{\mathrm{T}}\mathbf{u}, \qquad (6.32)$$

and $\chi^2(\mathbf{u})$ is given by Eq. (6.26). The connection with the MEM is now clarified by expanding the expression Eq. (6.27) for the cross-entropy around its peak at $u_p = 0$ as a power series in u_p, which reads

$$S(u_p, \mu_p) = -\frac{u_p^2}{4\mu_p} + \mathcal{O}(u_p^4). \qquad (6.33)$$

Since we have taken $\mu_p = \tfrac{1}{2}$ for all p, we thus find that the functions $F_{\mathrm{MEM}}(\mathbf{u}; \alpha)$ and $F_{\mathrm{WF}}(\mathbf{u})$, given in Eqs. (6.30) and (6.32) respectively, are *equal* to second-order in $\mathbf{u}$, provided $\alpha = 1$. Thus, the WF can be regarded as the quadratic approximation to the MEM in this case.

Determination of the regularization constant

So far, we have made no mention of how to choose the parameter α, which determines the amount of regularization on the reconstruction. It is clear that minimizing χ^2 only (by setting $\alpha = 0$) would lead to closer agreement with the data, and thus to noise fitting. On the other hand, maximizing the entropy alone (by setting $\alpha = \infty$) would lead to the reconstruction $\hat{\mathbf{u}} = \mathbf{0}$. Indeed, for every choice of α there is an estimated vector $\hat{\mathbf{u}}(\alpha)$ (and hence $\hat{\mathbf{s}}(\alpha)$) corresponding to the minimum of $F_{\mathrm{MEM}}(\mathbf{u}; \alpha)$ for that particular choice. The estimates $\hat{\mathbf{u}}(\alpha)$ vary along a trade-off curve (sometimes called the maximum-entropy trajectory) as α is varied.

There are several methods for assigning an optimal value for α. As we have shown above, setting $\alpha = 1$ recovers the WF to second order, and choosing this fixed value is often a reasonable approach. Nonetheless, in general, it is preferable to determine the appropriate value of α (and hence the level of regularization) from the data themselves. In early MEM applications, the 'optimal' value $\alpha = \hat{\alpha}$ was chosen such that at the estimate $\hat{\mathbf{u}}(\hat{\alpha})$ the value of χ^2 equalled its expectation value, i.e., $\chi^2 = N_d$, where N_d is the number of data values. However, it can be shown (Titterington 1985) that this choice leads to systematic under-fitting of the data. Therefore, an alternative method has been proposed (Gull 1989; Gull & Skilling 1990) in which a value for α is obtained within the Bayesian framework itself. Treating α as an additional (nuisance) parameter, one would ideally remove the dependence of the posterior on this parameter by marginalizing over α. Thus (omitting the other conditional parameters for brevity), one has

$$\Pr(\mathbf{u}|\mathbf{d}) = \int \Pr(\mathbf{u}|\mathbf{d}, \alpha) \Pr(\alpha|\mathbf{d}) \, \mathrm{d}\alpha. \qquad (6.34)$$

This integral is usually too computationally demanding to perform, and so one instead chooses an optimal value $\alpha = \hat{\alpha}$ by maximizing $\Pr(\alpha|\mathbf{d}) \propto \Pr(\mathbf{d}|\alpha)\Pr(\alpha)$. Moreover, since $\Pr(\mathbf{d}|\alpha)$ should overwhelm any prior on α, one in fact performs a maximization with respect to α of the evidence

$$\Pr(\mathbf{d}|\alpha) = \int \Pr(\mathbf{d}|\mathbf{u})\Pr(\mathbf{u}|\alpha)\,\mathrm{d}\mathbf{u} = \int L(\mathbf{u})\pi(\mathbf{u}|\alpha)\,\mathrm{d}\mathbf{u} \propto \int P(\mathbf{u}|\alpha)\,\mathrm{d}\mathbf{u}, \tag{6.35}$$

where $P(\mathbf{u}|\alpha)$ is the MEM posterior given in Eq. (6.29). This integral is usually performed by making a Gaussian approximation to the posterior about its peak $\hat{\mathbf{u}}(\alpha)$. Differentiating the resulting expression with respect to α leads to an implicit analytic expression for α, which may be solved numerically to yield the optimal value $\alpha = \hat{\alpha}$ (Gull & Skilling 1990; Hobson *et al.* 1998).

It is worth noting that an equivalent procedure can be used within the WF approach. In other words, one could replace the WF objective function $F_{\mathrm{WF}}(\mathbf{u})$, given in Eq. (6.32), by

$$F_{\mathrm{WF}}(\mathbf{u};\alpha) = \tfrac{1}{2}\chi^2(\mathbf{u}) + \tfrac{1}{2}\alpha\mathbf{u}^{\mathrm{T}}\mathbf{u}, \tag{6.36}$$

and determine the optimal value of the regularization constant α in an analogous way to that presented above.

Iterative updating of the signal covariance matrix

Aside from determining the optimal level of regularization on the solution, Hobson *et al.* (1998) found that the above approach for determining the regularization constant α allowed one to update iteratively the assumed signal covariance matrix $\mathbf{S}_0$. In this process, one uses the covariance matrix of the estimated signals as the input $\mathbf{S}_0$ to a new separation step, and repeats until convergence is obtained. This mitigates, to some extent, the reliance of the MEM approach on prior knowledge of the signal covariance matrix. This method was further demonstrated in Stolyarov *et al.* (2002). Figure 6.3 shows examples obtained in this way of all-sky reconstructions of the components in Figure 6.1, obtained by analyzing the frequency channel data shown in Figure 6.2. The MEM approach should be contrasted with the standard WF, which is well known to suppress power in high-k Fourier modes, where the signal-to-noise ratio becomes small. As a result, if the covariance matrix of the reconstructed signals is used as the input $\mathbf{S}_0$ to a subsequent WF separation, the solution gradually tends to zero as more iterations are performed.

Accommodation of spatially varying noise and spectral parameters

We conclude our discussion of the harmonic-space MEM separation with a brief account of how one can accommodate (at least approximately), the common

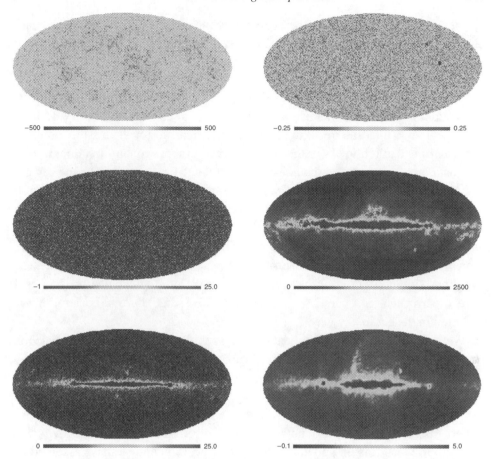

Fig. 6.3. All-sky reconstructions obtained using the harmonic-space MEM of the components in Figure 6.1, derived by analyzing the frequency channel data shown in Figure 6.2. (Reproduced from Stolyarov *et al.* 2002.)

complications of spatially varying noise properties and spectral parameters. We note that although these partial fixes have been developed in the context of the MEM (Stolyarov *et al.* 2005), they can equally well be applied to the WF separation method, which also neglects coupling between different k-modes.

Let us first consider the case of anisotropic instrument noise, which we represent as a non-stationary Gaussian random process in pixel space. This leads to correlations between different Fourier modes of the noise field. The elements of the noise covariance matrix in Fourier space are defined as

$$[N(\mathbf{k}, \mathbf{k}')]_{\nu\nu'} \equiv \langle \widetilde{n}_\nu(\mathbf{k}) \widetilde{n}_{\nu'}(\mathbf{k}') \rangle, \tag{6.37}$$

which are then, in general, non-zero for $\mathbf{k} \neq \mathbf{k}$ (and for $\nu \neq \nu'$, if there are correlations between the noise fields in different frequency channels). Stolyarov *et al.* (2005) advocate setting the elements of the *inverse* noise covariance matrix appearing in the χ^2 misfit function Eq. (6.26) directly, such that at each Fourier mode **k** one has

$$[N^{-1}]_{\nu\nu'} = [N^{-1}(\mathbf{k}, \mathbf{k})]_{\nu\nu'}. \tag{6.38}$$

Note that this is *not*, in general, equivalent simply to setting $[N]_{\nu\nu'} = [N(\mathbf{k}, \mathbf{k})]_{\nu\nu'}$ and then inverting the resulting matrix. The reason for using the definition Eq. (6.38) is that it allows for the straightforward (approximate) treatment of cut-sky data, particularly when the noise is uncorrelated between pixels. One can consider the missing areas in a cut-sky map as an extreme case of anisotropic noise in which the noise rms of the pixels in the cut is formally infinite. This leads to elements of the noise covariance matrix that are also formally infinite, which can cause computational problems in the analysis. By using the assignment Eq. (6.38), however, a formally infinite noise rms in the cut is easily accommodated: it corresponds simply to omitting the pixels in the cut from the calculation. One is, in effect, performing a separation over the entire sky patch, but not constraining the solution in the cut in any way.

We now turn to the accommodation of (weakly) spatially varying spectral parameters. In general, the sky emission from the signal component c in observing channel ν is

$$s_c(\mathbf{x}; \nu) = s_c(\mathbf{x}) F_{\nu c}(\phi_c(\mathbf{x})), \tag{6.39}$$

where (as previously) $s_c(\mathbf{x})$ denotes the signal at some reference frequency $\nu = \nu_0$. The mixing matrix is, in general, a function of position and we have assumed it is parameterized in terms of a set of physically motivated spectral parameters $\phi_c(\mathbf{x})$. As noted in Section 6.4.2, however, the choice of the Fourier domain as the data space (as used in WF and harmonic-space MEM) requires one to assume a fixed mixing matrix if one is to avoid a considerable additional computational burden. Nonetheless, Stolyarov *et al.* (2005) propose a (partial) solution to this problem by performing a Taylor expansion in the spectral parameters around their assumed mean values $\bar{\phi}_c$. Thus, one writes

$$s_c(\mathbf{x}; \nu) = s_c(\mathbf{x}) \left[F_{\nu c}(\bar{\phi}_c) + \sum_j \Delta\phi_{c,j}(\mathbf{x}) \left. \frac{\partial F_{\nu c}}{\partial \phi_{c,j}} \right|_{\phi_{c,j} = \bar{\phi}_{c,j}} + \cdots \right], \tag{6.40}$$

where j labels the spectral parameters in ϕ_c and $\Delta\phi_{c,j}(\mathbf{x}) = \phi_{c,j}(\mathbf{x}) - \bar{\phi}_{c,j}$. It is now possible to use the standard formalism to estimate several separate (but highly correlated) fields, namely $s_c(\mathbf{x})$ and $s_c(\mathbf{x})\Delta\phi_{c,j}(\mathbf{x})$ (for all j), from which one can then obtain $s_c(\mathbf{x})$ and $\Delta\phi_{c,j}(\mathbf{x})$ separately. If necessary, second- and higher-order terms in the Taylor expansion can be employed to obtain an increasingly accurate approximation to $s_c(\mathbf{x}, \nu)$. One must remember, however, that one is typically con-strained such that the total number of fields to be estimated does not exceed the number of frequency channels at which observations are made.

6.6.3 Mixed-space maximum-entropy method

As our final non-blind signal separation technique, we briefly consider the mixed-space MEM proposed by Barreiro *et al.* (2004), which circumvents many of the problems associated with wholly Fourier-domain algorithms such as WF and harmonic-space MEM, but at the cost of considerably larger computational require-ments. In fact, in this approach, one does ultimately estimate (spatially constant) spectral parameters, so it does go beyond a simple non-blind analysis, but it is best considered here, alongside non-blind techniques.

In this method, one again begins by assuming fixed, spatially constant values for the mixing matrix parameters, $\boldsymbol{\Phi} = \boldsymbol{\Phi}_0$. One also assumes that the covariance matrix of the signals is known, $\mathbf{S} = \mathbf{S}_0$, and works in terms of a-priori uncorre-lated variables. Thus, once again, the only free parameters (initially) in the hidden space are $\mathbf{h} = \mathbf{u}$, and these are represented in the Fourier domain, as above. More-over, the prior on these variables is again taken as the entropic form Eq. (6.28), namely,

$$\pi(\mathbf{u}|\mathbf{S}_0, \alpha) \propto \exp[\alpha S(\mathbf{u}, \boldsymbol{\mu})], \tag{6.41}$$

with $\mu_p = \frac{1}{2}$ for all p, although the WF quadratic approximation to the entropy is another possibility.

The key difference in the mixed-space MEM is that the data space is now chosen to be the pixel domain, rather than the Fourier domain. Thus, the χ^2 misfit statistic in Eq. (6.26) is replaced by

$$\chi^2(\mathbf{u}; \boldsymbol{\Phi}) = (\mathbf{d} - \mathcal{F}^{-1}\mathbf{R}\mathbf{L}\mathbf{u})^{\mathrm{T}}\mathbf{N}^{-1}(\mathbf{d} - \mathcal{F}^{-1}\mathbf{R}\mathbf{L}\mathbf{u}), \tag{6.42}$$

where $\mathbf{N}$ is now the noise covariance matrix in the pixel domain, and $\mathcal{F}^{-1}$ denotes a (real) inverse Fourier transform matrix (for data defined on the celestial sphere, the Fourier transform would be replaced by a spherical harmonic transform). For later convenience, we have also made explicit the dependence of χ^2 on the spectral parameters $\boldsymbol{\Phi}$.

By choosing the data space to be the pixel domain, anisotropic noise and sky-cuts can easily be accommodated in the analysis (although inter-pixel noise corre-lations are harder to incorporate). One could, in principle, also allow the mixing matrix parameters to vary with position, but this was not considered by Barreiro *et al.* (2004). On the other hand, by still representing the component fields in the

Fourier domain, one preserves the advantage of easily including a-priori (cross) power spectrum information in the algorithm.

Combining Eqs. (6.41) and (6.42), we see that the posterior again has the form Eq. (6.29) and its maximum is located by numerical minimization of Eq. (6.30), namely

$$F_{\mathrm{MEM}}(\mathbf{u}; \boldsymbol{\Phi}, \alpha) = \tfrac{1}{2}\chi^2(\mathbf{u}; \boldsymbol{\Phi}) - \alpha S(\mathbf{u}, \boldsymbol{\mu}), \qquad (6.43)$$

but with χ^2 now given by Eq. (6.42). Since the reconstruction is obtained in Fourier space, but the misfit to the observed data is calculated in pixel space, one is not able to perform a mode-by-mode optimization, and coupling between Fourier modes is now taken into account. Unfortunately, the computational price one has to pay is a *single* minimization for *all* the variables $\mathbf{u}$ simultaneously. Moreover, each evaluation of the objective function (and its derivatives) requires inverse Fourier transforms to be performed. The method is thus much slower than harmonic-space MEM, and is only applicable to low-resolution signal separation problems; Barreiro *et al.* (2004) apply the method to COBE data.

For a given value of α, the covariance matrix on the reconstruction residuals is again estimated as in Eq. (6.31), although this matrix is now much larger. The optimal value of the regularization parameter α is again determined using the method outlined in the previous section, but it is a more computationally demanding calculation in this case. Details are presented in Barreiro *et al.* (2004), but it is worth mentioning here that the use of optimal regularization again allows for the signal covariance matrix $\mathbf{S}_0$ of the components to be updated iteratively.

Thus, for any given set of the (assumed spatially constant) spectral parameters $\boldsymbol{\Phi}$, one iterates the signal covariance matrix to convergence, determining the optimal value $\hat{\alpha}$ of the regularization parameter at each iteration, to obtain an estimated signal separation $\hat{\mathbf{s}}(\hat{\alpha}) = \mathbf{L}\hat{\mathbf{u}}(\hat{\alpha})$. The estimation of the spectral parameters is then performed by repeating this entire procedure with different sets of spectral parameter values with the aim of minimizing an empirically defined selection function. In Barreiro *et al.* (2004), this function is a weighted linear combination of: the χ^2 value of the estimate $\hat{\mathbf{s}}(\hat{\alpha})$, its entropy, the cross-correlations between the estimates of the CMB and Galactic components and the dispersion of the estimated CMB component. The best set of weights to use in the linear combination were determined empirically from the analysis of simulated datasets.

A related, but more Bayesian, approach to the estimation of the spectral parameters (not explored by Barreiro *et al.*) would be to maximize the posterior distribution $\Pr(\boldsymbol{\Phi}|\mathbf{d}, \hat{\alpha}) \propto \Pr(\mathbf{d}|\boldsymbol{\Phi}, \hat{\alpha}) \Pr(\boldsymbol{\Phi})$, where the evidence factor $\Pr(\mathbf{d}|\boldsymbol{\Phi}, \hat{\alpha})$ is given by

$$\Pr(\mathbf{d}|\boldsymbol{\Phi}, \hat{\alpha}) = \int \Pr(\mathbf{d}|\mathbf{u}, \boldsymbol{\Phi}) \Pr(\mathbf{u}|\hat{\alpha}) \, d\mathbf{u} \propto \int P(\mathbf{u}|\hat{\alpha}, \boldsymbol{\Phi}) \, d\mathbf{u}. \qquad (6.44)$$

In this expression, $P(\mathbf{u}|\hat{\alpha}, \mathbf{\Phi}) \propto \exp[-F_{\mathrm{MEM}}(\mathbf{u}; \hat{\alpha}, \mathbf{\Phi})]$ is the MEM posterior given the current set of spectral parameters, and the integral may be evaluated by making a Gaussian approximation, as discussed in Section 6.6.2.

6.7 (Semi-)blind signal separation

In this section, we move on to consider methods for both semi-blind and blind signal separation methods. As mentioned in Section 6.3.1, the usual (somewhat arbitrary) distinction between semi-blind and blind methods is that the former parameterize the mixing matrix with some (physical) spectral parameters ϕ, whereas the latter use the mixing matrix elements $F_{\nu c}$ themselves as the parameters. Here we combine the discussion of these classes of methods, since they are so similar. In each case, there is a wide range of possibilities for the assumed (partial) knowledge of parameters or imposed constraints, which may be quantified by simply defining the prior probability assigned to the parameter set $\mathbf{\Phi}$ (see Section 6.5.1). Moreover, we begin by assuming the number of components N_c to be fixed in advance, although we will relax this condition in Section 6.7.4.

6.7.1 Pixel-domain parameter estimation

A direct approach to cosmological signal separation, first advocated by Brandt *et al.* (1994), is to treat it as a traditional parameter estimation problem in the pixel domain; this approach has since been refined by Eriksen *et al.* (2006) and further extended by Dunkley *et al.* (2009).

In this approach, one works with data convolved to a common resolution, as in Eq. (6.2), and takes both the data space and the parameterization of the component fields to be in the pixel domain. It is further assumed that the window function $W(\mathbf{x})$ in Eq. (6.2) is either unity or zero, and one simply ignores all the pixels in which it is zero. Thus, in general, the full data vector containing the values of all the pixels in the unmasked region in all the frequency channels is given by

$$\mathbf{d} = \mathbf{F}(\mathbf{\Phi})\mathbf{s} + \mathbf{n}, \tag{6.45}$$

where we have omitted the overbars used in Eq. (6.2) for the sake of notational simplicity. Expressing this in component form, we have

$$d_{\nu p} = \sum_{c=1}^{N_c} F_{\nu c}(\mathbf{\Phi}) s_{cp} + n_{\nu p}, \tag{6.46}$$

where p labels pixels. Typically the mixing matrix is expressed in terms of a set of physical parameters $\mathbf{\Phi} = \{\phi(x)\}$ which may, in general, depend on position (see Section 6.3.1).

Recall that our full hidden space of parameters is usually $\mathbf{h} = \{\boldsymbol{\Phi}, \mathbf{s}, \mathbf{S}\}$. In this semi-blind separation method, one usually attempts to make an inference on the mixing matrix parameters $\boldsymbol{\Phi}$ and the signal amplitudes $\mathbf{s}$ simultaneously, but not on the signal covariance matrix $\mathbf{S}$ (although the general methodology does allow for this). The assumed likelihood has the usual Gaussian form Eq. (6.11) and the prior has the general form Eq. (6.13).

In the opinion of the current authors, the best methodology for setting the prior Eq. (6.13) is first to write $\pi(\mathbf{s}, \mathbf{S}) = \pi(\mathbf{s}|\mathbf{S})\pi(\mathbf{S})$. If one does not wish to infer $\mathbf{S}$, then one can simply fix it to some known value $\mathbf{S}_0$, so that $\pi(\mathbf{S}) = \delta(\mathbf{S} - \mathbf{S}_0)$, and then assume the corresponding prior on the signal amplitudes to have the Gaussian form Eq. (6.14), namely,

$$\pi(\mathbf{s}|\mathbf{S}_0) = \frac{\exp(-\frac{1}{2}\mathbf{s}^{\mathrm{T}}\mathbf{S}_0^{-1}\mathbf{s})}{(2\pi)^{N_{\mathrm{pix}}N_c/2}|\mathbf{S}_0|^{1/2}}. \tag{6.47}$$

As outlined below, however, neither Eriksen *et al.* nor Dunkley *et al.* follow this prescription. The prior on the (physical) mixing matrix parameters $\pi(\boldsymbol{\Phi})$ will depend on the form of the model chosen, so we will not specify it further here.

Combining the priors and likelihood using Bayes' theorem, we obtain the joint posterior distribution $P(\mathbf{s}, \boldsymbol{\Phi}) \equiv \Pr(\mathbf{s}, \boldsymbol{\Phi}|\mathbf{d}, \mathbf{S}_0, h)$, which is given by

$$P(\mathbf{s}, \boldsymbol{\Phi}) = \frac{L(\mathbf{s}, \boldsymbol{\Phi})\pi(\mathbf{s}|\mathbf{S}_0)\pi(\boldsymbol{\Phi})}{E}. \tag{6.48}$$

In general, this posterior has a large number of parameters and we must find an efficient way of exploring it. We shall now look at how we can use this posterior distribution to make inferences under a number of assumptions about the properties of the signals and noise. As discussed in Section 6.3.3, the model $\mathbf{m}$ for the data that enters into the likelihood function is typically linear in the signal amplitudes $\mathbf{s}$, but non-linear in the physical mixing matrix parameters $\boldsymbol{\Phi}$.

Uncorrelated signals and noise

As mentioned in Section 6.4.1, when both the data and component fields are described in the pixel domain, the simplest approach is to ignore any spatial correlations in the components and noise fields (as advocated by Eriksen *et al.* 2006). This allows one to perform a pixel-by-pixel signal separation, where at each pixel one has the much lower dimensionality inference problem given by Eq. (6.4).

Formally, this assumption means that the likelihood can be factorized $L = \prod_{p=1}^{N_{\mathrm{pix}}} L_p$ into the individual likelihoods for each pixel. Thus, for the pth pixel, one has

$$L_p(\mathbf{s}_p, \boldsymbol{\Phi}_p) = \frac{\exp\{-\frac{1}{2}[\mathbf{d}_p - \mathbf{m}_p]^{\mathrm{T}}\mathbf{N}_p^{-1}[\mathbf{d}_p - \mathbf{m}_p]\}}{(2\pi)^{N_\nu/2}|\mathbf{N}_p|^{1/2}}, \tag{6.49}$$

where each vector and matrix contains only the relevant values corresponding to that pixel. Similarly, one could factorize the prior $\pi(\mathbf{s}|\mathbf{S}_0)$ on the signal amplitudes, such that for the pth pixel,

$$\pi(\mathbf{s}_p|\mathbf{S}_{p0}) = \frac{\exp(-\frac{1}{2}\mathbf{s}_p^{\mathrm{T}}\mathbf{S}_{0p}^{-1}\mathbf{s}_p)}{(2\pi)^{N_c/2}|\mathbf{S}_{0p}|^{1/2}}, \tag{6.50}$$

where $\mathbf{S}_{0p}$ could, in general, vary between pixels. We note that since both $\mathbf{N}_p$ and $\mathbf{S}_{0p}$ do not depend on the parameters $\mathbf{s}_p$ and $\boldsymbol{\Phi}_p$ to be inferred, then the denominators in Eqs. (6.49) and (6.50) can be safely ignored. In fact, Eriksen *et al.* do not impose a prior of the form Eq. (6.50). Rather, they make no mention of $\mathbf{S}_{0p}$ and simply impose a uniform positivity prior on the signal amplitudes (over a sufficiently wide range), such that the posterior is proportional to the likelihood.

Let us now turn to the prior on the (physical) mixing matrix parameters $\boldsymbol{\Phi}$. In principle, one could allow for these spectral parameters to vary from pixel to pixel by considering a separate $\boldsymbol{\Phi}_p$ vector at each pixel. In practice, however, one expects (hopes) that spatial variations in spectral parameters will occur only on much larger scales. To take account of this, and thereby reduce the number of free parameters in the model, Eriksen *et al.* propose a two-stage hierarchical approach to the signal separation, as follows. In the first stage, the mapped region is divided into large 'pixels' (or patches), in each of which the spectral parameters $\boldsymbol{\Phi}_P$ (and the signal amplitudes $\mathbf{s}_P$) are allowed to vary independently (where P labels large pixels); Eriksen *et al.* adopt a wide uniform prior on each spectral parameter, but this is easily modified. For each P, one then obtains estimates $\hat{\boldsymbol{\Phi}}_P$ of the spectral parameters by maximizing or (Gibbs) sampling from the corresponding posterior; alternatively, via sampling one can obtain a full one-dimensional marginal distribution for each of the spectral parameters $\boldsymbol{\Phi}_P$ (note that, in general, these are non-linear parameters). In the second stage, the mapped region is divided into small pixels. For each small pixel $p \in P$, the spectral parameters are either taken to be $\boldsymbol{\Phi}_p = \hat{\boldsymbol{\Phi}}_P$ or their values are sampled from the one-dimensional marginals of the $\boldsymbol{\Phi}_P$ found in the first stage. Once spectral parameters are fixed, estimates of the signal amplitudes $\mathbf{s}_p$ are obtained by maximizing or (Gibbs) sampling from the corresponding conditional posterior for that (small) pixel. We note that, in general, the signal amplitudes are linear parameters of the model, and so analytical maximization of the posterior and error estimation are often possible, although care must be paid to positivity priors on the signal amplitudes. Figure 6.4 shows the results of the above technique obtained by Eriksen *et al.* when applied to simulated WMAP observations.

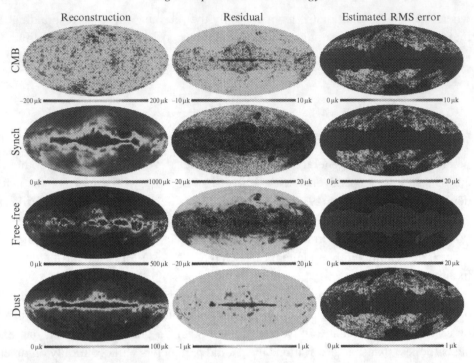

Fig. 6.4. Results obtained using the pixel-by-pixel separation method applied to all-sky simulated multi-frequency WMAP data smoothed to a common resolution of a 1-degree FHWM Gaussian beam. (Reproduced from Eriksen *et al.* 2006.)

Correlated signals and noise

It is clear that, in reality, there will exist signal and noise correlations between pixels, both intrinsic and induced by the convolution of the separate frequency channels to a common resolution. Dunkley *et al.* (2009) therefore extended the pixel-by-pixel approach outlined above to include noise correlations between pixels; thus the corresponding likelihood has the general form of Eq. (6.11). This means that one must estimate all the components of the full signal vector **s** (and spectral parameters vector $\mathbf{\Phi}$) simultaneously, which presents a far greater computational challenge than performing a pixel-by-pixel analysis, and hence limits the applicability of this approach to data convolved to a very low common resolution.

As above, Dunkley *et al.* apply only a uniform positivity prior to the signal amplitudes **s**. Similarly, they define larger 'pixels' (or patches) over which the spectral parameters are assumed constant, i.e., $\mathbf{\Phi}_p = \mathbf{\Phi}_P$ for all $p \in P$, and adopt a mixture of physically motivated Gaussian and uniform priors on the $\mathbf{\Phi}_P$. To explore the large joint $(\mathbf{s}, \boldsymbol{\Phi}_P)$ parameter space, Dunkley *et al.* have to employ a Metropolis-within-Gibbs sampling procedure to achieve a reasonably efficiency.

6.7.2 Independent component analysis (ICA)

An alternative approach to (semi-)blind signal separation is to make some assumptions about the statistical properties of the signals. In particular, a common assumption is that the signals are statistically independent, which leads to the important class of methods known collectively as Independent Component Analysis (ICA). In general, the assumption of statistical independence is much stronger than assuming the signals are merely uncorrelated; the latter leads to the Principal Component Analysis (PCA) and related methods. It is worth noting, however, that for Gaussian processes the notions of statistical independence and lack of correlation coincide.

Once one has assumed the signals to be statistically independent, there are many methods, both Bayesian and non-Bayesian, for performing a signal separation. Indeed, several techniques for cosmological signal separation have been proposed that are based on ICA and its variations, including FastICA, ensemble learning ICA and spectral matching ICA (Maino *et al.* 2002, 2003; Donzelli *et al.* 2006; Stivoli *et al.* 2006), and Correlated Component Analysis (CCA; Bonaldi *et al.* 2006, 2007). We illustrate the ideas behind ICA with an example based on spectral matching ICA (SMICA), which may be naturally considered in a Bayesian framework.

SMICA data model

The main idea underlying SMICA is to separate the signals on the basis of their different power spectra. Indeed, since SMICA models the component as Gaussian random fields, their power spectra and cross power spectra provide a complete statistical description. Moreover, the assumption of Gaussianity means that statistical independence between components is equivalent to them being uncorrelated. In fact, as we shall see, SMICA can allow for correlations between components, so overall the method's name is rather misleading! As mentioned in Section 6.6, SMICA may be considered as a 'pre-processing' step where (some of) both the mixing matrix parameters $\boldsymbol{\Phi}$ and the signal covariance matrix $\mathbf{S}$ are determined, which can then be used as input to non-blind separation methods such as the Wiener filter and the harmonic-space MEM. In fact, SMICA can also be used to estimate the noise covariance matrix $\mathbf{N}$, if it is not known in advance.

Like the Wiener filter and harmonic-space MEM, SMICA typically adopts a Fourier-domain data space, as discussed in Section 6.6 (although this is not necessary). Similarly, one assumes (at least formally) that no window (or cut) has been applied to the data, that the component fields and the noise fields in each channel are statistically homogeneous, and that the spectral parameters are independent of position. Hence, at each Fourier mode $\mathbf{k} = \mathbf{k}_p$ separately, one has the data model Eq. (6.7). In fact, although again not necessary, SMICA usually further assumes that the data in each frequency channel has been convolved to a common resolution. Since the effective observing beam is thus the same in each channel, its effect can be absorbed into the signals s, which now represent the convolved components. Hence, at each Fourier mode $\mathbf{k} = \mathbf{k}_p$ separately, one has the effective data model,

$$\mathbf{d}_p = \mathbf{F}(\mathbf{\Phi})\mathbf{s}_p + \mathbf{n}_p, \tag{6.51}$$

in which the mixing matrix $\mathbf{F}$ is the same for all modes (since the spectral behaviour is assumed to be spatially invariant), but we do not know it in advance. In its original form, SMICA parameterized spectral behaviour directly in terms of the elements of the mixing matrix, such that $\mathbf{\Phi} = \{F_{\nu c}\}$. It is straightforward, however, to use some physically motivated spectral parameterization $\mathbf{\Phi} = \phi$ (see Section 6.3.1).

SMICA likelihood

At each mode $\mathbf{k} = \mathbf{k}_p$, the signals are assumed to be drawn from a Gaussian distribution of zero mean and covariance

$$\langle \mathbf{s}_p \mathbf{s}_{p'} \rangle = \delta_{pp'} \mathbf{S}_p. \tag{6.52}$$

Since we have assumed that the signals are independent of one another (and hence uncorrelated), the matrix $\mathbf{S}_p$ is diagonal, and we denote its cth diagonal element as S_{pc}. Similarly, the noise is assumed drawn from a Gaussian distribution of zero mean and covariance

$$\langle \mathbf{n}_p \mathbf{n}_{p'} \rangle = \delta_{pp'} \mathbf{N}_p. \tag{6.53}$$

We can also assume that the noise covariance matrix is diagonal, since the noise is unlikely to be correlated between the instrumental channels. The νth diagonal element of the matrix is denoted as $N_{p\nu}$.

Combining the above, we find that the covariance matrix of the data $\mathbf{d}_p$ is

$$\langle \mathbf{d}_p \mathbf{d}_{p'} \rangle = \langle (\mathbf{F}\mathbf{s}_p + \mathbf{n}_p)(\mathbf{F}\mathbf{s}_{p'} + \mathbf{n}_{p'})^{\mathrm{T}} \rangle = \delta_{pp'}(\mathbf{F}\mathbf{S}_p\mathbf{F}^{\mathrm{T}} + \mathbf{N}_p) \equiv \delta_{pp'}\mathbf{D}_p, \tag{6.54}$$

where we have assumed that there is no correlation between the signals and the noise. We can use this to write the likelihood $L(\mathbf{S}_p, \mathbf{N}_p, \mathbf{F}) \equiv \Pr(\mathbf{d}_p|\mathbf{S}_p, \mathbf{N}_p, \mathbf{F})$ of the data in mode p given the signal covariance, noise covariance and mixing matrix,

$$L(\mathbf{S}_p, \mathbf{N}_p, \mathbf{F}) = \frac{\exp(-\frac{1}{2}\mathbf{d}_p^{\mathrm{T}}\mathbf{D}_p^{-1}\mathbf{d}_p)}{(2\pi)^{N_\nu/2}|\mathbf{D}_p|^{1/2}}, \tag{6.55}$$

where the dependence on the parameters is concealed in the data covariance matrix. Note that we are allowing, in general, for the noise covariance matrix also to be estimated, since this is straightforward within the SMICA approach; it is, however, simple to condition on a known noise covariance matrix if desired. In what follows, it is more convenient to work with the log-likelihood,

$$\ln L(\mathbf{S}_p, \mathbf{N}_p, \mathbf{F}) = -\tfrac{1}{2}(\mathbf{d}_p^{\mathrm{T}}\mathbf{D}_p^{-1}\mathbf{d}_p + \ln |\mathbf{D}_p|) + \text{const.}, \tag{6.56}$$

where we may ignore the irrelevant additive constant. The full log-likelihood $\ln L(\mathbf{S}, \mathbf{N}, \mathbf{F}) \equiv \ln \Pr(\mathbf{d}|\mathbf{S}, \mathbf{N}, \mathbf{F})$ for this problem is simply the sum of the log-likelihoods of the individual modes

$$\ln L(\mathbf{S}, \mathbf{N}, \mathbf{F}) = -\tfrac{1}{2}\sum_p(\mathbf{d}_p^{\mathrm{T}}\mathbf{D}_p^{-1}\mathbf{d}_p + \ln |\mathbf{D}_p|). \tag{6.57}$$

Note that the log-likelihood does not depend on the amplitudes of the signals, only their variance. By eliminating the signal variance, we have reduced the dimension of the hidden parameter space.

We further reduce the dimension of the parameter space by binning the signal and noise covariance matrices. We define a number of regions in Fourier space, $r = 1, \ldots, N_r$, where the signal and noise covariance matrices (and therefore the data covariance matrix) take the same value, that is,

$$\mathbf{S}_p = \mathbf{S}_r \tag{6.58}$$

$$\mathbf{N}_p = \mathbf{N}_r \tag{6.59}$$

$$\mathbf{D}_p = \mathbf{D}_r = \mathbf{F}\mathbf{S}_r\mathbf{F}^{\mathrm{T}} + \mathbf{N}_r \tag{6.60}$$

for all modes p that fall within region r ($p \in r$). We denote the number of modes that fall within region r as n_r. The usual procedure is to define the regions to be annuli in Fourier space centred on the origin, so that the they correspond to bins in the power spectrum. Thus the likelihood becomes

$$\ln L(\mathbf{S}, \mathbf{N}, \mathbf{F}) = \tfrac{1}{2}\sum_{r=1}^{N_r}\sum_{p\in r}(\mathbf{d}_p^{\mathrm{T}}\mathbf{D}_r^{-1}\mathbf{d}_p + \ln |\mathbf{D}_r|) \tag{6.61}$$

$$= \tfrac{1}{2}\sum_{r=1}^{N_r}\sum_{p\in r}\{\mathrm{Tr}\,(\mathbf{d}_p\mathbf{d}_p^{\mathrm{T}}\mathbf{D}_r^{-1}) + \ln |\mathbf{D}_r|\}, \tag{6.62}$$

where in Eq. (6.62) we have used a matrix identity to rearrange the term involving $\mathbf{d}_p$. Having compressed the hidden parameter space, we can also compress the data space by defining the 'measured' covariance matrix of the data in region r to be

$$\widetilde{\mathbf{D}}_r = \frac{1}{n_r}\sum_{p\in r}\mathbf{d}_p\mathbf{d}_p^{\mathrm{T}}. \tag{6.63}$$

Substituting this into the likelihood, we obtain

$$\ln L(\mathbf{S}, \mathbf{N}, \mathbf{F}) = -\tfrac{1}{2} \sum_{r=1}^{N_r} n_r \{ \mathrm{Tr}\, (\widetilde{\mathbf{D}}_r \mathbf{D}_r^{-1}) + \ln |\mathbf{D}_r| \}. \tag{6.64}$$

The problem has been transformed into one where the hidden space consists of the signal power spectra S_{rc}, the noise power spectra $N_{r\nu}$ and the mixing matrix F, which enter the likelihood through the data covariance matrices

$$\mathbf{D}_r = \mathbf{D}_r(\mathbf{S}_{rc}, \mathbf{N}_{r\nu}, \mathbf{F}), \tag{6.65}$$

and the data space consists of the measured covariance matrices Eq. (6.63). The measured covariance matrices contain $N_r N_\nu^2$ data which are used to constrain the parameters. There are $N_r N_c$ signal power spectrum parameters, $N_r N_\nu$ noise power spectrum parameters and $N_c N_\nu$ mixing matrix parameters.

The SMICA likelihood Eq. (6.64) has a number of degeneracies which we must be aware of when trying to make an inference about the sources. Since this is a blind method, the ordering of the sources is not set in advance. We may exchange any two sources, by exchanging their power spectra and the corresponding columns of the mixing matrix, and leave the likelihood unchanged. We may also change the sign of the elements of the mixing matrix for any component and leave the likelihood unchanged. There is one further degeneracy between the elements of the mixing matrix and the power spectrum: for any component, we may multiply its elements of the mixing matrix by a factor λ, and multiply its power spectrum by λ^{-2} and leave the likelihood unchanged. Some of these degeneracies can be avoided by imposing appropriate priors on the parameters, as discussed below.

SMICA priors

The signal and noise power spectra are positive quantities, so the priors must reflect this fact. If we do not know the scale of the power spectra parameters, then it is appropriate to use the Jeffreys prior for them. If, instead, we have some prior knowledge of how large the spectra parameters are expected to be, we could use an exponential prior or a top-hat prior. In the latter case we would set the minimum value allowed by the prior to be zero and the maximum value to be something larger than the anticipated value of the power spectrum.

Some of the degeneracies in the likelihood can be avoided by careful choice of the priors on the elements of the mixing matrix. In order to avoid the scaling degeneracy between the power spectrum of a signal and the corresponding elements of the mixing matrix, we can fix the sum of the squares of the elements to be unity,

$$\pi(F_{\nu c}) = \delta \left(\sum_{\nu=1}^{N_\nu} F_{\nu c}^2 - 1 \right). \tag{6.66}$$

To fix the degeneracy in the sign of the mixing matrix elements for a component, we can additionally constrain one of the elements to be positive or negative, depending on our knowledge of the components. The degeneracy involving the arbitrary ordering of the components cannot be solved unless we know what some of the components will be. This is addressed in Section 6.7.3.

SMICA posterior

Now we have the defined the likelihood and the prior distributions, we can use Bayes' theorem to find the posterior distribution $P(\mathbf{S}, \mathbf{N}, \mathbf{F}) \equiv \Pr(\mathbf{S}, \mathbf{N}, \mathbf{F} | \mathbf{d})$ of the parameters in the usual manner. This distribution is typically multi-modal, however, and so exploring it can be a challenge. The most common approach is to use numerical optimization to maximize the posterior, but one must be careful to avoid the degeneracies outlined above. Once the maximum a posteriori point is found, the uncertainties in the parameters can be found by using the curvature of the log-posterior, which can be evaluated either analytically or numerically. Since we have relatively few parameters, it may be possible to sample from the posterior using MCMC. If we chose the species of MCMC algorithm carefully, it could be used to explore the degeneracies between the parameters.

Once we have found the best-fit parameters from the posterior, we may wish to find a physical interpretation of the results. This can be challenging since there is much freedom in the model. We imposed priors on the parameters to fix degeneracies and to be able to perform the inference efficiently. These priors do not necessarily correspond to physical constraints on the sources, and so we can relax them when attempting to match the results to our physical understanding of the problem. We can, for example, change the sign of the mixing matrix parameters to match a known spectrum. The results of applying the SMICA approach to the analysis of simulated Planck data similar to that plotted in Figure 6.2 are shown in Figure 6.5.

Often the physical sources underlying the data are not statistically independent, even though we have assumed that in order to be able to make the separation. If this is the case, we can make linear combinations of the sources and their spectra in the search for a physical interpretation. Alternatively, one can introduce off-diagonal terms in the signal covariance matrices $\mathbf{S}_r$, thereby allowing for explicit correlations between components. One must be careful, however, not to introduce too many additional free parameters into the model.

6.7.3 Correlated component analysis (CCA)

We can make a connection between blind separation and semi-blind separation by considering an alternative parameterization of the likelihood in SMICA. Instead of

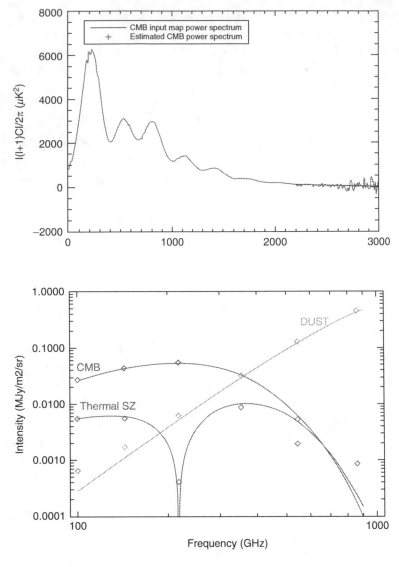

Fig. 6.5. Results obtained in applying the SMICA approach to all-sky simulated multi-frequency Planck similar to that shown in Figure 6.2. Top: the estimated CMB power spectrum (reproduced from Patanchon 2003). Bottom: the recovered mixing matrix elements for some components (diamonds) overplotted on the input frequency spectrum (solid lines). (Reproduced from Snoussi *et al.* 2002.)

using the elements of the mixing matrix as our parameters, we could instead use a physical parameterization. This provides a more prescriptive model that can break the degeneracies in the likelihood. Since the physical sources are not necessarily statistically independent, we can also again relax the requirement that the signal covariance matrices $\mathbf{S}_r$ are diagonal, although we must still be wary of having too many degrees of freedom in the model to constrain using the data.

A very similar approach is adopted in correlated component analysis (CCA; Bonaldi *et al.* 2006, 2007), except that CCA is formulated in the pixel domain. In CCA, the signals are modelled by their covariance matrices evaluated between points at a number of displacements in the pixel domain. The data consist of the covariance matrices of the observed images evaluated at the same displacements. Since this method is modelling the correlation structure of the signals, we can see that it is the pixel-space equivalent of SMICA which is doing the same thing in Fourier space. The list of displacements is therefore equivalent to the spectral bins of SMICA, and we may formulate the likelihood in an identical manner. The physical parameterization of the mixing matrix allows the assumption of independence to be relaxed, so the signals may be correlated, hence the name of the method.

6.7.4 Determining the optimal number of components

So far, we have assumed the number N_c of physical components is known. In general, however, it is desirable to infer N_c from the data themselves. The theoretically most desirable approach is to include N_c as an additional variable in the inference problem, on which one could then place a prior $\pi(N_c)$, such as a Poisson distribution with mean equal to the expected number of physical components. It is clear, however, that a crucial complication inherent to this approach is that the length of the (hidden space) parameter vector $\mathbf{h} = (\mathbf{s}, \mathbf{S}, \mathbf{\Phi}, N_c)$ is *variable*, since it depends on the unknown value N_c. If one were to explore the posterior distribution using MCMC methods, for example, the proposal distribution in the Metropolis–Hastings algorithm must be able to propose moves between spaces of differing dimension. In this case, the detailed balance conditions must be carefully considered (Green 1994; Phillips & Smith 1995), and can lead to algorithmic complications, but reliable implementations of variable-dimension samplers do exist, such as the BAYESYS sampler (Skilling 2004a).

The high computational cost of the above approach results largely from needing to sample from a parameter space of variable dimension. One therefore usually resorts to the 'poor man's' approach for achieving virtually the same result 'by hand', which is algorithmically much simpler and usually computationally faster. Instead of allowing N_c to vary during the sampling process, one instead considers a *series* of models H_{N_s}, each with a *fixed* number of components N_c, where N_c goes from $N_{\min}$ to $N_{\max}$. For each such model, one can then sample from (or find the maxima of) a posterior defined in a parameter space of fixed length to obtain parameter estimates (and errors) for the N_c components. Of equal importance, one can use any of the standard methods to obtain an estimate of the evidence E_{N_c}

for the model containing N_c components. One can then apply Bayesian model selection to determine the favoured number of sources by finding the model that maximizes $E_{N_c} \pi(N_c)$.

References

Barreiro, R. B., Hobson, M. P., Banday, A. J., Lasenby, A. N., Stolyarov V., Vielva P. and Górski, K. M. (2004). *Mon. Not. Roy. Astron. Soc.*, **351**, 515.

Bonaldi, A., Bedini, L., Salerno, E., Baccigalupi, C. and De Zotti, G. (2006). *Mon. Not. Roy. Astron. Soc.*, **373**, 271.

Bonaldi, A. *et al.* (2007). *Mon. Not. Roy. Astron. Soc.*, **382**, 1791.

Bouchet, F. R., Gispert, R., Boulanger, F., Puget, J. L. (1997). In F. R. Bouchet, R. Gispert, B. Guideroni and J. Tran Thanh Van, eds., *Proceedings of the 16th Moriond Astrophysics Meeting, Les Arcs, France*. Gif-sur-Yvette: Editions Frontière, p. 481.

Bouchet, F. R. and Gispert, R. (1999). *New Astron.*, **4**, 443.

Brandt, W. N., Lawrence, C. R., Readhead, A. C. S., Pakianathan, J. N. and Fiola T. M. (1994). *Astrophys. J.*, **424**, 1.

Donzelli, S. *et al.* (2006). *Mon. Not. Roy. Astron. Soc.*, **369**, 441.

Dunkley, J. *et al.* (2009). *Astrophys. J. Supp.*, **180**, 306.

Eriksen, H. K. *et al.* (2006). *Astrophys. J.*, **641**, 665.

Feroz, F. and Hobson, M. P. (2008). *Mon. Not. Roy. Astron. Soc.*, **384**, 449.

Feroz, F., Hobson, M. P. and Bridges, M. (2008). *Mon. Not. Roy. Astron. Soc.*, submitted (arXiv0809.3437).

Green, P. J. (1994). *J. R. Stat. Soc.*, **56**, 589.

Gull, S. F. (1989). In J. Skilling, ed., *Maximum Entropy and Bayesian Methods*. Dordrecht: Kluwer, p. 53.

Gull, S. F. and Skilling J. (1990). *The MEMSYS5 User's Manual*. Maximum Entropy Data Consultants Ltd, Royston, UK.

Hobson, M. P., Bridle, S. L. and Lahav, O. (2002). *Mon. Not. Roy. Astron. Soc.*, **335**, 377.

Hobson, M. P., Jones, A. W., Lasenby, A. N. and Bouchet, F. R. (1998). *Mon. Not. Roy. Astron. Soc.*, **300**, 1.

Hobson, M. P. and Lasenby, A. N. (1998). *Mon. Not. Roy. Astron. Soc.*, **298**, 905.

Kolmogorov, A. (1939). *C. R. Acad. Sci.*, **208**, 2043.

Levinson, N. (1947). *J. Math. Phys.*, **25**, 261.

Maino, D. *et al.* (2002). *Mon. Not. Roy. Astron. Soc.*, **334**, 53.

Maino, D., Banday, A. J., Baccigalupi, C., Perrotta, F. and Górski, K. M. (2003). *Mon. Not. Roy. Astron. Soc.*, **344**, 544.

Maisinger, K., Hobson, M. P. and Lasenby, A. N. (2004). *Mon. Not. Roy. Astron. Soc.*, **347**, 339.

O'Ruanaidh, J. J. K. and Fitzgerald, W. J. (1996). *Numerical Bayesian Methods Applied to Signal Processing*. New York: Springer-Verlag.

Patanchon, G. (2003). *New Astron. Rev.*, **47**, 871.

Phillips, D. B. and Smith, A. F. M. (1995). In W. R. Gilks, S. Richardson and D. J. Spiegelhalter, eds., *Markov Chain Monte Carlo in Practice*. London: Chapman & Hall.

Press, W. H., Teukolsky, S., Vetterling, W. T. and Flannery, B. P. (1997). *Numerical Recipes in C*. Cambridge: Cambridge University Press, p. 394.

Shannon, C. (1948). *Bell System Tech. J.*, **27**, 379.

Sivia, D. S. and Skilling, J. (2006). *Data Analysis: A Bayesian Tutorial*. Oxford: Oxford University Press.

Skilling, J. (1989). In J. Skilling, ed., *Maximum Entropy and Bayesian Methods*. Dordrecht: Kluwer, p. 45.

Skilling, J. (2004a). *AIP Conf. Proc.*, **707**, 388.

Skilling, J. (2004b). *AIP Conf. Proc.*, **735**, 395.

Snoussi, H., Patanchon, G., Macias-Pérez, J. F., Mohammad-Djafari, A. and Delabrouille, J. (2002). *AIP Conf. Proc.*, **617**, 125.

Stivoli, F., Baccigalupi, C., Maino, D. and Stompor, R. (2006). *Mon. Not. Roy. Astron. Soc.*, **372**, 615.

Stolyarov, V., Hobson, M. P., Ashdown, M. A. J. and Lasenby A. N. (2002). *Mon. Not. Roy. Astron. Soc.*, **336**, 97.

Stolyarov, V., Hobson, M. P., Lasenby, A. N. and Barreiro, R. B. (2005). *Mon. Not. Roy. Astron. Soc.*, **357**, 145.

Tegmark, M. and Efstathiou, G. (1996). *Mon. Not. Roy. Astron. Soc.*, **281**, 1297.

Titterington, D. M. (1985). *Astron. Astrophys.*, **144**, 381.

Wiener, N. (1949). *Extrapolation, Interpolation and Smoothing of Stationary Time Series, with Engineering Applications*. Cambridge, MA: MIT Press.

Zaroubi, S., Hoffman, Y., Fisher, K. B. and Lahav, O. (1995). *Astrophys. J.*, **449**, 446.

Part II

Applications

7

Bayesian source extraction

M. P. Hobson, Graça Rocha and Richard S. Savage

Source extraction is a generic problem in modern observational astrophysics and cosmology. Indeed, one of the major challenges in the analysis of astronomical observations is to identify and characterize a localized signal immersed in some general background. Typical one-dimensional examples include the extraction of point or extended sources from time-ordered scan data or the detection of absorption or emission lines in quasar spectra. In two dimensions, one often wishes to detect point or extended sources in astrophysical images that are dominated either by instrumental noise or contaminating diffuse emission. Similarly, in three dimensions, one might wish to detect galaxy clusters in large-scale structure surveys. Moreover, the ability to perform source extraction with reliable, automated methods has become vital with the advent of modern large-area surveys too large to be inspected in detail 'by eye'. Indeed, much of the science derived from the study of astronomical sources, or from the background in which they are immersed, proceeds directly from accurate source extraction.

In extracting sources from astronomical data, we typically face a number of challenges. Firstly, there is instrumental noise. Nonetheless, it is often possible to obtain an accurate statistical characterization of the instrumental noise, which can then be used to compensate for its effects to some extent. More problematic are any so-called 'backgrounds' to the observation. These can be astrophysical or cosmological in origin, such as Galactic emission, cosmological backgrounds, faint source confusion, or even simply emission from parts of the telescope itself. These backgrounds are often much harder to characterize and often constitute an in-depth study in themselves. A prime example is provided by high-resolution observations of the cosmic microwave background (CMB). In addition to the CMB emission, which varies on a characteristic scale of order ~ 10 arcmin, one is often interested in detecting emission from discrete objects such as extragalactic 'point' (i.e., beam-shaped) sources or the Sunyaev–Zel'dovich (SZ) effect in galaxy clusters, which have characteristic scales similar to that of the primordial CMB emission. More-over, the rms of the instrumental noise in CMB observations is often comparable to the amplitude of the discrete sources. We may also have to contend with systematic effects such as glitches that can be caused by cosmic ray hits on the detectors of space telescopes.

In this chapter, we will focus on the important specific example of extracting discrete sources from a diffuse background in a two-dimensional astronomical im-age, although our discussion will, in general, be applicable to datasets of arbitrary dimensionality.

7.1 Traditional approaches

Several excellent, general-purpose packages exist for performing source extrac-tion from two-dimensional images, such as DAOfind (Stetson 1992) and SExtrac-tor (Bertin & Arnouts 1996). In both cases, it is assumed that the background is smoothly varying, with a characteristic length scale much larger than that of the dis-crete source being sought. For example, SExtractor approximates the background emission by a low-order polynomial, which is subtracted from the image. Object detection is then performed by finding sets of connected pixels above some given threshold.

Nonetheless, when one has some prior knowledge of the properties of the sources, backgrounds or instrumental noise, it is clearly advantageous to make use of this information. Over the years, a number of methods have been devised that use various sets of information to perform 'optimal' source extraction (subject to certain assumptions). Many such techniques are based on the concept of filtering the data to enhance, relative to the background, the signal due to sources with a certain set of characteristics.

In these approaches, one typically applies a linear filter to the original image $d(\mathbf{x})$ and then analyses the resulting filtered field,

$$d_f(\mathbf{x}) = \int \psi(\mathbf{x} - \mathbf{y}) d(\mathbf{y}) \, \mathrm{d}^2 \mathbf{y}. \tag{7.1}$$

This process is usually performed by Fourier transforming the image $d(\mathbf{x})$ to ob-tain $\tilde{d}(\mathbf{k})$, multiplying by some filter function $\tilde{\psi}(\mathbf{k})$ and inverse Fourier transform-ing. The form of the filter function clearly determines which Fourier modes are suppressed and by what factor. In the simplest cases $\tilde{\psi}(\mathbf{k})$ may set to zero the am-plitudes of all $\mathbf{k}$-modes with $|\mathbf{k}|$ above (or below) some critical value k_c, and one obtains a simple low-pass (or high-pass) Fourier filter. Alternatively, one may re-tain only some specific set of Fourier modes by setting to zero all modes with $|\mathbf{k}|$ lying outside some range $k_{\min}$ to $k_{\max}$ (see, for example, Chiang *et al.* 2002).

More generally, suppose one is interested in detecting sources with some given spatial template $t(\mathbf{x})$ (normalized for convenience to unit peak amplitude). If the original image contains N_s sources at positions $\mathbf{X}_i$ with amplitudes A_i, together with contributions from backgrounds and additive instrumental noise, we may write

$$d(\mathbf{x}) \equiv s(\mathbf{x}) + b(\mathbf{x}) + n(\mathbf{x}) = \sum_{i=1}^{N_s} A_i t(\mathbf{x} - \mathbf{X}_i) + b(\mathbf{x}) + n(\mathbf{x}), \qquad (7.2)$$

where $s(\mathbf{x})$ is the signal of interest, $b(\mathbf{x})$ is the (diffuse) 'background' (or sum of backgrounds) in which the sources are embedded, and $n(\mathbf{x})$ is the instrumental noise. In many cases, the background $b(\mathbf{x})$ is well modelled as a (sum of) stochastic process(es), in which case it is common to consider the 'generalized noise' $g(\mathbf{x}) \equiv b(\mathbf{x}) + n(\mathbf{x})$, which is defined as all contributions to the image aside from the sources of interest.

In the latter case, it is well known (see, e.g., Haehnelt & Tegmark 1996) that, under the assumption that $g(\mathbf{x})$ is a statistically homogeneous random field, one can derive a 'matched filter' (MF) $\psi(\mathbf{x})$ such that the filtered field Eq. (7.1) has the following properties: (i) $d_f(\mathbf{X}_i)$ is an unbiassed estimator of A_i; and (ii) the variance of the filtered generalized noise field $g_f(\mathbf{x})$ is minimized. Various extensions to the MF have been proposed, in which further constraints on the filtered field are imposed but, in each case, one may consider the filtering process as 'optimally' boosting (in a linear sense and subject to certain constraints) the signal from discrete sources, with a given spatial template, and simultaneously suppressing emission from the background.

Most notable amongst the extensions to the MF is the scale-adaptive filter (SAF; Sanz, Herranz & Martinez-Gonzalez 2001), for which one additionally requires that (iii) the expected value of the filtered field at a source position, when considered as a function of the (unknown) spatial extent R of the source, has an extremum at some value $R = R_0$ that can be related to the source's true spatial extent. Thus the drop in gain of the SAF relative to the MF that results from imposing the additional condition (iii) is offset by the ability to determine the scale of the source, if this is unknown (see McEwen, Hobson and Lasenby (2006) for a full discussion). The SAF is thus particularly useful if the set of sources in the map do not all have the same spatial extent. In the case of the SZ effect, for example, one might model all clusters as having the same functional form of the template, but with an unknown 'core radius' that differs from one cluster to another. If one uses the MF in this case, one needs to repeat the filtering process at a number of spatial scales to obtain several filtered fields, each of which would optimally boost sources with that scale (Herranz *et al.* 2002a). Closely related to the SAF are the wavelet filters (see, e.g., Lopez-Caniego *et al.* 2005; Barreiro *et al.* 2003). More recently, Makovoz and Marleau (2005) have derived a filter of this type using the Bayesian formalism, thus allowing for the explicit inclusion of prior knowledge. It is also possible through the construction of directional filters (see, e.g., McEwen *et al.* 2006) to accommodate sources that are not circularly symmetric and have orientations that vary from one source to another.

Finally, it is unnecessary to restrict oneself to analyzing a single astronomical image at a particular observing frequency. In many cases, images at different frequencies may be available. Once again, the SZ effect provides a good example. CMB satellite missions, such as the Planck experiment, provide high-sensitivity, high-resolution observations of the whole sky at a number of different observing frequencies. Owing to the distinctive frequency dependence of the thermal SZ effect, it is better to use the maps at all the observed frequencies simultaneously when attempting to detect and characterize thermal SZ clusters hidden in the emission from other astrophysical components. The generalization of the above filter techniques to multi-frequency data is straightforward, and leads to the concept of multi-filters (Herranz *et al.* 2002b). We also note that an alternative approach to this problem, which relies only on the well-known frequency dependencies of the thermal SZ and CMB emission, has been proposed by Diego *et al.* (2002).

7.2 The Bayesian approach

The traditional approaches outlined above have been shown to produce good results, but the filtering process is only optimal among the rather limited class of linear filters and is logically separated from the subsequent source detection step performed on the filtered map(s). Moreover, the above approaches are not, in general, sufficiently flexible to take full account of the statistical characteristics of the (generalized) noise background, particularly when it is non-Gaussian. It is important, for example, to be able to take proper account of the Poissonian nature of many forms of data, most notably low-photon-count CCD output from X-ray observations. Also, the traditional techniques do not allow for the straightforward inclusion of prior information on the sources of interest, either in terms of their physical structure or their abundance.

As a result, Hobson and McLachlan (2003; hereinafter HM03) introduced a Bayesian approach to the detection and characterization of discrete sources in a diffuse background, which allows for the inclusion of all pertinent information; the general framework we present below follows this closely. As in the filtering techniques, the method assumes a parameterized form for the sources, but the optimal values of these parameters, and their associated errors, are obtained in a single step by evaluating their full posterior distribution. In principle, any statistical form for the (generalized) noise can be accommodated by defining an appropriate likelihood function; indeed the Bayesian methodology has also been applied in the Poisson noise regime (see, e.g., Guglielmetti *et al.* 2004). If available, one can also place physical priors on the parameters defining a source and on the number of sources present. Moreover, as discussed below, one can decide whether each peak found in the posterior distribution corresponds to a real source or a background fluctuation by performing a Bayesian model comparison using (an approximation to) the evidence. The approach therefore represents the theoretically optimal method for performing parameterized source extraction.

7.2.1 Discrete sources in a background

To keep our discussion as general as possible, let us denote the totality of our available data by the vector **d**. This may represent the pixel values in a single image, or in a collection of images, such as a multi-frequency dataset. Equally, **d** could represent the Fourier coefficients of the image(s), or coefficients in some other basis, but we will concentrate here on the pixel basis for simplicity.

We first consider the contribution to the data of the discrete sources of interest. Let us suppose that we wish to detect and characterize some set of (two-dimensional) discrete sources, each of which is described by a template $\tau(\mathbf{x}; \mathbf{p})$. In the analysis of single-frequency maps, one could take the template to be the (pixelized) shape of the source after convolution with the beam. For multi-frequency data, however, it is more natural to treat the intrinsic source shape and the beam profiles separately, in which case the template is taken to describe the former (perhaps at a specific observing frequency, if the intrinsic shape of the source is frequency dependent).

The template is defined in terms of a set of parameters **p** that might typically denote (collectively) the position (X, Y) of the source, its amplitude A and some measure R of its spatial extent. A particular example is the circularly symmetric Gaussian-shaped source defined by

$$\tau(\mathbf{x}; \mathbf{p}) = A \exp\left[-\frac{(x - X)^2 + (y - Y)^2}{2R^2} \right], \tag{7.3}$$

so that $\mathbf{p} = \{X, Y, A, R\}$. In what follows, we will consider only circularly symmetric sources, but our general approach accommodates templates that are, for example, elongated in one direction, at the expense of increasing the dimensionality of the parameter space.[1]

---

[1] Such an elongation can be caused by asymmetry of the beam, source or both. If our model assumes a circularly symmetric source but an asymmetric beam, the elongation will be constant across the map, but not if the sources are intrinsically asymmetric. In the latter case, one thus has to introduce the source position angle as a parameter, in addition to the elongation.

If N_s sources of interest are present in the map and the contribution of each source to the data is additive, and so too is the noise, we may write

$$\mathbf{d} = \sum_{k=1}^{N_s} \mathbf{s}(\mathbf{p}_k) + \mathbf{b}(\mathbf{q}) + \mathbf{n}(\mathbf{r}), \qquad (7.4)$$

where the 'signal vector' $\mathbf{s}(\mathbf{p}_k)$ denotes the (pixelized) contribution to the data vector from the kth discrete source, which includes, for example, the effect of the observing beam(s). The vector $\mathbf{b}(\mathbf{q})$ denotes the (pixelized) background contribution, which may itself be characterized in terms of a set of (potentially unknown) parameters $\mathbf{q}$; again $\mathbf{b}$ must include the effect of the observing beam(s). The vector $\mathbf{n}$ is the (pixelized) instrumental noise, the statistical description of which may be characterized by a number of (potentially unknown) parameters $\mathbf{r}$. As mentioned above, it is often useful to combine the background and instrumental noise contributions into the generalized noise vector $\mathbf{g}(\mathbf{q}, \mathbf{r})$. It is also convenient occasionally to denote the total contribution to the data by *all* the sources simply by $\mathbf{s}(\mathbf{p})$, where $\mathbf{p}$ denotes collectively the source parameters $\{\mathbf{p}_1, \mathbf{p}_2, \ldots, \mathbf{p}_{N_s}\}$.

Clearly, we wish to use the data $\mathbf{d}$ to place constraints on the values of the unknown source parameters N_s and $\mathbf{p}_k$ ($k = 1, \ldots, N_s$). If the values of any background and noise parameters $\mathbf{q}$ and $\mathbf{r}$ are unknown, then these are usually nuisance parameters and should ideally be marginalized over. In some cases, however, one may wish to estimate the values of (a subset of) these parameters as well.[2] In any case, we will denote the full parameter space by $\boldsymbol{\Theta} = \{N_s, \mathbf{p}, \mathbf{q}, \mathbf{r}\}$.

7.2.2 Bayesian inference

Bayesian inference methods provide a consistent approach to the estimation of a set of parameters $\boldsymbol{\Theta}$ in a model (or hypothesis) H for the data $\mathbf{d}$. Bayes' theorem states that

$$\mathrm{Pr}(\boldsymbol{\Theta}|\mathbf{d}, H) = \frac{\mathrm{Pr}(\mathbf{d}|\boldsymbol{\Theta}, H)\,\mathrm{Pr}(\boldsymbol{\Theta}|H)}{\mathrm{Pr}(\mathbf{d}|H)}, \qquad (7.5)$$

where $\mathrm{Pr}(\boldsymbol{\Theta}|\mathbf{d}, H) \equiv P(\boldsymbol{\Theta})$ is the posterior probability distribution of the parameters, $\mathrm{Pr}(\mathbf{d}|\boldsymbol{\Theta}, H) \equiv L(\boldsymbol{\Theta})$ is the likelihood, $\mathrm{Pr}(\boldsymbol{\Theta}|H) \equiv \pi(\boldsymbol{\Theta})$ is the prior, and $\mathrm{Pr}(\mathbf{d}|H) \equiv E$ is the Bayesian evidence.

In parameter estimation, the normalizing evidence factor is usually ignored, since it is independent of the parameters $\boldsymbol{\Theta}$. This (un-normalized) posterior constitutes the complete Bayesian inference of the parameter values. Inferences are

---

[2] It should be remembered, however, that our general expression Eq. (7.4) already assumes additive noise. This is often true, but an important case for which Eq. (7.4) is not valid is when the data are Poisson distributed; this is discussed further in Section 7.2.3.

usually obtained either by taking samples from the (un-normalized) posterior using Markov chain Monte Carlo (MCMC) methods, or by locating its maximum (or maxima) and approximating the shape around the peak(s) by a multivariate Gaussian.

In contrast to parameter estimation problems, in model selection the evidence takes the central role and is simply the factor required to normalize the posterior over $\mathbf{\Theta}$:

$$E = \int L(\mathbf{\Theta})\pi(\mathbf{\Theta})\mathrm{d}^D\mathbf{\Theta}, \tag{7.6}$$

where D is the dimensionality of the parameter space. As the average of the likelihood over the prior, the evidence is larger for a model if more of its parameter space is likely and smaller for a model with large areas in its parameter space having low likelihood values, even if the likelihood function is very highly peaked. Thus, the evidence automatically implements Occam's razor: a simpler theory with compact parameter space will have a larger evidence than a more complicated one, unless the latter is significantly better at explaining the data. The question of model selection between two models H_0 and H_1 can then be decided by comparing their respective posterior probabilities given the observed dataset $\mathbf{d}$, as follows:

$$\frac{\Pr(H_1|\mathbf{d})}{\Pr(H_0|\mathbf{d})} = \frac{\Pr(\mathbf{d}|H_1)\Pr(H_1)}{\Pr(\mathbf{d}|H_0)\Pr(H_0)} = \frac{E_1\Pr(H_1)}{E_2\Pr(H_0)}, \tag{7.7}$$

where $\Pr(H_1)/\Pr(H_0)$ is the a-priori probability ratio for the two models, which can often be set to unity but occasionally requires further consideration. Unfortunately, evaluation of the multidimensional integral (7.6) is a challenging numerical task, but sampling-based methods do exist, such as thermodynamic integration (see, e.g., Ó'Ruanaidh & Fitzgerald 1996) or, better, nested sampling (Skilling 2004a; Sivia & Skilling 2006). Moreover, a fast approximate method for evidence evaluation is to model the posterior as a multivariate Gaussian centred at its peak(s) (see, e.g., Hobson, Bridle & Lahav 2002).

7.2.3 Defining the posterior distribution

For any given model H of the source signal $\mathbf{s}(\mathbf{p})$, background $\mathbf{b}(\mathbf{q})$ and noise $\mathbf{n}(\mathbf{r})$, one can write down a likelihood function $L(\mathbf{\Theta}) \equiv \Pr(\mathbf{d}|\mathbf{\Theta}, H)$. As a typical example, suppose that the background is deterministic (rather than stochastic) and the instrumental noise is a (zero mean) statistically homogeneous Gaussian random field with covariance matrix $\mathbf{N} = \langle\mathbf{nn}^{\mathrm{T}}\rangle$. In this case, the likelihood function takes the form (including the parameter dependencies explicitly)

$$L(\boldsymbol{\Theta}) = \frac{\exp\left\{-\frac{1}{2}[\mathbf{d} - \mathbf{s}(\mathbf{p}) - \mathbf{b}(\mathbf{q})]^{\mathrm{T}}\mathbf{N}^{-1}(\mathbf{r})[\mathbf{d} - \mathbf{s}(\mathbf{p}) - \mathbf{b}(\mathbf{q})]\right\}}{(2\pi)^{N_{\mathrm{pix}}/2}|\mathbf{N}(\mathbf{r})|^{1/2}}. \tag{7.8}$$

If instead the background is also a (zero mean) statistically homogeneous Gaussian random field, then so too is the generalized noise $\mathbf{g} = \mathbf{b} + \mathbf{n}$. Denoting its covariance matrix by $\mathbf{G} = \langle \mathbf{g}\mathbf{g}^{\mathrm{T}}\rangle$, the likelihood function can then be written

$$L(\boldsymbol{\Theta}) = \frac{\exp\left\{-\frac{1}{2}[\mathbf{d} - \mathbf{s}(\mathbf{p})]^{\mathrm{T}}\mathbf{G}^{-1}(\mathbf{q}, \mathbf{r})[\mathbf{d} - \mathbf{s}(\mathbf{p})]\right\}}{(2\pi)^{N_{\mathrm{pix}}/2}|\mathbf{G}(\mathbf{q}, \mathbf{r})|^{1/2}}. \tag{7.9}$$

Another important example is Poisson-distributed data, such as those obtained using photon counting CCDs. As mentioned in Section 7.2.1, the ansatz (7.4) is not valid in this case, since the noise is not additive. Nonetheless, the likelihood can still be defined. For example, in the case where the background is taken as deterministic, this reads

$$L(\boldsymbol{\Theta}) = \prod_{i=1}^{N_{\mathrm{pix}}} \frac{e^{-\lambda_i}\lambda_i^{d_i}}{d_i!}, \tag{7.10}$$

where d_i (an integer) is the ith element of the data vector and we have defined $\lambda_i \equiv s_i(\mathbf{p}) + b_i(\mathbf{q})$. Note that for Poisson data the 'noise parameters' $\mathbf{r}$ are absent.

Since the (log-)likelihood is usually the computationally most expensive quantity to calculate, it is worth noting a simple method for reducing this burden considerably in the (frequently occurring) case that the background and noise parameters $\mathbf{q}$ and $\mathbf{r}$ are known. Consider, for example, the case Eq. (7.9), for which the log-likelihood function now reads

$$\ln L(\mathbf{p}) = c - \frac{1}{2}[\mathbf{d} - \mathbf{s}(\mathbf{p})]^{\mathrm{T}}\mathbf{G}^{-1}[\mathbf{d} - \mathbf{s}(\mathbf{p})], \tag{7.11}$$

where c is an unimportant constant. Following Carvalho, Rocha and Hobson (2008), we can rewrite the log-likelihood as

$$\ln L(\mathbf{p}) = c' - \frac{1}{2}\mathbf{s}(\mathbf{p})^{\mathrm{T}}\mathbf{G}^{-1}\mathbf{s}(\mathbf{p}) + \mathbf{d}^{\mathrm{T}}\mathbf{G}^{-1}\mathbf{s}(\mathbf{p}), \tag{7.12}$$

where $c' = c - \frac{1}{2}\mathbf{d}^{\mathrm{T}}\mathbf{G}^{-1}\mathbf{d}$ is again independent of the parameters $\mathbf{p}$. The advantage of this formulation is that the part of the log-likelihood dependent on the parameters $\mathbf{p}$ consists only of products involving the data and the signal. Since the signal from a (set of) discrete source(s) with parameters $\mathbf{p}$ is only (significantly) non-zero in a limited region centred on the putative position(s), one need only calculate the quadratic forms in Eq. (7.12) over a very limited number of pixels, whereas the quadratic form in the standard version Eq. (7.11) must be calculated over all the pixels in the image. Clearly, similar computational savings can be made for likelihoods of the form Eqs. (7.8) and (7.10).

Once the likelihood has been defined, all that remains is to specify the prior $\pi(\boldsymbol{\Theta}) \equiv \Pr(\boldsymbol{\Theta}|H)$ on the parameters for the model H. For most applications, it is natural to assume that the parameters $\mathbf{p}$, $\mathbf{q}$ and $\mathbf{r}$, characterizing the signal, background and noise respectively, are independent. Moreover, it is also reasonable to assume that the number of sources N_s and the parameters $\mathbf{p}_k$ for each source are mutually independent, so that

$$\pi(\boldsymbol{\Theta}) = \pi(N_s)\pi(\mathbf{p}_1)\pi(\mathbf{p}_2)\ldots\pi(\mathbf{p}_{N_s})\pi(\mathbf{q})\pi(\mathbf{r}). \tag{7.13}$$

As mentioned above, the parameters $\mathbf{p}_k$ that characterize the kth source will typically consist of its position X_k and Y_k, amplitude A_k and spatial extent R_k, and the priors imposed on these parameters will generally depend on the application. For example, one might impose uniform priors on X_k and Y_k within the borders of the image, whereas the priors on A_k and R_k may be provided by some physical model of the sources one wishes to detect. Similarly, one may impose a prior on the number of unknown sources N_s, which is clearly a discrete parameter. For example, if the sources of interest are not clustered on the sky and have a mean number density μ per image area, then one would set

$$\pi(N_s) = \frac{e^{-\mu}\mu^{N_s}}{N_s!}. \tag{7.14}$$

It is worth noting here that in source extraction problems the posterior distribution is typically an extremely complicated function of the parameters $\boldsymbol{\Theta}$, possessing many modes and elongated degeneracies. One therefore requires robust methods to explore (or maximize) the posterior distribution reliably.

7.3 Variable-source-number models

The theoretically most desirable approach is to attempt to detect and characterize all the sources in the image simultaneously by sampling from the (un-normalized) posterior distribution. In other words, the Bayesian purist would attempt to infer simultaneously the full set of parameters $\boldsymbol{\Theta} \equiv (N_s, \mathbf{p}_1, \mathbf{p}_2, \ldots, \mathbf{p}_{N_s}, \mathbf{q}, \mathbf{r})$. In particular, this allows one straightforwardly to include prior information regarding the number of sources expected in the image.

It is clear, however, that a crucial complication inherent to this approach is that the length of the parameter vector $\boldsymbol{\Theta}$ is *variable*, since it depends on the unknown value N_s. If one were to explore the posterior distribution using MCMC methods, for example, the proposal distribution in the Metropolis–Hastings algorithm must be able to propose moves between spaces of differing dimension. In this case, the

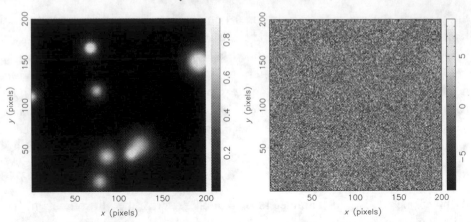

Fig. 7.1. The toy problem discussed in HM03. The 200×200 pixel test image (left panel) contains eight discrete Gaussian-shaped sources of varying widths and amplitudes. The corresponding data map (right panel) has independent Gaussian pixel noise added with an rms of 2 units.

detailed balance conditions must be carefully considered (Green 1994; Phillips & Smith 1995), and can lead to algorithmic complications, but reliable implementations of variable-dimension samplers do exist, such as the BAYESYS sampler (Skilling 2004b).

Nonetheless, further complications occur that are particular to the source extraction problem. Most notable is the source-swapping degeneracy. If one is sampling from a parameter space with $N_s > 1$, then one can swap the values of the parameters $\mathbf{p}_k$ and $\mathbf{p}_{k'}$ of any two sources without affecting the value of the posterior. One must therefore take care when assigning the source parameter vectors in each output sample to particular sources identified in the image. Another practical difficulty results if the prior $\pi(N_s)$ on the number of sources remains non-zero at large values of N_s, since then the size of the corresponding parameter space to be sampled becomes very large. Consequently, the algorithm can be slow to burn in and typically requires a lot of CPU time.

This variable-source-number approach has been illustrated by application to a simple toy source extraction problem by HM03. In Figure 7.1, the left panel shows a 200×200 pixel test image containing eight Gaussian sources defined by Eq. (7.3), with different values of the parameters X_k, Y_k A_k and R_k ($k = 1, \ldots, 8$). The X and Y position coordinates are drawn independently from the uniform distribution $\mathcal{U}(0, 200)$, whereas the amplitude A and size R of each source are drawn independently from the uniform distributions $\mathcal{U}(0.5, 1)$ and $\mathcal{U}(5, 10)$, respectively. The right panel of Figure 7.1 shows the corresponding data map, which has independent ('white') Gaussian pixel noise added, with an rms of 2 units. This corresponds

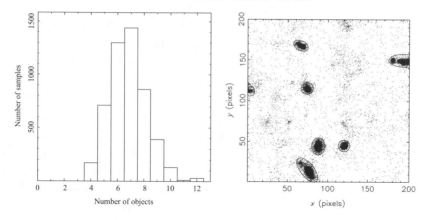

Fig. 7.2. Left: histogram of the number of post burn-in samples obtained in each subspace corresponding to a different number of sources N_s. Right: the samples obtained for $N_s = 7$, projected into the (X, Y) subspace; the ellipses indicate the samples used in the calculation of the properties of the sources.

to a signal-to-noise ratio of 0.25–0.5 as compared with the peak emission in each source; we note that no sources are visible to the naked eye.

In the analysis of this toy problem, we have no unknown background parameters $\mathbf{q}$ or noise parameters $\mathbf{r}$. On the number of sources N_s, we assume the Poisson prior Eq. (7.14) with a mean of $\mu = 4$, which is purposely chosen to be somewhat smaller than the actual number of sources $N_s = 8$. Since the Poisson prior imposes no upper limit on the possible number of sources, the overall parameter space under consideration is formally the countably infinite union of subspaces $\boldsymbol{\Theta} = \bigcup_{N_s=0}^{\infty} \boldsymbol{\Theta}_{N_s}$, where $\boldsymbol{\Theta}_{N_s} = \{\mathbf{p}_1, \ldots, \mathbf{p}_{N_s}\}$ denotes the $4N_s$-dimensional space corresponding to the model with N_s sources. The parameters of the kth source are $\mathbf{p}_k = \{X_k, Y_k, A_k, R_k\}$. The priors on X_k and Y_k $(k = 1, \ldots, N_s)$ are (correctly) assumed to be $\mathcal{U}(0, 200)$, whereas the priors for A_k and R_k are taken as $\mathcal{U}(0, 2)$ and $\mathcal{U}(3, 12)$, respectively.

The BAYESYS sampler was used, running ten interacting Markov chains, to obtain 5000 post burn-in samples from the variable-length parameter space $\boldsymbol{\Theta}$. The left panel of Figure 7.2 shows a histogram of the number of samples obtained in each subspace of different dimension, showing the most favoured number of sources is $N_s = 7$. One is free to use the 5000 post burn-in samples in a variety of ways to place limits on the parameters $\boldsymbol{\Theta}$. For illustration only, in the right panel of Figure 7.2 we plot the samples obtained for the case $N_s = 7$, projected into the (X, Y) subspace. We see that there exist seven main areas in which the samples are concentrated, which we highlight with ellipses. Comparison with Figure 7.1

(left panel) shows that each of these areas corresponds to a real source. The mean and standard deviation of the parameters $\{X_k, Y_k, A_k, R_k\}$ $(k = 1, \ldots, 7)$ for each detected source were calculated from the samples contained in each ellipse, and were found to recover the input parameters to reasonable accuracy. It must be noted, however, that the two overlapping sources in Figure 7.1 have been confused, and yield a single detected 'source'.

Clearly more optimal strategies exist for using the samples to characterize the sources and distinguish between real and spurious detections. We shall not pursue them further here, however, owing to the rather computationally intensive nature of the above approach; the above analysis required ~17 hours of CPU time on a standard desktop.

7.4 Fixed-source-number models

The high computational cost of the above approach results largely from needing to sample from a parameter space of variable dimension. Fortunately, there exists a 'poor man's' approach for achieving virtually the same result 'by hand', which is algorithmically much simpler and usually computationally faster.

Instead of allowing N_s to vary during the sampling process, one instead considers a *series* of models H_{N_s}, each with a *fixed* number of sources N_s, where N_s goes from (say) zero to some predefined maximum value $N_{\max}$. For each such model, one can then sample from (or find the maxima of) a posterior defined in a parameter space of fixed length to obtain parameter estimates (and errors) for the N_s sources. Of equal importance, one can use any of the standard methods to obtain an estimate of the evidence E_{N_s} for the model containing N_s sources. One can then apply Bayesian model selection to determine the favoured number of sources by finding the model that maximizes $E_{N_s}\pi(N_s)$ (see, e.g., Marshall 2006). It should be noted, however, that, for $N_s > 1$, this approach still exhibits the source-swapping degeneracy mentioned above. Moreover, the computational cost is still typically rather large, albeit less than the variable-source-number approach.

7.5 Single-source models

A very special case of the fixed-source-number approach is simply to set $N_s = 1$ throughout the analysis; this is equivalent to imposing the prior $\pi(N_s) = 1$ if $N_s = 1$ and zero otherwise. Since the model consists of just a single source, the full source parameter is $\mathbf{p} = \{X, Y, A, R\}$ (say), which is fixed and only 4-dimensional. Although one has fixed $N_s = 1$, it is important to understand that this does *not* restrict one to detecting just a single source in the map. Indeed, by modelling the data in this way, one would expect the posterior distribution to

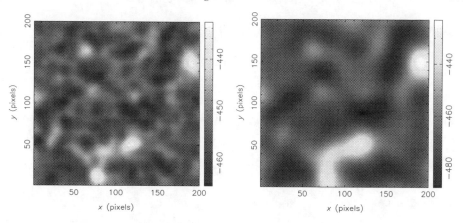

Fig. 7.3. The 2-dimensional conditional log-posterior distributions in the (X, Y) subspace for the toy problem illustrated in Figure 7.1, where the model contains a single source parameterized by $\mathbf{p} = \{X, Y, A, R\}$. The amplitude A and size R are conditioned at $A = 0.75$, $R = 5$ (left panel) and $A = 0.75$, $R = 10$ (right panel).

possess numerous local maxima in the 4-dimensional parameter space, some of which correspond to the location in this space of sources present in the image. This is illustrated in Figure 7.3, where we see that maxima do indeed occur at positions corresponding to each of the input sources in Figure 7.1, but there exist numerous other maxima that do not coincide with the position of a real source, but instead occur because the background noise in some areas has 'conspired' to give the impression that a source might be present.

HM03 show that the vastly simplified approach of setting $N_s = 1$ is both fast and reliable when the sources of interest are spatially well separated (but can encounter problems when sources overlap). In what follows, we will assume for simplicity that the background and noise parameters $\mathbf{q}$ and $\mathbf{r}$ are known, and so we are interested only in determining the source parameters $\mathbf{p}$. The process of source extraction then reduces to locating the local maxima of the posterior distribution in the parameter space $\mathbf{p}$; we consider various strategies for performing this task in the following subsections, and for assigning uncertainties to the derived parameter values associated with each posterior peak.

First, however, we discuss how one may decide whether a detected maximum corresponds to a real source by evaluating the evidence associated with that peak for two competing models for the data. The precise definition of the models H_0 and H_1 can vary, but we shall adopt here a formulation in which each model is parameterized by the same parameters, for example $\mathbf{p} = \{X, Y, A, R\}$, but with different priors.

In this case the evidence for each model is

$$E_i = \int L(\mathbf{p})\pi_i(\mathbf{p})\mathrm{d}\mathbf{p}, \tag{7.15}$$

where the likelihood function $L(\mathbf{p})$ is the same for both models. Although not required by the method, we will assume for illustration that for each model the prior is separable, so that for $i = 0, 1$,

$$\pi_i(\mathbf{p}) = \pi_i(\mathbf{X})\pi_i(A)\pi_i(R), \tag{7.16}$$

where $\mathbf{X} = (X, Y)$ is the vector position of the source.

It is convenient to consider some general spatial region S in the image, and take our two models for the data to be

$$H_0 = \text{'there is no source with its centre lying in the region } S\text{'},$$
$$H_1 = \text{'there is one source with its centre lying in the region } S\text{'}.$$

In this case, the priors on the source position in the two models are uniform: $\pi_0(\mathbf{X}) = \pi_1(\mathbf{X}) \equiv \pi(\mathbf{X}) = 1/|S|$ if $\mathbf{X} \in S$, and zero otherwise, where $|S|$ is the area of the region S.[3] We also assume the prior on R is the same for both models, so $\pi_0(R) = \pi_1(R) \equiv \pi(R)$, but the priors on A are substantially different. Guided by the forms for H_0 and H_1 given above, we take $\pi_0(A) = \delta(A)$ (which forces $A = 0$) and $\pi_1(A)$ to be an appropriate prior probability distribution on the amplitude of the sources of interest, which is non-zero in some range $[0, A_{\max}]$. One could, of course, consider alternative definitions of these hypotheses, such as setting H_0: $A \leq A_{\lim}$ and H_1: $A > A_{\lim}$, where $A_{\lim}$ is some (non-zero) cut-off value below which one is not interested in the identified source. We shall not, however, pursue this further here.

Assuming the above priors, the evidence for each model can be written

$$E_i(S) = \int L(\mathbf{X}, A, R)\,\pi(\mathbf{X})\pi_i(A)\pi(R)\,\mathrm{d}\mathbf{X}\,\mathrm{d}A\,\mathrm{d}R \equiv \frac{1}{|S|}\int_S \bar{P}_i(\mathbf{X})\,\mathrm{d}\mathbf{X}, \tag{7.17}$$

where we have written explicitly the dependence of the evidence on the chosen spatial region S and we have defined the (un-normalized) two-dimensional marginal posterior

$$\bar{P}_i(\mathbf{X}) = \int \pi_i(A)\pi(R)L(\mathbf{X}, A, R)\,\mathrm{d}A\,\mathrm{d}R. \tag{7.18}$$

---

[3] Non-uniform priors on the source position are often useful and can be easily accommodated with little modification of our discussion. For example, one might impose a Gaussian prior on $\mathbf{X}$ centred at the putative source position with a width indicative of one's uncertainty in its location.

In particular, using $\pi_0(A) = \delta(A)$, the evidence for model H_0 is then simply

$$E_0 = \frac{1}{|S|} \int_S L_0 \, \mathrm{d}\mathbf{X} = L_0, \tag{7.19}$$

since $L_0 \equiv L(\mathbf{X}, A = 0, R)$ is, in fact, independent of the source parameters and the priors are normalized. Note that E_0 is independent of S.

So far we have not addressed the prior ratio $\Pr(H_1)/\Pr(H_0)$ in Eq. (7.7). Although this factor is often set to unity in model selection problems, one must be more careful in the current setting. For the sake of illustration and simplicity, let us assume that the sources we seek are randomly distributed in spatial position, i.e., they are not clustered. In that case, if μ_S is the (in general, non-integer) expected number of sources (centred) in a region of size $|S|$, then the probability of there being N sources in such a region is Poisson distributed:

$$\Pr(N|\mu_S) = \frac{\mathrm{e}^{-\mu_S} \mu_S^N}{N!}. \tag{7.20}$$

Thus, bearing in mind the above definitions of H_0 and H_1, we have

$$\frac{\Pr(H_1)}{\Pr(H_0)} = \mu_S. \tag{7.21}$$

Hence, the key equation (7.7) for model selection becomes

$$\rho \equiv \frac{\Pr(H_1|\mathbf{D})}{\Pr(H_0|\mathbf{D})} = \mu_S \frac{E_1(S)}{L_0} = \mu \frac{\int_S \bar{P}_1(\mathbf{X}) \, \mathrm{d}\mathbf{X}}{L_0}, \tag{7.22}$$

where $\mu = \mu_S/|S|$ is the expected number of sources (centres) per unit area. One accepts the posterior peak as corresponding to a real source if $\rho > 1$ and rejects it otherwise; Carvalho *et al.* (2008) call this the 'symmetric loss' criterion, since it implies that a rejected real source is as undesirable as an accepted spurious one.

The only conceptual issue remaining is the choice of the region S, and various alternatives are possible. If one takes S to be the entire map, then the corresponding 'global' evidence $E_1(S)$ has contributions from the entire multi-modal posterior distribution. With this choice of S, the model H_1 essentially becomes 'there is one source somewhere in the map'. In source detection, however, one is usually more interested in whether an individual posterior peak is associated with a real source. One therefore wishes to evaluate separately the 'local' evidence associated with each posterior peak (of interest). Thus, for each posterior peak, S is taken to be a region (just) enclosing it entirely (to a good approximation) in the $\mathbf{X}$ subspace. Another choice for S, explored further in Section 7.5.5, is to set it equal to the region contained in a single pixel of interest. One can even take S to be a single point, so that $\pi(\mathbf{X}) = \delta(\mathbf{X} - \mathbf{X}_0)$, thereby fixing a priori the (centre) position of the source.

It is worth mentioning that, for the last two choices of S, one must exercise some caution. If one *knows* absolutely that, if a source is present, then it *must* be centred in some pixel or at some precise location, then one can use the ratio ρ in Eq. (7.22) as described above. This is not usually the case, however. More often, one might, for example, fix the source position $\mathbf{X}_0$ to lie 'close' to where one supposes a true source might be located, in order to save on computation by reducing the dimensionality of the parameter space. In this case, to decide if a source is really present, one should still calculate the 'local' evidence E_1 associated with the entire corresponding posterior peak. Nonetheless, an available approximation to the model selection ratio (7.22) in this case is given by

$$\rho = \mu \frac{\int_{\text{peak}} \bar{P}_1(\mathbf{X})\,\mathrm{d}\mathbf{X}}{L_0} \approx \mu \mathcal{A} \frac{\bar{P}_1(\mathbf{X}_0)}{L_0} \approx f_{\text{s}} \frac{\bar{P}_1(\mathbf{X}_0)}{L_0}, \qquad (7.23)$$

where $\mathcal{A}$ is the 'typical' projected area of a source of interest and hence f_{s} is the expected fraction of the image covered by such sources, assuming no sources overlap (see Cruz *et al.* 2007).

Although the approach outlined above for Bayesian source detection and validation is conceptually straightforward, the practical task of locating the multiple peaks of the posterior distribution, estimating uncertainties on derived parameter values and evaluating the required evidences for model selection can be extremely computationally demanding. In the remainder of this section, we therefore consider some approaches to these problems, roughly in order of increasing theoretical desirability and, typically, computational complexity. Our final Section 7.5.5 presents an alternative, pixel-by-pixel, method that takes a somewhat different approach.

7.5.1 Analytic source extraction and the matched filter

It is worth noting that, in a particular special case, the approach outlined above leads to the matched filter. Consider the case in which the non-source contribution to the image can be considered as a generalized noise with covariance matrix $\mathbf{G} = \langle \mathbf{g}\mathbf{g}^{\mathrm{T}} \rangle$ (again assuming the parameters $\mathbf{q}$ and $\mathbf{r}$ characterizing the generalized noise are known). Also let us write the template Eq. (7.3) as

$$\tau(\mathbf{x}; \mathbf{p}) = At(\mathbf{x} - \mathbf{X}; R), \qquad (7.24)$$

where A, $\mathbf{X}$ and R are, respectively, the amplitude, vector position and spatial extent of the source, and $t(\mathbf{x}; R)$ is the spatial profile of a unit amplitude reference source of size R, centred on the origin. Then the signal vector in the likelihood Eq. (7.9) is simply $\mathbf{s}(\mathbf{p}) = A\mathbf{t}(\mathbf{X}, R)$, where the vector $\mathbf{t}$ has components $t_i = t(\mathbf{x}_i - \mathbf{X}; R)$, in which $\mathbf{x}_i$ is the position of the centre of the ith pixel in the image.

Substituting this expression into the log-likelihood, assuming uniform priors on the source parameters in some (wide) range, differentiating the log-posterior with respect to A and setting the result to zero yields an analytic estimate for the source amplitude,

$$\widehat{A}(\mathbf{X}, R) = \frac{\mathbf{t}^{\mathrm{T}}(\mathbf{X}, R)\,\mathbf{G}^{-1}\mathbf{d}}{\mathbf{t}^{\mathrm{T}}(\mathbf{X}, R)\,\mathbf{G}^{-1}\mathbf{t}(\mathbf{X}, R)}. \tag{7.25}$$

Note that, under the assumption of statistical homogeneity, the denominator in the above expression does not in fact depend on the source position $\mathbf{X}$, and thus need only be evaluated once (for a source at the origin, say) for any given value of R.

More importantly, the estimator Eq. (7.25) is precisely the filtered field produced by the standard matched filter, in which one assumes a value for R. Moreover, the corresponding log-posterior can be written:

$$\ln P(\mathbf{X}, \widehat{A}, R) = \text{constant} + \tfrac{1}{2}\alpha(R)[\widehat{A}(\mathbf{X}, R)]^2. \tag{7.26}$$

Thus, for a given value of R, peaks in the filtered field correspond to peaks on the log-posterior considered as a function of $\mathbf{X}$. If the sources sought all have the same size R and this is known in advance, then the local maxima of the now 3-dimensional posterior in the space (X, Y, A) are all identified by this method. This scenario corresponds, for example, to point sources observed by an experiment with a known, invariant beam profile.

In this approach, uncertainties in the (source) parameter values derived from each posterior peak and the associated 'local' evidences required for source validation can be obtained simply by evaluating the curvature (or Hessian) matrix at each maximum and using it to construct a Gaussian approximation to each posterior peak (see, e.g., Hobson & McLachlan 2003; Carvalho *et al.* 2008), although this is rarely performed in matched filter analyses.

7.5.2 *Iterative source extraction: local maximization*

In the matched filter method, when the source size R is not known in advance, or the sources have different sizes, one typically adopts the rather ad hoc approach of filtering the field at several predetermined values of R and combining the results in some way to obtain estimates of the parameters $\mathbf{p} = \{X, Y, A, R\}$ for each identified source.

Carvalho *et al.* (2008) adopt a different strategy based on locating the 'substantial' local maxima of the posterior using multiple downhill minimizations. They show that the resulting POWELLSNAKES algorithm outperforms linear filters, while retaining their speed. The algorithm is technically quite complicated, but the basic approach can be summarized as follows:

- First, a pre-filtering step is applied in which the original image is replaced by the field Eq. (7.25), where R is taken equal to the mid-point of its prior range. The advantage of this step is that, even in the presence of sources of different sizes, the filtered field often has pronounced local maxima near many real sources.
- In the next step, one launches a large number (typically a few hundred) of downhill minimizations in the two-dimensional (X, Y) subspace, using the Powell algorithm (Press *et al.* 1994). The (X, Y) starting points for the 'snakes' are dithered randomly about a uniform rectangular grid, with the total number of snakes equal to the ratio of the total number of pixels to the smallest anticipated spatial extent of a source (in pixel units). The Powell minimization algorithm is also modified to include a 'jump procedure', inspired by quantum-mechanical barrier tunnelling, designed to avoid 'small' local maxima in the posterior, such as those resulting from the background and noise. The purpose of the 2-dimensional minimization step is simply to obtain a good estimate of the (X, Y) positions of the sources, although one may then use Eq. (7.25) also to obtain an approximate estimate of the putative source amplitude.
- In the final step, one performs multiple Powell minimizations (with no jump procedure) in the full 4-dimensional space (X, Y, A, R), starting each minimization from the X, Y, A values found in the previous step (and setting the initial value of R to the mid-point of its prior range). The posterior peak located at the end of each minimization is then either accepted as a real source using an evidence-based criterion (see below) or rejected. If accepted, the best-fit source template is subtracted from the map before the next minimization is launched.

As in the previous subsection, uncertainties in the (source) parameter values derived from each posterior peak and the associated 'local' evidences required for source validation are obtained by constructing a Gaussian approximation to each located posterior maximum by evaluating its Hessian matrix. Nonetheless, Carvalho *et al.* (2008) point out that the estimation of evidences via the Gaussian approximation does not take into account the fact that the prior might abruptly truncate a posterior peak before it falls (largely) to zero, and hence may lead to an overestimate of the evidence. They go on to suggest ways in which this effect might be mitigated. In particular, they compute an upper bound on the quality of a detection, and show that a quantity termed the 'normalized integrated signal-to-noise ratio' (NISNR) plays an important role in establishing a limit on the source amplitudes that can be reliably detected. They also consider the pixel-based prior mentioned in Section 7.5, where the region S is chosen as the pixel containing the relevant posterior peak.

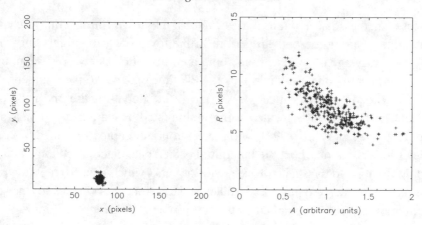

Fig. 7.4. The 500 samples from the posterior obtained on the third iteration of the MCCLEAN algorithm when applied to the toy problem in Figure 7.1. The samples are projected into the 2-dimensional subspaces (X, Y) (left panel) and (A, R) (right panel).

7.5.3 Iterative source extraction: global maximization

Rather than performing iterative source extraction based on local maximization, HM03 originally advocated locating the global maximum of the posterior at each iteration, subtracting the corresponding source if accepted as real based on the evidence criterion, and repeating the process until the first rejected detection. This approach has a stronger theoretical motivation than those presented above, since it concentrates at each stage on the most prominent source remaining in the map, but it is far more computationally expensive. HM03 suggest two different ways of performing this iterative approach.

- In the first method, called MCCLEAN, the posterior peak centred on the global maximum is located and explored using MCMC sampling. HM03 found that this could be achieved relatively straightforwardly by using just a few interacting MCMC chains and employing a relatively fast annealing schedule during the burn-in phase. The latter allows the chains initially to sample from remote regions of the posterior and locate the global posterior peak, before sampling from it during the post burn-in phase. Note that if two (or more) of the dominant posterior peaks are well explored by the MCMC chains, then one simply characterizes and subtracts more than one source at that iteration. HM03 applied this method to the toy problem illustrated in Figure 7.1, using the BAYESYS sampler with five chains and, at each iteration, taking 500 post burn-in samples. For illustration, the samples taken during the third iteration are shown in Figure 7.4. These samples can then be used to determine best-fit values and uncertainties for the source parameters. For each putative source, the evidence value required to assess whether to accept it as real is calculated by performing a thermodynamic integration during the burn-in period. In the toy model, the algorithm stopped after identifying six sources (one being a combination of two overlapping real sources), but required ~40 minutes of CPU time on a desktop computer.

- In the second method, called MAXCLEAN, the global maximum of the posterior at each iteration is located using a simulated annealing downhill simplex optimizer (Press *et al.* 1994), with an appropriate choice of annealing schedule. The Hessian matrix at the maximum is then calculated to obtain a Gaussian approximation to the shape of the posterior peak, which is used to obtain uncertainties on the derived parameter values and to estimate evidences for the source validation step. When applied to the toy problem illustrated in Figure 7.1, the results were similar to those obtained using the MCCLEAN method, identifying the same six sources before terminating. In this approach, however, the entire analysis required only ~8 minutes of CPU time.

7.5.4 Simultaneous source extraction

The iterative approach outlined in the previous subsection was in fact originally born out of necessity, since the computation required to sample reliably from the full multi-modal posterior (see Figure 7.3) is prohibitive using standard MCMC methods based on the Metropolis–Hastings algorithm. Indeed, straightforward implementations of this approach find it very difficult, in general, to transition between widely separated, narrow modes of a posterior, and can easily give spurious results. A state-of-the-art MCMC sampler, such as BAYESYS, which employs multiple interacting chains and compound proposals, can cope with such distributions, but requires over 10 hours of CPU time to explore the full posterior adequately when applied to the toy problem illustrated in Figure 7.1.

Fortunately, this problem has recently been overcome by the multi-modal nested sampling algorithm of Feroz and Hobson (2008), which can efficiently produce posterior samples and calculate global and local evidences for posteriors that are highly multi-modal. It is therefore ideally suited to performing Bayesian source detection and validation. The technique combines the basic nested sampling approach of Skilling (2004a) with an automatic clustering algorithm, and builds on the work of Mukherjee, Parkinson and Liddle (2006) and Shaw, Bridges and Hobson (2007). When applied to the toy problem, this method produces the samples shown in Figure 7.5, when projected into the (X, Y) subspace. Not only has the sampler clearly explored the full posterior distribution, but it has also automatically identified which samples belong to each posterior peak. In total, 11 posterior peaks are located, seven of which were correctly identified as real sources by evaluation of their local evidence (although one peak again corresponds to the

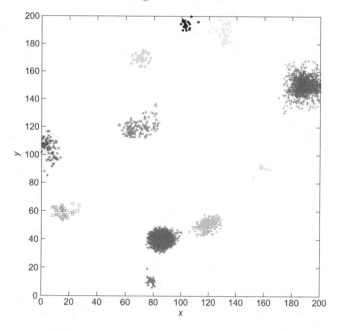

Fig. 7.5. The samples obtained from the toy model posterior using the multi-modal nested sampling algorithm, projected into the (X, Y) subspace. The different tints correspond to different posterior peaks.

combination of two overlapping real sources). The remaining four peaks were correctly rejected as sources using the local evidence criterion. The entire analysis took ~ 5 minutes of CPU time.

7.5.5 Pixel-by-pixel source extraction

We conclude our summary of Bayesian source extraction within the single-source model with a method based on that suggested by Savage and Oliver (2007), which takes a rather different approach to those outlined above. In this approach, one retains the classical source extraction methodology of dividing the process into two distinct stages: source detection and source photometry. This is computationally much quicker than the single, combined approach used thus far, although at the cost of losing some theoretical rigour. We take the opportunity here to make some straightforward generalizations to the basic method originally presented.

In the source detection step, the key idea is to consider in turn each pixel in the map separately. Suppose one is currently concerned with the kth pixel at position $\mathbf{x}_k$. One first reduces the computational burden by taking the corresponding data vector $\mathbf{d}$ to be a small portion of the map centred on this pixel, rather than the whole map. Clearly the size of this patch should be chosen by considering the range of spatial sizes of sources of interest. Following Eq. (7.4) the (reduced) data vector is then considered to be of the form

$$\mathbf{d} = \mathbf{s}(\mathbf{p}) + \mathbf{b}(\mathbf{q}) + \mathbf{n}(\mathbf{r}), \tag{7.27}$$

where, as we have assumed throughout this section, the signal vector is assumed to result from just a single source (i.e., $N_s = 1$). In the current approach, however, one additionally assumes that $\mathbf{X} = \mathbf{x}_k$, i.e., that the source is centred in the current pixel. Moreover, one assumes a value for the source size R, typically the mid-point of its allowed range (note that no assumption is necessary in the case where all the sources have the same, known size, for example when one is considering populations of point-like sources, such as distant galaxies or stars). Thus the only remaining source parameter to be determined is the amplitude A. To simplify matters further, the background is assumed to be uniform across the patch, so the only unknown parameter $\mathbf{q}$ is the overall background level B. It is also assumed that the noise parameters $\mathbf{r}$ are known and the noise is statistically homogeneous with a covariance matrix $\mathbf{N}$. Hence the model contains only the two free parameters A and B.

One then considers two models for the data:

$$H_0 = \text{'there is no source centred in the } k\text{th pixel'},$$

$$H_1 = \text{'there is one source centred in the } k\text{th pixel'},$$

where, as above, the former model is simply obtained from the latter by setting $A = 0$. Thus the log-posteriors for H_0 and H_1 read

$$\ln P_0(B) = c - \tfrac{1}{2}(\mathbf{d} - B\mathbf{1})^{\mathrm{T}}\mathbf{N}^{-1}(\mathbf{d} - B\mathbf{1}) + \ln \pi(B), \tag{7.28}$$

$$\ln P_1(A, B) = c - \tfrac{1}{2}(\mathbf{d} - A\mathbf{t} - B\mathbf{1})^{\mathrm{T}}\mathbf{N}^{-1}(\mathbf{d} - A\mathbf{t} - B\mathbf{1})$$
$$+ \ln \pi(A) + \ln \pi(B), \tag{7.29}$$

where c is an unimportant constant, $\mathbf{1}$ is the vector with all elements equal to unity, and the vector $\mathbf{t}$ has components $t_i = t(\mathbf{x}_i - \mathbf{x}_k)$ and corresponds to a unit amplitude reference source centred in the kth pixel.

To simplify subsequent results, it is useful to define the quantities

$$\alpha \equiv \mathbf{t}^{\mathrm{T}}\mathbf{N}^{-1}\mathbf{t}, \quad \beta \equiv \mathbf{1}^{\mathrm{T}}\mathbf{N}^{-1}\mathbf{1}, \quad \gamma \equiv \mathbf{t}^{\mathrm{T}}\mathbf{N}^{-1}\mathbf{d},$$
$$\delta \equiv \mathbf{1}^{\mathrm{T}}\mathbf{N}^{-1}\mathbf{d}, \quad \epsilon \equiv \mathbf{t}^{\mathrm{T}}\mathbf{N}^{-1}\mathbf{1}, \tag{7.30}$$

where it is worth noting that the quantities α, γ and ϵ must be calculated separately for each pixel k, whereas β and δ need only be calculated once. It is then easy to show that the optimal value for B in model H_0 must satisfy

$$\beta \hat{B}_0 - \left.\frac{\partial \ln \pi(B)}{\partial B}\right|_{\hat{B}_0} = \delta. \tag{7.31}$$

Approximating the posterior peak as a Gaussian, one finds that the rms uncertainty in $\hat{B}_0$ is

$$\sigma_{\hat{B}_0} = \left(\beta - \frac{\partial^2 \ln \pi(B)}{\partial B^2} \bigg|_{\hat{B}_0} \right)^{-1/2} \tag{7.32}$$

and the corresponding estimate of the log-evidence is

$$\ln E_0 = \tfrac{1}{2} \ln(2\pi) + \ln \sigma_{\hat{B}_0} + c - \tfrac{1}{2} \mathbf{d}^{\mathrm{T}} \mathbf{N}^{-1} \mathbf{d} - \tfrac{1}{2} \hat{B}_0^2 \beta + \hat{B}_0 \delta + \ln \pi(\hat{B}_0). \tag{7.33}$$

Similarly, for the model H_1, the optimal values of A and B satisfy

$$\alpha \hat{A}_1 - \frac{\partial \ln \pi(A)}{\partial A} \bigg|_{\hat{A}_1} + \epsilon \hat{B}_1 = \gamma, \tag{7.34}$$

$$\epsilon \hat{A}_1 + \beta \hat{B}_1 - \frac{\partial \ln \pi(B)}{\partial B} \bigg|_{\hat{B}_1} = \delta. \tag{7.35}$$

The Hessian matrix at the posterior peak reads

$$\mathbf{H} = \begin{pmatrix} -\alpha + \dfrac{\partial^2 \ln \pi(A)}{\partial A^2} \bigg|_{\hat{A}_1} & -\epsilon \\ -\epsilon & -\beta + \dfrac{\partial^2 \ln \pi(B)}{\partial B^2} \bigg|_{\hat{B}_1} \end{pmatrix} \tag{7.36}$$

and hence defines a Gaussian approximation leading to the log-evidence estimate

$$\ln E_1 = \ln(2\pi) - \tfrac{1}{2} \ln |-\mathbf{H}| + c - \tfrac{1}{2} \mathbf{d}^{\mathrm{T}} \mathbf{N}^{-1} \mathbf{d} - \tfrac{1}{2} \hat{A}_1^2 \alpha + \hat{A}_1 \gamma - \hat{A}_1 \hat{B}_1 \epsilon$$
$$+ \hat{B}_1 \delta - \tfrac{1}{2} \hat{B}_1^2 \beta + \ln \pi(\hat{A}_1) + \ln \pi(\hat{B}_1). \tag{7.37}$$

Detection of the sources then proceeds by comparing the models H_0 and H_1 at each pixel in the map. At each pixel, the logarithm of the model selection ratio Eq. (7.22) is

$$\ln \rho = \ln E_1 - \ln E_0 + \ln\langle n \rangle$$
$$= \tfrac{1}{2} \ln(2\pi) - \tfrac{1}{2} \ln |-\mathbf{H}| - \ln \sigma_{\hat{B}_0} - \tfrac{1}{2} \hat{A}_1^2 \alpha + \hat{A}_1 \gamma$$
$$- \hat{A}_1 \hat{B}_1 \epsilon + (\hat{B}_1 - \hat{B}_0)\delta - \tfrac{1}{2}(\hat{B}_1^2 - \hat{B}_0^2)\beta$$
$$+ \ln \pi(\hat{A}_1) + \ln \pi(\hat{B}_1) - \ln \pi(\hat{B}_0) + \ln\langle n \rangle, \tag{7.38}$$

where we have used the fact that $\Pr(H_1)/\Pr(H_0) = \langle n \rangle$, i.e., the (in general, non-integer) expectation value of the number of sources per pixel (which is likely to be $\ll 1$); in the second equality we note that the quantities c and $\mathbf{d}^{\mathrm{T}} \mathbf{N}^{-1} \mathbf{d}$ have cancelled out of the final expression. By cycling through the pixels in turn, one creates a 'map' of $\ln \rho$. Each peak in this map with $\ln \rho > 0$ is identified as the position of a real source, with the corresponding $\hat{A}_1$ giving its flux. An illustration of the process is shown in Figure 7.6, assuming uniform priors on A and B.

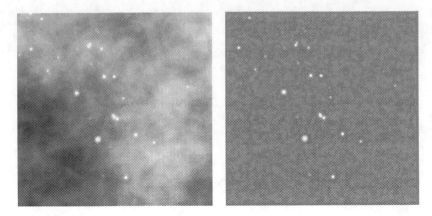

Fig. 7.6. Input map (left) and $\ln \rho$ map (right) produced by the pixel-by-pixel source detection method.

If all the sources have the same known size (such as point sources in a map with a fixed beam), the source detection step above can, in fact, be the complete analysis. In more general problems, however, the detection step is followed by a source photometry step. For each source identified in the detection step, an MCMC analysis is performed in the full source parameter space $\{X, Y, A, R\}$, but with narrow priors imposed on X and Y, centred on the identified source position. Thermodynamic integration can then be used to calculate a more accurate evidence value to decide whether to accept the source as real.

The motivation for the two-stage approach adopted in this method is speed of analysis, which can be very important in the early stages of modern, large astronomical surveys (when the production of early results and metrics of survey depth, and so on, are desirable as quickly as possible). To this end, it is noted that because of the simplicity of models we use here (i.e., single sources, centred on pixel positions), one can implement the above calculations using Fast Fourier Transforms, making the method virtually as fast (and having the same algorithmic dependence on data size) as even the simplest implementations of the matched filter. In addition, the requirement for a fast, *analytic* solution still leaves some freedom in the choice of prior on the source amplitude A and background B. In particular, one can choose a prior on A of the form

$$\pi(A) \propto A^{n}, \tag{7.39}$$

which includes the following special cases: $n = 0$ is a uniform prior; $n = -1$ is a (non-informative) Jeffreys prior; and $n < -1$ is a good description of observed (power-law) source number counts, which is useful when, for example, one is extracting distant galaxies from survey data.

7.6 Conclusions

We have presented the Bayesian approach to the detection and characterization of discrete sources in a diffuse background, which allows for the inclusion of all pertinent prior information available. The method assumes a parameterized form for the sources, and the optimal values of these parameters, and their associated errors, are usually obtained in a single step by evaluating their full posterior distribution. In principle, any statistical form for the (generalized) noise can be accommodated by defining an appropriate likelihood function. If available, one can also place physical priors on the parameters defining a source and on the number of sources present. Moreover, one can decide whether each peak found in the posterior distribution corresponds to a real source or a background fluctuation by performing a Bayesian model comparison using (an approximation to) the evidence. The approach therefore represents the theoretically optimal method for performing parameterized source extraction.

Acknowledgments

We thank Farhan Feroz for useful discussions and for providing Figure 7.5.

References

Barreiro, R. B., Sanz, J. L., Herranz, D. and Martinez-Gonzalez, E. (2003). *Mon. Not. Roy. Astron. Soc.*, **342**, 119.

Bertin, E. and Arnouts, S. (1996). *Astron. Astrophys. Supp.*, **117**, 393.

Carvalho, P., Rocha, G. and Hobson, M. P. (2008). *Mon. Not. Roy. Astron. Soc.*, submitted.

Chiang, L.-Y., Jorgensen, H. E., Naselsky, I. P., Naselsky, P. D., Novikov, I. D. and Christensen, P. R. (2002). *Mon. Not. Roy. Astron. Soc.*, **335**, 1054.

Cruz, M., Turok, N., Vielva, P., Martinez-Gonzalez, E. and Hobson, M. P. (2007). *Science*, **318**, 1612.

Diego, J. M., Vielva, P., Martinez-Gonzalez, E., Silk, J. and Sanz, J. L. (2002). *Mon. Not. Roy. Astron. Soc.*, **336**, 1351.

Feroz, F. and Hobson, M. P. (2008). *Mon. Not. Roy. Astron. Soc.*, **384**, 449.

Green, P. J. (1994). *J. R. Stat. Soc.*, **56**, 589.

Guglielmetti, F., Fischer, R., Voges, W., Boese, G. and Dose, V. (2004). In V. Schoenfelder, G. Lichti and C. Winkler, eds., *5th INTEGRAL Workshop*. ESA **SP-552**.

Haehnelt, M. G. and Tegmark, M. (1996). *Mon. Not. Roy. Astron. Soc.*, **279**, 545.

Herranz, D., Sanz, J. L., Barreiro, R. B. and Martinez-Gonzalez, E. (2002a). *Astrophys. J.*, **580**, 610.

Herranz, D., Sanz, J. L., Hobson, M. P., Barreiro, R. B., Diego, J. M., Martinez-Gonzalez, E. and Lasenby, A. N. (2002b). *Mon. Not. Roy. Astron. Soc.*, **336**, 1057.

Hobson, M. P., Bridle, S. L. and Lahav, O. (2002). *Mon. Not. Roy. Astron. Soc.*, **335**, 377.

Hobson, M. P. and McLachlan, C. (2003). *Mon. Not. Roy. Astron. Soc.*, **338**, 765.

López-Caniego, M., Herranz, D., Barreiro, R. B. and Sanz, J. L. (2005). *Mon. Not. Roy. Astron. Soc.*, **359**, 993.

Makovoz, D. and Marleau, F. R. (2005). *Proc. Astron. Soc. Pac.*, **117**, 1113.

Marshall, P. J. (2006). *Mon. Not. Roy. Astron. Soc.*, **372**, 1289.

McEwen, J. D., Hobson, M. P. and Lasenby, A. N. (2008). *IEEE Trans. Signal Process.*, **56**, 3813, available as astro-ph/0612688.

Mukherjee, P., Parkinson, D. and Liddle, A. R. (2006). *Astrophys. J. Lett.*, **638**, L51.

Ó'Ruanaidh, J. J. K. and Fitzgerald, W. J. (1996). *Numerical Bayesian Methods Applied to Signal Processing*. New York: Springer-Verlag.

Phillips, D. B. and Smith, A. F. M. (1995). In W. R. Gilks, S. Richardson and D. J. Spiegelhalter, eds., *Markov Chain Monte Carlo in Practice*. London: Chapman & Hall.

Press, W. H., Teukolsky, S. A., Vetterling, W. T. and Flannery, B. P. (1994). *Numerical Recipes in Fortran*. Cambridge: Cambridge University Press.

Sanz, J. L., Herranz, D. and Martinez-Gonzalez, E. (2001). *Astrophys. J.*, **552**, 484.

Savage, R. and Oliver, S. (2007). *Astrophys. J.*, **661**, 1339.

Shaw, R., Bridges, M. and Hobson, M. P. (2007). *Mon. Not. Roy. Astron. Soc.*, **378**, 1365.

Sivia, D. S. and Skilling, J. (2006). *Data Analysis: A Bayesian Tutorial*. Cambridge: Cambridge University Press.

Skilling, J. (2004a). In E. Erickson, J. T. Rychert and C. R. Smith, eds., *AIP Conf. Proc.*, **735**, 395.

Skilling, J. (2004b). BayeSys and MassInf, online at http://www.inference.phy.cam. ac.uk/bayesys/

Stetson, P. B. (1992). User's Manual for Daophot II, online at http://www.star.bris. ac.uk/~mbt/daophot.

8

Flux measurement

Daniel Mortlock

8.1 Introduction

The measurement of the flux of an astronomical source is a classic parameter estimation problem in which the quantity of interest must be inferred from noisy data. The Bayesian methods described in this book provide a clear, unambiguous, self-consistent and optimal method for answering this type of question, but the vast majority of flux measurements are made by applying a heuristic classical estimator to the data. This raises some immediate questions: Why is the estimator-based approach adopted in most cases? How do the resultant flux measurements differ? What is the relationship between the two techniques?

To answer these questions first requires an understanding of the astronomical measurement process itself (Section 8.2), which leads very naturally to the definition of the standard flux estimator (Section 8.3). Using a model for the source population (Section 8.4) as a prior, it is also possible to apply Bayesian inference to the problem (Section 8.5), although care is required to avoid some potential inconsistencies (Section 8.6). Even with the full Bayesian result in hand, however, the existence of databases containing billions of classically estimated fluxes and errors leads to a number of practical considerations which argue against simply reporting posterior distributions for astronomical fluxes (Section 8.7).

8.2 Photometric measurements

How is the flux of an astronomical source measured? 'With great difficulty' is one possible answer, especially given a history of photographic plates, dipole antennae, microdensitometers and other arcane equipment. A more pertinent answer would be 'with technology', as almost all astronomical measurements now rely on some combination of large telescopes and highly efficient electronic detectors. Many modern detectors (e.g., optical or X-ray charge-coupled devices) are close to 100% efficient, and even the registration of individual photons is a mature technique (see, e.g., Law *et al.* 2006; Murphy *et al.* 2008). As photon emission is usually an independent stochastic process, an immediate implication is that the probability[1] of registering N_γ source photons is Poisson distributed. Hence

$$\mathrm{P}(N_\gamma|\bar{N}_\gamma) = \frac{\bar{N}_\gamma^{N_\gamma} \, e^{-\bar{N}_\gamma}}{\Gamma(N_\gamma + 1)}, \tag{8.1}$$

where the average number of photons expected, $\bar{N}_\gamma$, depends not only on the source's flux, but also on the details of the observation. For a source of flux F observed for integration time T_{obs} using a telescope of collecting area A,

$$\bar{N}_\gamma = \frac{F \, T_{\mathrm{obs}} A}{\bar{E}_\gamma}, \tag{8.2}$$

where $\bar{E}_\gamma$ is the average energy[2] of the photons. Thus, for example, an average of $\bar{N}_\gamma \simeq 10^3$ photons would be registered from a typical redshift ~ 0.1 galaxy in a routine 1-minute observation on the 2.5 m diameter telescope used for the Sloan Digital Sky Survey (SDSS; York *et al.* 2000), whereas even a 100-hour integration on a faint active galactic nucleus (AGN) with the $0.04 \, \mathrm{m}^2$ Chandra X-Ray Observatory might have $\bar{N}_\gamma \simeq 5$ (cf. the example in Chapter 12).

There are, of course, a great many real-world complexities (e.g., atmospheric absorption, non-ideal detectors and terrestrial interference) which combine to make accurate photometry such a difficult task, but most of these effects simply modify the conversion between flux and counts. Thus it is useful to combine these factors with the basic observational parameters described above to give a single calibration constant, C, which relates the expected counts to the flux by $C = F/\bar{N}_\gamma$, and reduces to $C = \bar{E}_\gamma/T_{\mathrm{obs}}A$ in the ideal case. The count distribution given in Eq. (8.1) can then be rewritten as

$$\mathrm{P}(N_\gamma|F) = \frac{(F/C)^{N_\gamma} \, e^{-F/C}}{\Gamma(N_\gamma + 1)}, \tag{8.3}$$

where just one photon would be expected from a source of flux C in this observation.

---

[1] Here, 'probability' is used strictly in the logical sense adopted by, e.g., Jaynes (2005). Hence all expressions of the form $\mathrm{P}(A|B)$ should be interpreted as the degree to which (the truth of) proposition B implies the truth of proposition A. As such $\mathrm{P}(A|B)$ is not a mathematical function in the usual sense, although if A and B are mathematical in nature then the more formal $\mathrm{P}(x = x_0|y = y_0)$ is generally replaced by the less cumbersome, if potentially ambiguous, shorthand $\mathrm{P}(x|y)$.

[2] The spectral response, or passband, of an instrument is characterized in terms of energy in X-ray and gamma-ray astronomy, by wavelength in optical and infrared astronomy, and by frequency in radio and microwave astronomy. Further, it is almost always the flux density averaged over a finite energy range that is obtained in actual astrophysical measurements. Some treatments of flux estimation (e.g., Murdoch *et al.* 1973) work in terms of flux density; however, this complication is irrelevant to the statistical aspects of the problem, and so only total fluxes are considered here.

Implicit in the above results is that the field of view of the measurement, of angular area Ω_{obs}, is greater than that of the observational point-spread function.[3] If this were not the case then an 'aperture correction' would be required, but a more important implication of a finite aperture size is that source photons might not be the only contribution to the data.

Unfortunately, the majority of astronomical measurements includes contamination from actual astrophysical backgrounds (e.g., the cosmic microwave background in the submillimetre), terrestrial foregrounds (e.g., sky-glow in the near-infrared) or instrumental imperfections (e.g., the read-out noise from solid state detectors or thermal noise in microwave bolometers). These physical processes might differ greatly, but the effect is the same: a reasonably uniform but noisy contribution to the data which must be accounted for when measuring a source's flux. Expressed in flux units, the raw data take the form

$$d = CN_\gamma + b, \tag{8.4}$$

where b is the total background in the aperture (as opposed to the background surface brightness).

The presence of a smooth, perfectly known, background would not affect flux measurements, but any uncertainty in the value of b will result in increased errors. Instrumental noise, intrinsic variations across the background (e.g., due to clouds in the atmosphere or Poisson fluctuations in astrophysical backgrounds) and contamination by other sources in the field can all lead to an error in the estimation of b. It might seem these processes should be treated separately, even if only at the level of formalism, but the differences between the nature of these processes are largely immaterial, and it is only the state of knowledge that matters, something which is particularly clear within the Bayesian framework. If the background signal is estimated (by whatever means) to be $\hat{b}$ with uncertainty σ_b, its sampling distribution is taken to be

$$P(\hat{b}|\text{data}) = \frac{1}{(2\pi)^{1/2}\sigma_b} \exp\left[-\frac{1}{2}\left(\frac{\hat{b}-b}{\sigma_b}\right)^2\right], \tag{8.5}$$

where 'data' refers to all the information used to determine the background level. The assumption of Gaussianity may be justified either because of the central limit theorem or due to maximum entropy principles (e.g., Jaynes 2003).

With the stochastic nature of the observation process fully described by the sampling distributions given in Eqs. (8.3) and (8.5), the task of flux measurement is, then, to use these relationships to constrain a source's flux from whatever data

---

[3] In optical and infrared observations this is characterized by the seeing, usually defined as the full-width at half-maximum of the point-spread function of the instrument. In the microwave and radio regime the term 'beam' is usually used for the same quantity.

have been obtained. Two very different approaches are considered here: the construction of a classical unbiased estimator (Section 8.3); and the calculation of the full Bayesian posterior (Section 8.5). Despite the latter method being the main focus of this book, it is the former which is treated first, partly because it has been used to generate almost all reported astronomical flux measurements, but mainly because it is simpler.

8.3 Classical flux estimation

Having made a photometric observation of a source (as described in Section 8.2), the most common method of flux measurement is to construct an unbiased estimator simply by subtracting the measured background, $\hat{b}$, from the data, d, to give

$$\hat{F} = d - \hat{b} = C N_\gamma - (\hat{b} - b). \tag{8.6}$$

Aside from being a reasonable estimator of the flux, $\hat{F}$ can also function as a more intuitive representation of the data, being a sufficient statistic provided that $\hat{b}$ is known. However, it is also immediately clear from Eq. (8.6) that $\hat{F}$ can be negative if $d < \hat{b}$, one of several shortcomings of the estimator-based approach to flux estimation that are discussed at the end of this section.

The sampling distribution of $\hat{F}$ can be found by combining Eqs. (8.1) and (8.5) to give

$$P(\hat{F}|F) = \sum_{N_\gamma=0}^{\infty} \int_{-\infty}^{\infty} P(N_\gamma|F)\, P(\hat{b}|\text{data})\, \delta\left\{\hat{F} - \left[C N_\gamma - (\hat{b} - b)\right]\right\} d\hat{b},$$

$$= \sum_{N_\gamma=0}^{\infty} \frac{(F/C)^{N_\gamma} e^{-F/C}}{\Gamma(N_\gamma + 1)} \frac{1}{(2\pi)^{1/2}\sigma_{\rm b}} \exp\left[-\frac{1}{2}\left(\frac{\hat{F} - C N_\gamma}{\sigma_{\rm b}}\right)^2\right]. \tag{8.7}$$

The sum effectively marginalizes over the unknown background, although it is revealing to note that the sampling distribution is independent of its actual (or estimated) value. Whilst Eq. (8.7) cannot be simplified in general, one of two approximations that result in analytic forms for $P(\hat{F}|F)$ can be adopted in almost all cases of practical interest.

In X-ray astronomy, for example, there is often no significant background, which is equivalent to it being perfectly known [i.e., $\sigma_{\rm b} = 0$ and $P(\hat{b}|\text{data}) = \delta(\hat{b} - b)$], in which case

$$P(\hat{F}|F) = \frac{(F/C)^{\hat{F}/C} e^{F/C}}{\Gamma(\hat{F}/C + 1)}. \tag{8.8}$$

Flux estimation in this regime is discussed in more detail in Chapter 12.

It is more common – and almost always the case in optical and infrared astronomy – that the background is appreciable but the source photon counts are high as well, $\bar{N}_\gamma \gg 1$. In this case it is legitimate to take the continuum limit in which the discrete photon distribution given in Eq. (8.3) becomes a Gaussian. The sum in Eq. (8.7) then simplifies to

$$
\begin{aligned}
\mathrm{P}(\hat{F}|F) &\simeq \int_{-\infty}^{\infty} \frac{1}{(2\pi)^{1/2}(CF)^{1/2}} \exp\left[-\frac{1}{2}\frac{(F'-F)^2}{CF}\right] \\
&\quad \times \frac{1}{(2\pi)^{1/2}\sigma_{\mathrm{b}}} \exp\left[-\frac{1}{2}\frac{(\hat{F}-F')^2}{\sigma_{\mathrm{b}}^2}\right] \, \mathrm{d}F' \\
&= \frac{1}{(2\pi)^{1/2}(\sigma_{\mathrm{b}}^2+CF)^{1/2}} \exp\left[-\frac{1}{2}\frac{(\hat{F}-F)^2}{\sigma_{\mathrm{b}}^2+CF}\right],
\end{aligned}
\tag{8.9}
$$

where $CF = C^2 F/C$ is the Poisson variance, as exactly one photon is expected from a source of flux C in such an observation. The simple Gaussian form of the sampling distribution makes clear the separate influences of the Poisson and background terms, and is used as the basis for all the results that follow (and, even when applied in the low-N_γ limit, produces the same mean and variance as the full expression in Eq. (8.7)).

The most important, if obvious, implication of Eq. (8.9) is that $\hat{F}$ is an unbiased estimator of the source flux, with an expectation value over noise realizations of

$$
\langle \hat{F} \rangle = F
\tag{8.10}
$$

and a variance of

$$
\langle (\hat{F}-F)^2 \rangle = \sigma_{\mathrm{b}}^2 + CF.
\tag{8.11}
$$

Whilst care was taken to derive these results rigorously, the same expression for the variance could have been arrived at using more heuristic arguments, by adding the variances from the source and background terms in quadrature. Often one of the two terms in Eq. (8.11) dominates: the low σ_{b}^2 case appropriate to some X-ray observations was already mentioned, while the other extreme, characterized by $\langle (\hat{F}-F)^2 \rangle \simeq \sigma_{\mathrm{b}}^2$, is appropriate to all but the brightest sources in most optical or infrared surveys.

It is not possible, however, to calculate $\langle (\hat{F}-F)^2 \rangle$ from the data, as the source's actual flux is unknown. The commonly used estimate for the measurement error is, instead, obtained by simply replacing F with $|\hat{F}|$ in Eq. (8.11) to obtain

$$
\Delta \hat{F} = \left(\sigma_{\mathrm{b}}^2 + C|\hat{F}| \right)^{1/2}.
\tag{8.12}
$$

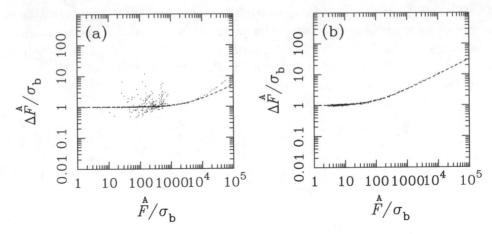

Fig. 8.1. Scatter plots of the estimated error, $\Delta\hat{F}$, as a function of the estimated flux, $\hat{F}$, from sources taken from the SDSS (York *et al.* 2000) in (a) and the UKIRT Infrared Deep Sky Survey (UKIDSS; Lawrence *et al.* 2000) in (b). Both the estimated fluxes and errors are normalized to the background uncertainty, $\sigma_{\rm b}$, and the dashed lines show the expected form $\Delta\hat{F} = (\sigma_{\rm b}^2 + C|\hat{F}|)^{1/2}$ of the data model adopted here (Eq. 8.12). The SDSS data model deviates from this in two ways: there is a scatter from deblended close pairs of sources; and a minimum magnitude error (i.e. $\Delta\hat{F} \propto \hat{F}$ for bright sources) is applied to account for overall calibration uncertainties.

The relative influences of the constant and flux-dependent contributions to $\Delta\hat{F}$ are shown in Figure 8.1. Whilst $\Delta\hat{F} \simeq \langle(\hat{F} - F)^2\rangle^{1/2}$ for bright sources, the necessity of using $|\hat{F}|$ in calculating the error term is a serious shortcoming close to the detection limit (e.g., Lupton *et al.* 1999).

This problem is just one of the limitations of an estimator-based approach to parameter inference. No method which ignores prior information can do more than summarize the properties of the data, including even the most rigorous frequentist confidence limits which account for the measurement process but ignore the underlying source population. Whilst it is true by construction that, for a source of flux F, $\hat{F}$ will be drawn from the sampling distribution given in Eq. (8.7), the apparent implication that the true flux has an 'effective posterior',

$$P(F|\hat{F}) = \frac{1}{(2\pi)^{1/2}(\sigma_{\rm b}^2 + C|\hat{F}|)^{1/2}} \exp\left[-\frac{1}{2}\frac{(F - \hat{F})^2}{\sigma_{\rm b}^2 + C|\hat{F}|}\right], \qquad (8.13)$$

does not necessarily follow. Even though this distribution is not explicitly implied by the estimator-based approach to flux measurement, it is the natural distribution in F to construct from $\hat{F}$ and $\Delta\hat{F}$, and even what many non-Bayesians might think of subconsciously when presented with a measurement and its error. Moreover, an effective posterior like this facilitates comparison with the Bayesian results derived in Section 8.5, although these also require the explicit inclusion of prior information. In the case of flux measurement, this comes from the astronomical source population, which is described in Section 8.4.

8.4 The source population

Astronomical sources, from the faintest sub-stellar objects to the most luminous galaxies and quasars, are seen at distances from tens to billions of light years, resulting in observed fluxes that span more than ten orders of magnitude. Fortunately, the incredible variety of evolving astronomical populations is irrelevant to the problem of flux estimation, and it is only the distribution of source fluxes that is important for an observer. Under the assumption that the sources are spatially uncorrelated (e.g., Tsutomu & Takako 2004) this can be characterized by $d\Sigma/dF$, the differential number of sources of flux F per steradian on the sky.

The form of $d\Sigma/dF$ is largely determined by geometry: the quadratic dependence of the volume of a spherical shell on its radius ensures that distant sources are more common than those nearby, and hence that there are many more sources of low flux than high flux, independent of their intrinsic luminosity distributions. In an unevolving, eternal, transparent and spatially flat Universe, the source number counts are determined purely by this effect, yielding the classic result that $d\Sigma/dF \propto F^{-5/2}$ (e.g., Peebles 1993). In reality, cosmological evolution (of extra-Galactic populations) and the finite size of the Galaxy (for stars) combine to ensure that most faint source populations actually have decidedly flatter slopes than this (e.g., Hopkins *et al.* 1998; Beckwith *et al.* 2006). Not that this result is surprising: the darkness of the night sky (the subject of Olbers' paradox) can only be reconciled with power-law source counts if they rise less steeply than $F^{-3/2}$ for low F. Further, the finite volume of the observable Universe means that the number counts must drop to zero below some minimum flux, F_{min}, corresponding roughly to the faintest source at the greatest possible distance. The actual value of F_{min} is somewhat uncertain but, critically, is considerably fainter than current observational limits in all passbands. At optical wavelengths, for example, the Hubble Ultra Deep Field (HUDF; Beckwith *et al.* 2006) observations have revealed that galaxies have a still-rising surface density of $\Sigma \simeq 10^6 \, \text{deg}^{-2}$ at the survey's detection limit, as shown in Figure 8.2.

Despite the considerable complexity of, and uncertainty about, astronomical source populations, most of the relevant properties of the combined flux distribution can be captured by a simple bounded power law of the form

$$\frac{d\Sigma}{dF} = \Theta(F - F_{min})\frac{(\alpha - 1)\Sigma_*}{F_*}\left(\frac{F}{F_*}\right)^{-\alpha}, \tag{8.14}$$

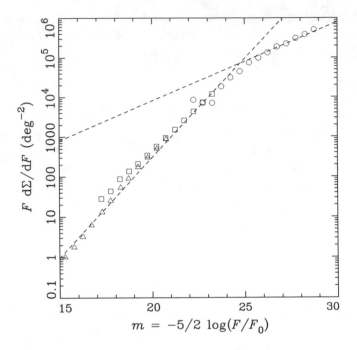

Fig. 8.2. Observed optical source counts taken from the Automatic Plate Measuring (APM) Galaxy Survey (Maddox *et al.* 1990; triangles), the Jones *et al.* (1991) sample (squares), and the HUDF (Beckwith *et al.* 2006; circles). The dashed lines illustrate the degree to which a simple power-law model can fit the observed counts over flux ranges of $\gtrsim 100$. The slightly arcane abscissa is chosen to approximately match standard astronomical magnitudes, m; by way of comparison, the faintest stars visible to the naked eye have $m \simeq 5$.

where $\Theta(x)$ is the Heaviside step function, α is the logarithmic slope, and $F_{\min}$ is the low-flux cut-off. Provided that $\alpha > 1$ and $F_* \geq F_{\min}$, the corresponding cumulative distribution is

$$\Sigma(> F) = \Sigma_* \left[\frac{\max(F, F_{\min})}{F_*} \right]^{-(\alpha-1)} . \qquad (8.15)$$

This gives the normalization condition that there are Σ_* sources per steradian brighter than F_* on average.

Aside from being a reasonable approximation to the observed source counts over a large range of fluxes (as shown in Figure 8.2), the use of a power-law model also facilitates comparison with a number of previous statistical studies of flux measurement (e.g., Murdoch *et al.* 1973; Condon 1974). In most cases, however, the low-flux cut-off has been treated only qualitatively, if at all. Whilst it is possible to pursue a Bayesian approach using the improper prior that results from taking

$F_{\min} \to 0$ (e.g., Hogg & Turner 1998), it is preferable to work with a correctly normalized posterior, as in Section 8.5. Moreover, the value of $F_{\min}$ plays a surprisingly important role in the problem of flux estimation, as shown in Section 8.6.

8.5 Bayesian flux inference

Given a full statistical description of the observation process (Section 8.2) and a model for the source population (Section 8.4), Bayesian inference is the obvious method of using the available data to estimate a source's flux, F. The resultant posterior incorporates all the available information in a self-consistent manner and thus gives the tightest legitimate parameter constraints. Using the classical flux estimator, $\hat{F}$, defined in Eq. (8.6), as a convenient data statistic, the posterior is given by

$$P(F|\hat{F}) = \frac{\pi(F)\,L(\hat{F}|F)}{\int_0^\infty \pi(F')\,L(\hat{F}|F')\,\mathrm{d}F'}. \qquad (8.16)$$

In this expression $\pi(F) = P(F|\text{source})$ is the prior, $L(\hat{F}|F) = P(\hat{F}|F, \text{source})$ is the likelihood, and the evidence integral in the denominator ensures that $P(F|\hat{F}) = P(F|\hat{F}, \text{source})$ is a correctly normalized probability distribution in F. The explicit use of 'source' emphasizes that all these results are conditional on there actually being a single source in the aperture, an issue that is considered further in Section 8.6.

The prior is obtained by applying unit normalization to the source number counts, $\mathrm{d}\Sigma/\mathrm{d}F$, to give

$$\pi(F) = \frac{\mathrm{d}\Sigma/\mathrm{d}F}{\int_0^\infty \mathrm{d}\Sigma/\mathrm{d}F'\,\mathrm{d}F'}$$

$$= \Theta(F - F_{\min})\frac{\alpha - 1}{F_{\min}}\left(\frac{F}{F_{\min}}\right)^{-\alpha}, \qquad (8.17)$$

where the second expression assumes the power-law form given in Eq. (8.14), provided that $F_{\min} > 0$ and $\alpha > 1$.

The use of $\hat{F}$ as a data surrogate means that the likelihood is identical to the sampling distribution given in Eq. (8.9). This immediately gives

$$L(\hat{F}|F) = \frac{1}{(2\pi)^{1/2}(\sigma_{\mathrm{b}}^2 + CF)^{1/2}}\exp\left[-\frac{1}{2}\frac{(\hat{F} - F)^2}{\sigma_{\mathrm{b}}^2 + CF}\right], \qquad (8.18)$$

although it is worth noting that whilst this is a simple normalized Gaussian distribution in $\hat{F}$, it is a considerably more complicated function of F.

Multiplying the prior from Eq. (8.17) with the likelihood in Eq. (8.18) then gives the posterior for the source flux as

$$P(F|\hat{F}) \propto \Theta(F - F_{\min})\frac{d\Sigma/dF}{(\sigma_b^2 + CF)^{1/2}} \exp\left[-\frac{1}{2}\frac{(F - \hat{F})^2}{\sigma_b^2 + CF}\right]$$

$$\propto \Theta(F - F_{\min})\frac{F^{-\alpha}}{(\sigma_b^2 + CF)^{1/2}} \exp\left[-\frac{1}{2}\frac{(F - \hat{F})^2}{\sigma_b^2 + CF}\right], \qquad (8.19)$$

where the second expression again assumes the power-law source counts. Note that the posterior is only given up to a normalization constant, which must be computed numerically.

Having calculated the full posterior, the Bayesian inference task is complete; all the available information about this source's flux is summarized in this one expression. Examples of this distribution are shown in Figure 8.3, revealing a variety of distinct forms which can best be understood by examining a number of special cases.

Before looking at the influence of the source population, it is illustrative to examine a uniform prior (i.e., $\alpha = 0$, with the added assumption that there is some high maximum flux to normalize the prior). Under this assumption, the posterior in Eq. (8.19) simplifies to

$$P(F|\hat{F}) \propto \Theta(F - F_{\min})\frac{1}{(\sigma_b^2 + CF)^{1/2}} \exp\left[-\frac{1}{2}\frac{(F - \hat{F})^2}{\sigma_b^2 + CF}\right], \qquad (8.20)$$

examples of which are shown in Figure 8.3(a) and (b). Although unrealistic for astronomical sources, this brings the Bayesian result closer to the 'effective posterior' defined in Eq. (8.13). The two distributions would be identical if not for the flux-dependent Poisson term, which causes the peak of the Bayesian posterior to be skewed towards lower fluxes and results in a longer tail at high fluxes. Thus, even in a hypothetical universe with sources drawn from a flat flux distribution, the Bayesian posterior and the classical flux estimator give very different results.

A more realistic special case than a uniform flux distribution is the situation where there is no significant background, astrophysical or otherwise, and hence no uncertainty in its (non-)subtraction. With $\sigma_b = 0$, the posterior given in Eq. (8.19) simplifies to (cf. Chapter 12)

$$P(F|\hat{F}) \propto \Theta(F - F_{\min})F^{-(\alpha+1/2)} \exp\left[-\frac{1}{2}\frac{(F - \hat{F})^2}{CF}\right]. \qquad (8.21)$$

As can be seen from Figure 8.3(c) and (d), the peak of this posterior, which is not Gaussian or even symmetric, is skewed to lower fluxes than $\hat{F}$, due to the sharply

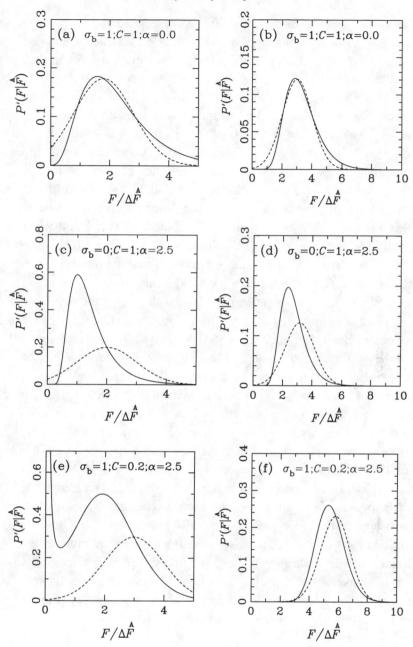

Fig. 8.3. Un-normalized posterior probability distributions, $P'(F|\hat{F})$, of the flux, F, scaled by the classical estimated noise, $\Delta\hat{F}$ (as this is always non-zero, unlike σ_b or C). In each panel the solid line shows the Bayesian posterior and the dashed line shows the 'effective posterior' derived from the classical flux estimator.

rising prior. In classical flux estimation, the increased number of fainter sources scattered bright is a well-known bias (Eddington 1913) which has to be accounted for post hoc, rather than being a natural part of the formalism. A corollary of this is that any natural estimator (e.g., the mean or the median) that might be constructed from the skewed Bayesian posterior would, in a classical sense, be 'biased'. This is often cited as a flaw in the Bayesian approach to parameter estimation, but as the bias is only defined in terms of a long-run frequency of identical trials, it is inappropriate to apply it to a single measurement. Another important aspect of the posterior given in Eq. (8.21) is that the huge prior probability at $F \simeq F_{\min}$ is overwhelmed by the even more extreme drop in the likelihood resulting from the flux dependence of the variance. Hence the posterior does not depend on $F_{\min}$, provided it is well below both C and $\hat{F}$. Combined with the fact that the posterior is normalizable, this is a sufficient condition for taking the limit $F_{\min} \to 0$.[4] Given the lack of knowledge about $F_{\min}$, this is a desirable state of affairs, although this result relies on perfect knowledge of the background.

Turning finally to the most general problem, with $\sigma_{\rm b} > 0$ and steep source counts with $\alpha \simeq 5/2$, it is immediately clear from Figure 8.3(e) and (f) that the situation is more complicated than in either of the special cases considered above. That the form of the peak at $F \simeq \hat{F}$ is skewed by both the signal-dependent error term and the prevalence of fainter sources is as expected; but the rising source counts and the background noise combine to have the far more drastic effect of producing a dominant second peak in the posterior at $F = F_{\min}$. For faint sources with $\hat{F} \simeq \sigma_{\rm b}$ the posterior increases monotonically to $F_{\min}$, which reflects the fact that the posterior is merely a perturbation of the prior, with the data only providing an upper limit on the flux. However, the existence of a second peak at $F = F_{\min}$ even for bright, well-measured sources – including the example given in Figure 8.3(f), even though the peak is not visible – is counter-intuitive at best. Can it really be the case that most apparently bright objects are really just undetectable faint sources that have been subject to extreme positive noise fluctuations? In more concrete terms, can it really be true that, upon glancing up into the night sky and noticing a faint star, it is at all likely to be one of the distant galaxies of the type seen in the HUDF? The answer to this question, and the resolution of this apparent paradox, requires a more careful treatment of the faintest sources, which is the subject of the next section.

8.6 The faintest sources

The seemingly straightforward Bayesian analysis described in Section 8.5 seems to imply that the numerous faint sources in the Universe can combine with back-

---

[4] Note that taking the limit at this stage is a rigorous mathematical operation, unlike taking $F_{\min} \to 0$ initially and adopting an improper prior in the hope that the resultant posterior will be normalizable. The common success of the latter approach hides the fact that, as emphasized by Jaynes (1991), taking the limit of a ratio and evaluating the ratio of limits are not equivalent processes.

ground noise to dominate the numbers of bright detections. A corollary of this would be that many apparently luminous objects are actually orders of magnitude fainter than they appear. At very least this is counter-intuitive and, if correct, would imply that most astrophysical knowledge is severely compromised. Before discarding several centuries' worth of hard-won understanding, however, it is worth re-examining the assumptions and subsequent reasoning which led to the double-peaked posterior of Eq. (8.19).

An obvious starting point is the source population that provides the prior for the flux measurement. It is always tempting to assume a simple mathematical form like the power law introduced in Eq. (8.14), but is it valid to apply it over such a large range of fluxes? In short, yes, as can be seen from Figure 8.2, which shows measured source counts rising steadily over ten orders of magnitude in flux. Given the empirical existence of such faint sources, the low-flux peak in Eq. (8.19) is clearly not caused by an unrealistic model of the source population.

Another common source of error in statistical analyses is the unwitting contradiction of a hidden assumption, such as that implicit in the first sentence of Section 8.2, that it is a single source under consideration. That would be uncontentious if it were possible to travel to the object in question and investigate it in isolation; but astronomy is an observational, not an experimental, science, and so sources can only be observed projected along a line-of-sight onto the plane of the sky. In the context of photometric measurements, a source can only be considered isolated if there are no others within the observational aperture introduced in Section 8.2.

This assumption is very obviously broken in some situations (e.g., low angular resolution gamma-ray observations, or near-infrared imaging of the dense star fields near the Galactic centre) where the average angular separation between detectable sources is comparable to, or less than, the angular resolution of the instrument. In such crowded fields, such as in Figure 8.4(a), the model must include the possibility of multiple sources in the aperture, and blindly applying a single source model would result in over estimating the flux. Of course, the classical estimator described in Section 8.3 would be similarly biased; no statistical method can produce correct results if the underlying model is wrong. Reliable photometry in crowded fields is possible, but only by using more complicated image analysis algorithms than are considered here (e.g., Stetson 1987; Irwin 1990).

Whether or not a field is crowded can be judged directly from an image, if available, but this can also be determined directly from the source counts. A conservative 'rule of thumb' in image analysis is that a field is crowded if there is, on average, more than one detectable source per $\sim$30 beam areas (e.g., Wall & Jenkins 2003). In the context of the power-law model given in Eq. (8.14) this reduces to $\Sigma(> F_{\mathrm{det}}) \gtrsim 1/30$, where F_{det} is the flux of the faintest individually detectable source. As expected, this depends on both the source counts and the observational

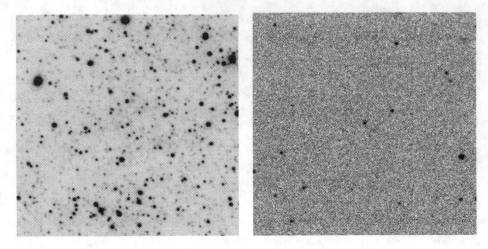

Fig. 8.4. Examples of a crowded field (left) and non-crowded field (right) observed in the near-infrared by UKIDSS (Lawrence *et al.* 2007).

parameters, and so the possibility of crowding must be assessed on a case-by-case basis.

As important as crowding is, it is not the explanation for the low-flux peak in the posterior given in Eq. (8.19). In the context of the rather extreme example invoked at the end of Section 8.5, bright, naked eye stars are clearly well separated (even for an observer who, like the author, is short-sighted). This situation is also illustrated in Figure 8.4(b), which emphasizes that any flux measurement of the central target source is unlikely to be completely unaffected by any detectable sources nearby. If crowding of bright sources is not the problem, then, what about the much higher surface density of fainter undetectable sources? They are so numerous that there are inevitably many per beam in any astronomical observation, more than satisfy-ing the above crowding criterion; but, rather than being identifiable as discrete, if possibly overlapping, sources, they are so dense on the sky that the net result is essentially that of a noisy background. Termed source confusion (as distinct from source crowding at detectable fluxes), this phenomenon has now been observed in most passbands, although it will probably always be most famous as the cause of the discrepancy between the source counts inferred by two early radio surveys that were attempting to test the Big Bang hypothesis (see, e.g., Wall & Jenkins 2003).

This dispute was settled when Scheuer (1957) developed a method for calculat-ing the distribution of total faint source flux, $F_{\text{faint,tot}}$, in an aperture of solid angle Ω_{obs}. The distribution is analytic for power-law source counts (Condon 1974), but simplifies further in the confusion limit relevant to this problem. The expected number of faint sources in any aperture is huge, but subject to Poisson fluctuations; the result is that the distribution of $F_{\text{faint,tot}}$ tends to a Gaussian. More importantly, there are so many undetectable sources that they combine to mimic a uniform, if noisy, background.

The presence of an additional noisy background would inevitably increase flux uncertainties, but the fact that the background contribution is estimated from the plentiful data ought to ensure that it does not induce any systematic bias. Indeed, the standard flux estimator is insensitive to the faint source population for this reason, but a problem does arise in the Bayesian approach because these sources are accounted for initially in the estimated background and then again in the prior which extends to fluxes well below the detection limit. The measured background, $\hat{b}$, implicitly includes all contributions, and cannot distinguish these faint sources from other putative contaminants. To then admit the possibility that a detected, isolated source might actually be one of the background sources whose flux has been absorbed into $\hat{b}$ is a clear contradiction, and it is this double-counting which is, finally, the source of the spurious low-flux peak in the posterior given in Eq. (8.19).

To obtain a self-consistent Bayesian flux estimate, the prior should be modified in such a way that the numerous ultra-faint sources that have already been accounted for during background subtraction do not enter into the calculation of the posterior. For non-crowded fields this can be accomplished by truncating the posterior at the confusion flux, $F_{\text{confusion}}$, above which only ~ 1 source is expected in the aperture. In the case of power-law source counts this is given from Eq. (8.15) as $F_{\text{confusion}} \simeq F_*(\Omega_{\text{obs}}\Sigma_*)^{1/(\alpha-1)}$. The correct posterior for well-detected sources is thus obtained by modifying Eq. (8.19) to give

$$P(F|\hat{F}) \propto \Theta(F - F_{\text{confusion}}) \frac{d\Sigma/dF}{(\sigma_{\text{b}}^2 + CF)^{1/2}} \exp\left[-\frac{1}{2}\frac{(F - \hat{F})^2}{\sigma_{\text{b}}^2 + CF}\right]$$

$$\propto \Theta(F - F_{\text{confusion}}) \frac{F^{-\alpha}}{(\sigma_{\text{b}}^2 + CF)^{1/2}} \exp\left[-\frac{1}{2}\frac{(F - \hat{F})^2}{\sigma_{\text{b}}^2 + CF}\right], \qquad (8.22)$$

where, as before, the second expression assumes power-law source counts. This form is also what would have been expected from qualitative arguments (Hogg & Turner 1998): a posterior with a single peak near $F \simeq \hat{F}$ that both the rising source counts and Poisson noise skew towards lower flux.

For well-detected sources, the absence of the low-flux peak does not depend on the exact value of $F_{\text{confusion}}$ adopted, as long as it falls in the trough between the two peaks of the posterior shown in Figure 8.5. Critically, in non-crowded fields, $F_{\text{confusion}}$ is comfortably below the detection limit, F_{det}, and so this condition is met by definition. The main effect of truncating the prior at $F_{\text{confusion}}$ is simply to remove the anomalous low-flux peak in the posterior, leaving only the likelihood-dominated peak at $F \simeq \hat{F}$.

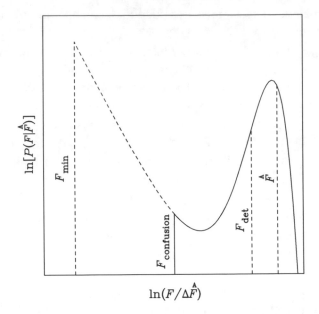

Fig. 8.5. Schematic diagram showing the various fluxes that enter into the problem of aperture flux measurement: the minimum flux of the population, F_{min}; the confusion flux, $F_{confusion}$; the detection limit, F_{det}; and the classical estimated flux, $\hat{F}$. The solid line is the (un-normalized) posterior truncated at $F_{confusion}$; the dashed line is the naive posterior that extends all the way down to F_{min} and is dominated by the spurious low-flux peak. Note that both axes are logarithmic.

For fainter sources (with $F \simeq \sigma_b$), however, there may not be any trough in the posterior, with the result that the truncated distribution still rises to $F_{confusion}$ without any significant peak at $F \simeq \hat{F}$. This result immediately makes sense once it is realized that such a 'source' may not even be real; if the data are consistent with a three- or four-'sigma' deviation in the noise then this prior, which is conditional on there actually being a source there in the first place, is not necessarily valid. The posterior could then be understood in terms of the data merely applying an upper limit to the power-law prior. In this situation, the identification and classification of the source are not separable – something that is always true in principle (e.g., Hobson & McLachlan 2003), even if performing the two tasks independently is generally reasonable in practice. The validity of the truncated posterior in this low-flux case is less clear-cut, but such difficulties are a direct consequence of the non-Bayesian treatment of the background described in Section 8.2. This could be seen as just punishment for violating Bayesian principles, by subtracting the background rather than modelling it, even if fitting for all the contributions to real astronomical datasets is impractical at best (e.g., Irwin 1990).

Whatever the appropriate prior, the full Bayesian answer to any parameter inference problem is the posterior distribution, in this case $P(F|\hat{F})$, as given in Eq. (8.22). Even with the statistical problem solved, however, there are a number of practical issues that must also be considered when working with, or generating, catalogues of flux measurements.

8.7 Practical flux measurement

With the correct handling of undetectable faint sources it is possible to obtain a well-defined Bayesian posterior for the flux of a source by combining prior knowledge about the population with the data from a photometric observation (Section 8.5 and Section 8.6). In practice, however, it is often necessary to characterize a source's flux and uncertainty with a few numbers (e.g., in a catalogue or database). For the fairly simple posteriors given in Eq. (8.22), the mean and variance of the posterior contain most of the information, although these can differ significantly from the classically estimated mean, $\hat{F}$, and variance, $\Delta\hat{F}$, defined in Section 8.3.

Nonetheless, it is inevitably $\hat{F}$ and $\Delta\hat{F}$ which are reported in survey databases, and, without reprocessing the raw data, it might seem there is no option but to 'make do' with these quantities, maybe constructing the effective posterior defined in Eq. (8.13) to at least encode this information in a probabilistic form. However, it should also be possible to reconstruct likelihoods from the catalogued flux estimates and, given that the likelihood represents the most complete statistical description of the data, it could be argued that this is a necessary condition if a database is to live up to its name.

Clearly it is possible to ensure this is true in future surveys, but what about the billions of sources that are already catalogued? For bright sources, specifically those for which the background uncertainty can be neglected, $\hat{F}$ and $\Delta\hat{F}$ are sufficient to reconstruct the likelihood by setting $\sigma_b = 0$ and $C = (\Delta\hat{F})^2/\hat{F}$ in Eq. (8.18). Similarly, if the Poisson contribution to the error from finite photon counts is negligible (as is the case for most faint sources), the likelihood can be obtained from $\hat{F}$ and $\Delta\hat{F}$ by setting $\sigma_b = \Delta\hat{F}$ and $C = 0$ in Eq. (8.18). It is also possible to solve the more general problem of reconstructing the variance term $\sigma_b^2 + CF$ in Eq. (8.18), either by directly obtaining the background noise, σ_b, and the flux calibration, C, or, failing that, by utilizing the detection limit of the survey, F_{det}, as introduced in Section 8.6.

The flux limit is defined in many different ways, but mostly it is a signal-to-noise based criterion, in which a source must have $\hat{F}/\Delta\hat{F} \gtrsim (S/N)_{min}$ to be reliably detected. Working with this definition, Eqs. (8.6) and (8.12) can then be combined to give

$$C = F_{\text{det}} \frac{(\Delta\hat{F}/F_{\text{det}})^2/(S/N)^2_{\min} - 1}{\hat{F}/F_{\text{det}}/(S/N)^2_{\min} - 1} \tag{8.23}$$

and

$$\sigma_{\text{b}} = \frac{F_{\text{det}}/\hat{F}}{(S/N)^2_{\min}} \frac{F_{\text{det}}/\hat{F} - (\Delta\hat{F}/\hat{F})^2}{1 - F_{\text{det}}/\hat{F}/(S/N)^2_{\min}}. \tag{8.24}$$

Inserting these expressions into Eq. (8.18) then gives the full likelihood for a source from the estimated quantities $\hat{F}$ and $\Delta\hat{F}$ and the survey parameters F_{det} and $(S/N)_{\min}$. For different types of surveys an equivalent calculation should yield the desired quantities, the key point being that it must be possible to reconstruct the likelihood in order to extract all the information contained in the data.

It is possible to take this approach one step further by reconstructing not just the likelihood, but the posterior, from the catalogued data. Indeed, Hogg and Turner (1998) advocate that their maximum posterior corrections, which account for the slope of the source counts, 'should in principle be applied to all flux measurements in a flux-limited sample'. Although Hogg and Turner (1998) only treat the low-flux peak in the posterior qualitatively, their corrections are the same as would be obtained from Eq. (8.22), so why apply them only 'in principle'? And why has this suggestion largely been ignored (but see Laird *et al.* 2009), with survey databases still containing only classical flux estimates?

The immediate reaction of many Bayesians would be to put this down to a misguided faith in classical methods, but there are actually several good reasons why posteriors are not, and should not, be used in survey catalogues and databases. As appealing an idea as that might be in principle, it is flawed in practice because there is no definitive flux prior that can be applied to a source that will always be appropriate. A generic model like the power law used in Section 8.4 is probably adequate for a newly discovered source about which little else is known, but as more and more information is obtained (e.g., through follow-up spectroscopy or a proper motion measurement) the prior – or at least external – information will evolve. Similarly, a good measurement of a source's flux in one passband is likely to imply a very tight prior given by the previously observed distribution of flux ratios between the two bands (Hogg & Turner 1998): the range of X-ray hardness ratios or optical colours is much smaller than the full range of fluxes allowed by the prior given in Eq. (8.17). Essentially, it is impossible to combine measurements given only the corresponding posterior distributions; there is no general way of disentangling the information provided by each measurement from whatever priors have been used. The same problem could arise in the example given in Chapter 11, in which the source population is being modelled: a Bayesian approach to this problem requires the likelihood for each source, not the posterior obtained given some model for the population.

It might seem strange to be advocating against the use of prior information in a book on Bayesian methods, but this is the difference between the optimal recording of data and attempting to answer scientific questions from that data. The prior information is critical to any questions about the properties of the real world for which the data do not completely determine the answer. But with so many different possible constraints available, what prior knowledge should be applied? The answer, inevitably, is all of it; whatever relevant information is available should be utilized. And if more information becomes available subsequently then it is perfectly legitimate to apply that as well. As noted by Jaynes (1991), 'every Bayesian problem is open-ended: no matter how much analysis you have completed, this only suggests still other kinds of prior information that you might have had, and therefore still more interesting calculations that need to be done, to get still deeper insight into the problem'. While the likelihood for a given measurement should be fixed, other measurements – or insights – will always occur. One of the great beauties of Bayesian inference is that this evolution of knowledge is naturally incorporated into what is, essentially, a formalism for learning.

References

Beckwith, S. V. W. *et al.* (2006). *Astron. J.*, **132**, 1729.

Condon, J. J. (1974). *Astrophys. J.*, **188**, 279.

Eddington, A. S. (1913). *Mon. Not. Roy. Astron. Soc.*, **73**, 359.

Hobson, M. P. and McLachlan C. I. (2003). *Mon. Not. Roy. Astron. Soc.*, **338**, 765.

Hogg, D. W. and Turner, E. L. (1998). *Proc. Astron. Soc. Pac.*, **110**, 727.

Hopkins, A. M., Mobasher, B., Cram, L. and Rowan-Robinson, M. (1998). *Mon. Not. Roy. Astron. Soc.*, **296**, 839.

Irwin, M. J. (1990). *Mon. Not. Roy. Astron. Soc.*, **214**, 575.

Jaynes, E. T. (1991). In W. T. Grandy and L. Schick, eds., *Proceedings of the Tenth Annual MAXENT Workshop*. Dordrecht: Kluwer Academic Press.

Jaynes, E. T. (2003). *Probability Theory: The Logic of Science*. Cambridge: Cambridge University Press.

Jones, L. R., Fong, R., Shanks, T., Ellis, R. S. and Peterson, B. A. (1991). *Mon. Not. Roy. Astron. Soc.*, **249**, 481.

Laird, E. S. *et al.* (2009). *Astrophys. J. Supp.*, **180**, 102.

Law, N. M., Mackay, C. D. and Baldwin, J. E. (2006). *Astron. Astrophys.*, **446**, 739.

Lawrence, A. *et al.* (2007). *Mon. Not. Roy. Astron. Soc.*, **381**, 1400.

Lupton, R. H., Gunn, J. E. and Szalay, A. S. (1999). *Astron. J.*, **118**, L1406.

Maddox, S. J., Efstathiou, G., Sutherland, W. J. and Loveday, J. (1990). *Mon. Not. Roy. Astron. Soc.*, **243**, 692.

Murdoch, H. S., Crawford, D. F. and Jauncey, D. L. (1973). *Astrophys. J.*, **183**, 1.

Murphy, T. W. *et al.* (2008). *Proc. Astron. Soc. Pac.*, **120**, 20.

Peebles, P. J. E. (1993). *The Principles of Physical Cosmology*. Princeton, NJ: Princeton University Press.

Scheuer, P. A. G. (1957). *Proc. Camb. Phil. Soc.*, **53**, 764.

Stetson, P. B. (1987). *Proc. Astron. Soc. Pac.*, **99**, 191.

Tsutomu, T. and Takako, T. I. (2004). *Astrophys. J.*, **604**, 40.

Wall, J. V. and Jenkins, C. R. (2003). *Practical Statistics for Astronomers*. Cambridge: Cambridge University Press.

York, D. G. *et al.* (2000). *Astron. J.*, **120**, 1579.

9

Gravitational wave astronomy

Neil Cornish

9.1 A new spectrum

Just a century ago, our view of the cosmos was limited to the single window provided by optical telescopes. In the intervening years, new windows opened as cosmic rays, neutrinos, radio waves, X-rays, microwaves and gamma rays were enlisted in our quest to understand the Universe. These new messengers have revolutionized astronomy and profoundly altered our perception of the cosmos.

Despite these impressive advances, almost everything we know about the Universe beyond our own galaxy comes from observing light of various energies. Hopefully this will soon change as a new spectrum is opened by the direct detection of gravitational waves. Just as our sense of hearing complements our sense of sight, gravitational wave astronomy can extend and enrich the picture provided by electromagnetic astronomy (Hughes 2003).

Astrophysical and cosmological sources of gravitational waves are expected to produce signals that span over twenty decades in frequency, ranging from primordial signals with frequencies as low as 10^{-18} Hz, to supernovae explosions that reach frequencies of 10^4 Hz. A suite of detection techniques have been proposed to detect these signals, from acoustic (bar) detectors for narrow band detection of high-frequency waves, through to polarization maps of the cosmic microwave background radiation to look for imprints of the lowest frequency waves. Pulsar timing arrays are a promising technique for detecting waves in the 10^{-8} Hz range, and there is an outside chance that this technique might yield the first direct detection. Most attention in gravitational wave astronomy is focused on ground and space-based interferometric detectors, which together will span the frequency range between 10^{-5} and 10^4 Hz. The first generation of ground-based detectors, GEO, TAMA, LIGO and Virgo, have completed their initial science runs, including a full-year triple coincidence run at design sensitivity for the US-based LIGO detectors. Planning for the joint ESA–NASA space-based LISA mission is well advanced, and development will proceed rapidly once funding is made available.

The science potential for gravitational wave astronomy is as wide as its spectral range and as varied as the detector technologies. Any asymmetric movement of mass/energy will produce gravitational waves, but the extreme stiffness of space-time limits the detectable phenomena to those that are either extremely violent or nearby. The most promising sources involve large amounts of high-density material moving at close to the speed of light, such as occurs in tight binary systems composed of two compact stellar remnants, or core collapse supernovae. The final inspiral and merger of two black holes of the type found at our Galactic centre could be seen out to $z = 10$ or greater by the LISA observatory, while stellar remnant black-hole binaries could be seen out to $z = 1$ or greater by Advanced LIGO. These observations can be used to probe fundamental physics by testing general relativity in the dynamical, strong field regime, and to study star and galaxy formation throughout cosmic history.

For a more in-depth introduction to gravitational wave astronomy, see the recent review articles by Hughes (Hughes 2003) and Camp and Cornish (2004), and the overviews produced by LIGO (LIGO Scientific Collaboration 2007) and LISA (LISA International Science Team 2007) science collaborations.

9.2 Gravitational wave data analysis

The development of practical methods for applying Bayesian inference to scientific data largely predates the development of gravitational wave data analysis but, until recently, traditional frequentist approaches have held sway. The historical reasons for this choice appear to be a combination of the inertia of existing paradigms, and a highly influential paper by Thorne (1987) that set the direction for gravitational wave data analysis. Thorne's paper described the technique of Wiener matched filtering for constructing a frequentist detection statistic. Today, the majority of papers that describe the analysis of data from gravitational wave detectors employ frequentist techniques and, where applicable, the matched filtering statistic suggested by Thorne.

Given this state of affairs it is rather surprising to find that many of the seminal papers (Finn 1992; Finn & Chernoff 1993; Flanagan & Hughes 1998a,b; Anderson *et al.* 2001) on gravitational wave data analysis employ Bayesian probability theory. These analyses are generally of a hybrid type, where Bayesian reasoning is used to motivate the form of a particular frequentist statistic. This hybrid style of analysis is particularly evident in recent studies of stochastic backgrounds

(Allen *et al.* 2003), unmodeled bursts (Searle *et al.* 2008) and methods for setting upper limits (Biswas *et al.* 2007).

A short time after Thorne's article appeared, Davis (1989) suggested that Bayes estimators be used in gravitational wave data analysis, but he also noted that the computational cost may be prohibitive. The possibility of using Bayesian inference outside the confines of abstract theoretical analyses did not take root for another decade. The turning point came when Christensen and Meyer (1998) applied MCMC techniques to simulated gravitational wave data. This pioneering work was followed by a trickle, and later a flood, of papers that explored the use of practical Bayesian inference techniques in a vast array of gravitational wave data analysis problems. At the present time there is a wealth of data to analyze from the worldwide network of gravitational wave detectors, and MCMC pipelines are being developed to perform follow-ups to the triggers generated by the frequentist analyses. The application of Bayesian inference to parameter estimation is now fairly well accepted within the gravitational wave community, but there is currently little enthusiasm for taking a Bayesian approach to the detection problem.

Even without a first detection, gravitational wave astronomy has grown into a large field that involves many types of detectors and many different source classes. Here we will focus our attention on laser interferometers, and sources for which theoretical waveforms have been derived, though most of the discussion also applies to acoustic detectors, pulsar timing arrays, and deterministic yet unmodelled sources.

9.2.1 The traditional approach

The output of an idealized gravitational wave detector can be described by a time series $s(t)$ that is a linear combination of possible signal(s) $h(t)$ and noise $n(t)$:

$$s(t) = h(t) + n(t).\tag{9.1}$$

For much of the discussion to follow it is more convenient to work in the Fourier domain where

$$\tilde{s}(f) = \int_{-\infty}^{\infty} e^{2\pi i f t} s(t) \mathrm{d}t.\tag{9.2}$$

In what follows we will simplify our notation by writing $s(f)$ for the output of the Fourier transform.

For an ideal detector the response $h(t)$ to a gravitational wave signal with polarization components $h_+(t)$ and $h_\times(t)$ is linear, and described by

$$h(t) = F^+(t; \theta, \phi, \psi, \tau)h_+(t) + F^\times(t; \theta, \phi, \psi, \tau)h_\times(t).\tag{9.3}$$

Here F^+ and $F^\times$ are the antenna beam patterns of the detector, which depend on the sky location (θ, ϕ) and polarization angle ψ of the source in some fixed reference system, and the orientation of the detector in that reference system at time t. When more than one detector is available, or the signal duration is long compared with the orbital timescale of the detector, the response must also include the time delay τ between the wave striking the detector and the wave arriving at some fixed reference location (such as the solar barycenter). Additional complications arise when the wavelength is comparable to, or shorter than, the size of the detector, as then the antenna beam functions have to be replaced by linear time delay operators.

The output of our ideal detector includes instrument noise of the stationary and Gaussian variety favoured by theorists. Such noise is fully specified by the expectation values:

$$\langle n(f) \rangle = 0$$

$$\langle n(f) n^*(f') \rangle = \frac{T}{2} \delta_{ff'} S_n(f), \tag{9.4}$$

where $S_n(f)$ is the one-sided noise spectral density and T is the observation time. The statistical properties of the noise motivate a natural inner product on the vector space of signals, defined by

$$(a|b) = 2 \int_0^\infty \frac{a(f)b^*(f) + a^*(f)b(f)}{S_n(f)} \mathrm{d}f. \tag{9.5}$$

For several important classes of gravitational wave signal there exist analytic theoretical models for the gravitational waveforms that can be used to construct parameterized models $h(\vec{\lambda})$ that predict the instrument response to a source described by parameters $\vec{\lambda}$. Examples include the quasi-circular inspiral of two spinning black holes ($D = \dim(\vec{\lambda}) = 15$), and the post-merger ringdown of a single distorted black hole ($D = 6$). Standard matched-filter searches for modelled signals employ the Wiener filter statistic

$$\rho(\vec{\lambda}) = (s|\hat{h}(\vec{\lambda})), \tag{9.6}$$

where the filter functions are scaled to unit norm: $(\hat{h}|\hat{h}) = 1$. The filter is optimal in the sense that it maximizes the ratio of signal power to noise power for signals described by $h(\vec{\lambda})$ in the presence of stationary Gaussian noise. When a signal $h(\vec{\lambda})$ is present in the data, the expectation value of ρ is equal to the optimal (amplitude) signal-to-noise ratio (SNR) of the signal:

$$\langle \rho \rangle = \mathrm{SNR} = \sqrt{(h(\vec{\lambda})|h(\vec{\lambda}))}. \tag{9.7}$$

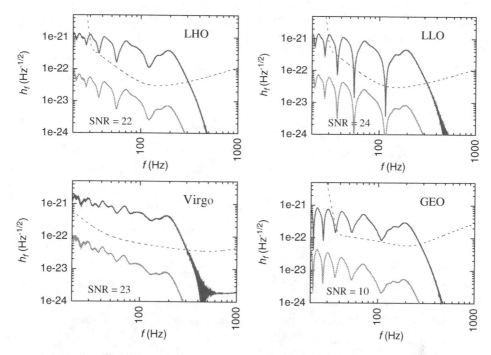

Fig. 9.1. Simulated strain spectral densities for a binary black-hole inspiral signal incident on the LIGO Hanford (LHO), LIGO Livingston (LLO), Virgo and GEO network of detectors. The dashed line corresponds to the design level instrument noise for the first generation of detectors. The lower line in each panel is the raw strain, and the upper line is the effective strain after matched filtering. The single detector SNR is listed in each panel. The simulated signal is for a pair of black holes with masses M of $10M_\odot$ and $5M_\odot$ and spin magnitudes S/M^2 of 0.7 and 0.5, at a luminosity distance of 10 Mpc.

The utility of the matched filtering approach can be seen by considering the contribution to the SNR from a logarithmic frequency interval with central frequency f:

$$\frac{d\,\mathrm{SNR}^2}{d\ln f} = 2\left(\frac{S_h(f)}{S_n(f)}\right) N(f)\,. \tag{9.8}$$

Here $S_h(f)/S_n(f)$ is the raw (power) signal-to-noise ratio, and $N(f) = fT$ is the number of wave cycles observed. By coherently matching the waveform over many cycles, the raw signal-to-noise is enhanced by $N(f)$, and it becomes possible to detect signals that are buried deep below the instrument noise. The utility of the matched filtering approach is illustrated in Figure 9.1.

The search for the best-fit signal parameters is performed by laying out a uniformly spaced template grid and filtering the signal against each template (Owen 1996). The grid spacing is then determined by using the overlap (or match) function $M(\vec{\theta}, \vec{\lambda}) = (\hat{h}(\vec{\theta})|\hat{h}(\vec{\lambda}))$, which defines a natural Riemannian metric $g_{ij}(\vec{\lambda})$ on the space of signals:

$$M(\vec{\lambda} + \Delta\vec{\lambda}, \vec{\lambda}) = 1 - g_{ij}(\vec{\lambda})\Delta\lambda^i \Delta\lambda^j + \cdots, \tag{9.9}$$

where

$$g_{ij}(\vec{\lambda}) = -\frac{1}{2}\frac{\partial^2 M}{\partial\lambda^i \partial\lambda^j}. \tag{9.10}$$

The choice of grid spacing involves a trade-off between computational cost and signal coverage. Adopting a minimum match, $M_{\min}$, between templates sets the allowed fractional loss in signal-to-noise, and determines the number of templates needed to cover the search space. For a hypercubic lattice, the number of templates is given by $\mathcal{N} = V/\Delta V$, where $V = \int d^D\lambda \sqrt{g}$ is the volume of the parameter space and $\Delta V = (2\sqrt{(1 - M_{\min})/D})^D$ is the volume covered by each template.

At first sight it is not obvious how a matched-filter analysis is related to the methods used in Bayesian inference. The relationship becomes less mysterious when optimal matched filtering is recast in terms of maximum likelihood estimation. To establish this link, consider the probability that the noise in a given frequency bin is equal to $n(f)$. For data $s_0(f)$ with no gravitational wave signal present, the conditional probability of observing $n(f)$ is, for Gaussian instrument noise,

$$pr(s_0|n) = \frac{1}{2\pi TS_n(f)}\exp\left(-\frac{2n(f)n(f)^*}{TS_n(f)}\right). \tag{9.11}$$

Now, the conditional probability of measuring $s(f)$ when a signal $h(f)$ is present, $pr(s|h)$, is equal to the conditional probability of measuring $s_0 = s - h$ assuming that no signal is present, $pr(s_0|n)$. Thus,

$$pr(s|h) = \frac{1}{2\pi TS_n(f)}\exp\left(-\frac{2(s(f) - h(f))(s^*(f) - h^*(f))}{TS_n(f)}\right). \tag{9.12}$$

Combining the contributions from all the frequency bins yields the likelihood

$$pr(s|h) = pr(s|\vec{\lambda}) = C\exp\left(-\frac{(s - h|s - h)}{2}\right). \tag{9.13}$$

The normalization constant $C \sim \exp(-\int \log S_n(f)df)$ is often ignored since it does not depend on the signal $h(\vec{\lambda})$. The generalization to a network of detectors is straightforward: simply multiply together the likelihoods for each detector. To evaluate the maximum likelihood, first maximize with respect to the template amplitude $A = (h|h)^{1/2}$, which yields the solution $A_{\mathrm{ML}} = (s|\hat{h}) = \rho(\vec{\lambda})$ and the partial maximization

$$\left[pr(s|\vec{\lambda})\right]_{\max_A} = C'\exp\frac{\rho^2(\vec{\lambda})}{2}. \tag{9.14}$$

Maximizing the likelihood with respect to the remaining signal parameters is then equivalent to maximizing the matched-filter statistic $\rho(\vec{\lambda})$.

In the frequentist setting, the detection problem is usually addressed by applying the Neyman–Pearson criterion, which seeks to minimize the false dismissal probability for a given false alarm probability. Here false dismissal corresponds to the conclusion that $h = 0$ when in fact a signal is present, and false alarm corresponds to the conclusion that $h \neq 0$ when no signal is present. For Gaussian noise, the likelihood ratio

$$\Lambda = \frac{pr(s|h)}{pr(s|0)} = \exp\left(-(s|h) + \frac{1}{2}(h|h)\right) \tag{9.15}$$

provides an optimal decision statistic for the Neyman–Pearson test. The space of possible observations is partitioned by setting a threshold $\Lambda_0 > 0$ that yields the desired false alarm probability. In practical applications, where the noise is neither stationary nor Gaussian, the likelihood ratio Λ, or the closely related matched-filter statistic ρ, are adopted as a detection statistic, and Monte Carlo simulations are used to estimate the false alarm and false dismissal probabilities. The false dismissal probability is estimated using hardware and software injections of simulated gravitational wave signals. Estimating the false alarm probability is more difficult since there is no way to know in advance if a signal is present in the data or not, as it is impossible to shield the detector from gravitational waves. The closest alternative is to artificially time-shift the data between detectors in a network, thereby destroying the coherent response to a gravitational wave, while leaving the noise largely unaffected.

While waiting for the first detection, gravitational wave astronomers amuse themselves by predicting what can be learned from observing various sources of gravitational waves. A key element of these investigations are estimates of how accurately the source parameters can be recovered. These estimates employ the Gaussian approximation to the likelihood in the neighbourhood of maximum likelihood:

$$pr(s|\vec{\lambda}_{\mathrm{ML}} + \Delta\vec{\lambda}) = \left(\det\left(\frac{\Gamma}{2\pi}\right)\right)^{1/2} e^{-\frac{1}{2}\Gamma_{ij}\Delta\lambda^i\Delta\lambda^j}, \tag{9.16}$$

where

$$\Gamma_{ij} = -\left\langle \frac{\partial^2 \log p(s|\vec{\lambda})}{\partial\lambda^i\partial\lambda^j} \right\rangle\Bigg|_{\mathrm{ML}} = (h_{,i}|h_{,j}). \tag{9.17}$$

At this level of approximation, the parameter error variance–covariance matrix is given by

$$\langle \Delta\lambda^i\Delta\lambda^j \rangle = \Gamma_{ij}^{-1}. \tag{9.18}$$

In the gravitational wave literature the matrix Γ_{ij} is generally referred to as the Fisher Information Matrix, even though this terminology is only strictly correct if the priors are uniform.

9.3 The Bayesian approach

The availability of fast computers and the development of efficient algorithms for mechanizing Bayesian inference has transformed data analysis in many fields. The fledgling field of gravitational wave astronomy has been slow to adopt these new techniques, in part because of history, and in part because of several technical challenges that will be described shortly. At present there is no end-to-end Bayesian inference pipeline for the analysis of data collected by the current generation of gravitational wave observatories, and it may be many years before such a pipeline is developed. The situation is more promising for the future space-based detector LISA, where Bayesian methods have taken root early in the data analysis development effort, and portions of a comprehensive Bayesian inference pipeline have been implemented and tested on simulated data. In the near term, the main application of Bayesian analysis will be to follow up detection candidates from the standard search pipelines. The goal is to improve upon the point estimate for the signal parameters that comes from a matched-filter or maximum likelihood search, by providing full posterior distributions. In other words, the Bayesian contribution will be to put 'error bars' on the parameters.

Two technical challenges have hindered the adoption of practical Bayesian inference in gravitational wave astronomy. The first is the complexity of the detectors, which makes it difficult to come up with an appropriate form for the likelihood function, and the second is the tight concentration of the posterior weight within a high-dimensional search space (in other words, the signals are hard to find).

The likelihood function quoted in Eq. (9.13) is deficient in several respects: it considers just the gravitational wave data channel, it assumes that the noise is stationary and Gaussian, and it assumes that the instrument response is linear. The state of the art gravitational wave interferometers at the LIGO and Virgo sites are complex machines with many thousands of components, controlled by dozens of non-linear feedback loops, and covered with sensors that record tens of thousands of channels of data in addition to the 'gravitational wave' channel that monitors the differential motion of the test masses. It is known that certain disturbances recorded in the monitoring channels can correlate with disturbances in the gravitational wave channel, and this should be taken into account in the likelihood function. However, despite heroic efforts by the detector characterization teams, the full spectrum of cross couplings is far from understood, and a multi-channel likelihood function remains a distant dream. Instead, a list of data quality flags are produced that tag sections of data with descriptions of possible problems and a numerical assessment of their severity. The highest-level flags are used to veto sections of data, while the lower-level flags are considered when following up potential detections. The instruments also exhibit various glitches and excursions that are not flagged by any monitoring channel, and the overall noise spectrum changes over time. The measurement process is further complicated by the feedback loop that is used to keep the interferometers locked: the length of the arms is adjusted so that the laser light from the two arms interferes destructively at the so-called anti-symmetric port. If a passing gravitational wave, or some other disturbance, acts to change the relative length of the two arms, forces are applied to the test masses (mirrors) so as to keep the anti-symmetric port dark. Thus, the gravitational wave signal is encoded in the response of the non-linear control loop that keeps the interferometers in lock. The response of the control loop is continuously measured by applying sinusoidally varying forces to the test masses at various frequencies. The amplitude of these calibration lines are used to determine the transfer function for the length sensing control loop, and from this the strain $h(t)$ is reconstructed. This procedure leads to a non-linear coupling of the gravitational wave signal and the instrument noise, and the instrument noise with itself, which violates two of the key assumptions used to derive the simple form for the likelihood in Eq. (9.13).

Despite all these difficulties, something very close to the expression in (9.13) can provide a good approximation to the likelihood function. The first modification is to introduce a Heaviside step function that vetoes data segments tagged by the data quality flags. The second modification is to treat the noise level and spectral shape as unknowns to be determined from the data, and to further allow these quantities to depend on time (either by working with shorter segments of data or using some other time-frequency decomposition such as wavelets). The final modification is to allow for a non-Gaussian tail in the noise distribution. One method for doing this is to model the noise as the sum of two Gaussians of different weight and variance. A non-Gaussian tail in the instrument noise will result in the noise fit identifying a second component with larger variance. On the other hand, if the noise is Gaussian, and the number of parameters in the noise fit is itself a parameter, the second component will be discarded. It has been shown (Allen *et al.* 2003) that these modifications lead to a robust frequentist statistic that is near optimal for weak signals buried in non-Gaussian noise.

The second technical challenge relates to the extreme concentration of the posterior mass. Define ΔV to be the smallest volume of parameter space to encompass 95% of the posterior weight, and define V to be the total volume of the parameter space. In many cases, the ratio $\Delta V/V$ is tiny – akin to finding a cork bobbing about somewhere in the Pacific Ocean – and the volume ΔV is often spread over several disjoint regions surrounding secondary maxima of the posterior distribution function. The smallness of this ratio is reflected in the large number of templates needed to perform a frequentist analysis, and the long burn-in times for an MCMC style analysis. A variety of techniques can be used to help mitigate this problem, such as hierarchical searches, adaptive grid refinement, simulated annealing and parallel tempering, but the bottom line is that the signals are hard to find. Much of the current research on Bayesian inference in gravitational wave astronomy is devoted to developing strategies for minimizing the burn-in time of MCMC style analyses.

9.3.1 Parameter estimation

In the happy event of a gravitational wave detection it will be desirable to quote more than just the best-fit values for the source parameters. Electromagnetic astronomers will want to know the shape and extent of the 'error ellipse' that describes the uncertainty in the sky location, and astrophysicists will want to know the level of uncertainty in quantities such as the component masses and spins. When a grid search is used to make the detection there will be some crude information available from the distribution of likelihood values on surrounding grid points, but the searches typically maximize, rather than marginalize, over quantities such as time of arrival and gravitational wave phase, and prior information is not accounted for. A far more informative analysis of the parameter distributions can be derived using standard MCMC techniques.

Metropolis and Hastings provided a simple recipe for performing MCMC analyses that requires just three ingredients: likelihood, prior and proposal. The basic form for the likelihood function (9.13) and its generalizations have already been discussed. Appropriate priors for the signal parameters are fairly well understood for most source types, and are arrived at using a combination of astrophysical modelling and electromagnetic observations. For example, distant sources are expected to be uniformly distributed on the sky, and the initial orientation of a binary's orbital plane should be uniformly distributed in the cosine of the inclination angle. The distribution of other parameters, such as the mass M and spin S of a black hole, are not as well understood, but theory does provide constraints, such as $S/M^2 \leq 1$.

The final ingredient in the MCMC recipe – the form of the proposal distribution – is where the MCMC practitioner gets to add a creative touch. In principle, any non-trivial proposal distribution will yield a Markov chain with stationary distribution equal to the posterior distribution function being sought. In practice, poorly chosen proposal distributions will lead to chains that take longer than the age of the Universe to reach stationarity. Proposal distributions that closely approximate the posterior distribution yield well-mixed chains that rapidly approach stationarity, but constructing such distributions seems to imply advance knowledge of the end result. The MCMC literature describes many strategies for choosing proposal distributions, and several of these strategies have been applied in the context of gravitational wave parameter estimation (Christensen & Meyer 2001; Cornish & Crowder 2005; Cornish & Porter 2006; Rover, Meyer & Christensen 2007). The main methods studied so far are delayed rejection, Gaussian approximation, and parallel tempering.

It has been found that a hybrid algorithm that uses parallel tempering and a Gaussian approximation to the posterior as the proposal distribution delivers good all-round performance. The method assumes that the dominant modes of the posterior distribution function have been identified by the original search algorithm. The Hessian of the log posterior, H_{ij}, is then computed at each posterior mode, either by direct numerical computation, or by using the Fisher matrix approximation of (9.17). The proposal distribution is then written as

$$q(\vec{x}|\vec{y}) = q(\vec{x}) = \sum_n A_n e^{-(\vec{x} - \vec{\lambda}_n) \cdot \overleftrightarrow{H}_n \cdot (\vec{x} - \vec{\lambda}_n)/2}, \qquad (9.19)$$

where $\vec{\lambda}_n$ is the position of the nth mode and $\overleftrightarrow{H}_n$ is the Hessian at this mode. The amplitudes A_n describe the relative weight of each mode, and are scaled to yield a properly normalized probability distribution. A distribution of this type can be drawn from by first randomly selecting the mode n (suitably weighted), then making Gaussian draws along each eigendirection of the Hessian, with variances given by the respective eigenvalues. The transitions between modes is aided by the technique of parallel tempering, whereby several chains with different 'temperatures' T are run in parallel, and Metropolis–Hastings transitions can occur between chains. The likelihood function for a chain with temperature T is given by $[pr(s|\vec{\lambda})]^{1/T}$, and the width of the peaks in the proposal distribution are similarly widened by replacing $H_{ij} \rightarrow H_{ij}/T$. While only the samples from the $T = 1$ chain can be used to construct the posterior, the higher-temperature chains help facilitate mixing and transitions between modes. There is always the danger that a particular choice of proposal distribution will restrict the movement of the chain, and with insufficient samples, produce a posterior distribution that follows the proposal distribution. The parallel tempering technique helps to alleviate this tendency, as does the use of a mixture of proposal distributions. For example, ten jumps drawn from Eq. (9.19) might be followed by one jump drawn from a proposal distribution of the form

$$q(\vec{x}|\vec{y}) = \vec{y} + \delta\vec{y}, \qquad (9.20)$$

where the ith component of $\delta\vec{y}$ is drawn from a uniform distribution of width ϵ^i. Figure 9.2 shows the marginalized posterior distributions for the 15 parameters

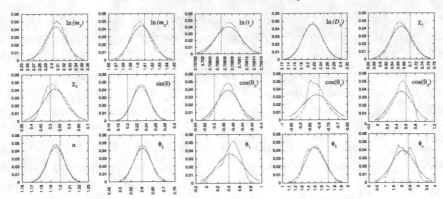

Fig. 9.2. Marginalized posterior distributions for the 15 parameters that describe the LIGO–Virgo–GEO response to a binary system of spinning black holes. The MCMC derived posterior distributions (the noisy curves with a higher peak) are compared with the predictions of the Fisher information matrix (the smoother, Gaussian curve). The vertical dashed line indicates the injected values for the source parameters.

that describe the response of the LIGO–Virgo–GEO network to the inspiral signal from a pair of spinning black holes (shown in Figure 9.1). The simulation employs synthetic stationary and Gaussian instrument noise and uses the MCMC algorithm described above.

9.3.2 Search strategies

As described in Section 9.2.1, the standard technique for searching for signals of a known type employs matched filtering against a uniformly spaced bank of templates. It is often possible to reduce the size of the search space by analytically maximizing over some of the signal parameters, such as time of arrival, distance to the source and the initial phase of the waveform. The computational cost of the search can be reduced further by adopting a hierarchical approach that starts with a coarse grid, then follows up with a finer grid in the region surrounding templates that gave an SNR above some threshold. An alternative hierarchical approach is to start with some subset of the available data – for example, some fraction of the total observation time or bandwidth – and progressively incorporate more of the data.

While hierarchical grid searches work well for simple signals and low-dimensional search spaces, the computational cost can become prohibitive when the complexity of the signal or the dimensionality of the search space becomes large. Some of the most challenging examples are the signals from pairs of spinning black holes, the combined signals from tens of millions of Galactic white dwarf binaries, and signals from extreme or intermediate mass ratio inspirals (EMRIs or IMRIs). While various hierarchical procedures have been proposed to deal with these sources, the most successful algorithms developed to date abandon the template grid approach and employ MCMC inspired search strategies.

The gridless search techniques that have been developed in gravitational wave astronomy are extensions of the methods used to speed up the 'burn-in' phase of MCMC posterior studies. Since the samples from the burn-in portion of the chain are discarded, the usual rules concerning reversibility and hysteresis can be ignored. Some of the more effective algorithms (Cornish & Porter 2007; Crowder & Cornish 2007) combine deterministic hill-climbing moves with stochastic Metropolis–Hastings sampling and simulated annealing. These hybrid algorithms are generally superior to purely deterministic approaches, such as the Nelder–Mead SIMPLEX (amoeba) algorithm, which tend to get stuck at local maxima, and they also out-perform standard random walk Metropolis–Hastings algorithms due to their superior hill-climbing abilities.

These directed-stochastic searches share many features in common with genetic algorithms (Crowder, Cornish & Reddinger 2006), and their success relies on the posterior landscape offering 'partial credit'. If the landscape resembles a flat plain with a single sharp spike at the posterior mode, then a directed-stochastic search would take longer to locate the mode than a methodical grid search. On the other hand, if the posterior landscape more closely resembled Colorado, where the eastern plains gradually rise up until they reach the foothills of the Rocky Mountains, then a directed-stochastic search would quickly home in on the posterior mode, while a simple grid search would waste time out in the great plains. An intelligently designed hierarchical grid search could also exploit the large-scale structure of the posterior to achieve comparable performance, but the advantage of the directed-stochastic searches is that they arrive at a near optimal strategy automatically.

9.3.3 Model selection

One of the most exciting areas that is currently undergoing rapid development in gravitational wave astronomy is the application of Bayesian model selection to important questions, such as 'Do the data contain a gravitational wave signal?' For the detection question the models are: Model 0 – 'only noise is present' and Model 1 – 'there is a gravitational wave signal and noise present'. As mentioned in Section 9.2.1, the traditional approach to the detection question employs the Neyman–Pearson test, calibrated with extensive Monte Carlo simulations of injected signals and scrambled data.

Bayesian inference provides a well-defined answer to the detection question. Bayes' theorem states that the posterior probability of model M_i, $pr(M_i|s)$, is given in terms of the prior probability of the model, $pr(M_i)$, and the marginal likelihood or *evidence* for M_i, $pr(s|M_i)$, by

$$pr(M_i|s) = \frac{pr(M_i)pr(s|M_i)}{pr(s)}. \tag{9.21}$$

The normalization factor $pr(s)$ is unimportant here since we are only interested in the relative probability, or odds ratio of model M_i against model M_j:

$$O_{ij} = \frac{pr(M_i|s)}{pr(M_j|s)} = \frac{pr(M_i)}{pr(M_j)}\frac{pr(s|M_i)}{pr(s|M_j)} = P_{ij}\,B_{ij}. \tag{9.22}$$

The odds ratio is given by the product of the prior belief, $P_{ij} = pr(M_i)/pr(M_j)$, and the Bayes factor $B_{ij} = pr(s|M_i)/pr(s|M_j)$. The Bayes factor is a measure of how the data have informed our degree of belief in the two models.

The evaluation of the model evidence involves an integral of the likelihood weighted by the priors on the model parameters $\vec{\lambda}$:

$$pr(M_i|s) = \int d\vec{\lambda}\,pr(\vec{\lambda}|M_i)\,pr(s|\vec{\lambda}, M_i). \tag{9.23}$$

In most cases of interest in gravitational wave astronomy, the parameter space dimension is large, and a direct numerical evaluation of the evidence integral is impractical. One exception is the search for gravitational wave signals associated with neutron star spin glitches. When treated as a power spectrum search, the glitch signal depends on just three parameters: amplitude, decay time, and characteristic frequency, and the evidence can be computed by brute force (Clark *et al.* 2008). Very recently the powerful technique of nested sampling (Skilling 2006) has been used to compute the model evidence for simulated black-hole inspiral signals embedded in synthetic noise (Veitch & Vecchio 2008). It is hoped that this technique will eventually be implemented as part of the detection follow-up procedure for the LIGO–Virgo–GEO network. Careful thought will have to be given to the assignment of prior belief in this implementation, as the events being considered will have been pre-selected as promising detection candidates.

The Reverse Jump Markov chain Monte Carlo (RJMCMC) (Green 1995) algorithm provides an alternative computational framework for performing Bayesian model selection that neatly side-steps the need to compute the model evidence. The RJMCMC approach extends the usual MCMC technique to include transitions between models. The Bayes factor B_{ij} is then given by the ratio of the time the chain spends exploring model M_i to the time spent exploring model M_j. The advantage of this approach is that it combines parameter estimation and model selection within a single framework. The RJMCMC approach is particularly well suited to the analysis of data from the future space-based LISA detector, where one has to contend with a Galactic foreground produced by tens of millions of white dwarf

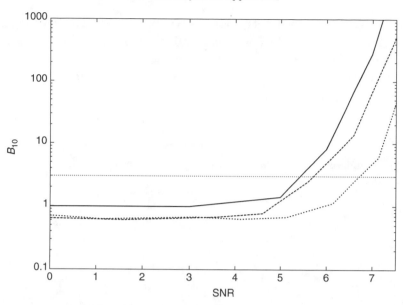

Fig. 9.3. Bayes factor as a function of signal-to-noise for model 0, pure instrument noise, and model 1, instrument noise and the signal from a white dwarf binary with frequency ~3 mHz. The three lines correspond to different noise realizations, with Bayes factors that cross the detection threshold for different signal-to-noise ratios. This illustrates that the detectability of a signal depends on the detailed structure of the noise, in addition to the relative strength of the signal and noise.

binaries in our galaxy. To avoid biasing the analysis of the more interesting extra-galactic signals from massive black holes and extreme mass ratio inspirals it will be necessary to simultaneously solve for all the signals present in the data. Since the number and nature of the resolvable signals is not known a priori, model selection will take centre stage in LISA data analysis. While MCMC techniques have been employed to isolate the signals from ~20 000 white dwarf binaries in simulated data and to explore the joint posterior (Crowder & Cornish 2007), the model selection in these large-scale simulations has relied on approximations to the evidence. The RJMCMC technique has been applied to the simpler model selection problem of deciding if a single white dwarf binary is better described as monochromatic or chirping (Cornish & Littenberg 2007), and there have been some pilot studies that have looked at determining the number of white dwarf signals present in sim-ulated data. The first of these studies considered the signals from two overlapping white dwarf binaries, and compared models with 1, 2 or 3 signals present (Stroeer, Gair & Vecchio 2006). This work has recently been extended to include the de-tection problem, where simulated data with 0 or 1 white dwarf signals present is used to compare models with 0 or 1 signals present (Littenberg & Cornish 2009). Figure 9.3 shows how the Bayes factor B_{10} varies with the signal-to-noise of a simulated white dwarf signal. The noise model employed eight parameters, while the signal plus noise model requires an additional eight parameters to describe the signal from a chirping white dwarf binary. Work is currently in progress to apply the RJMCMC approach to larger numbers of white dwarf signals, and to the binary inspiral detection problem for ground-based detectors.

References

Allen, B., Creighton, J. D. E., Flanagan, E. E. and Romano, J. D. (2003). *Phys. Rev. D*, **67**, 122002.

Anderson, W. G., Brady, P. R., Creighton, J. D. E. and Flanagan, E. E. (2001). *Phys. Rev. D*, **63**, 042003.

Biswas, R., Brady, P. R., Creighton, J. D. E. and Fairhurst, S. (2007). arXiv:0710.0465 [gr-qc].

Camp, J. B. and Cornish, N. J. (2004). *Annu. Rev. Nucl. Part. Sci.*, **54**, 525.

Christensen, N. and Meyer, R. (1998). *Phys. Rev. D*, **58**, 082001.

Christensen, N. and Meyer, R. (2001). *Phys. Rev. D*, **64**, 022001.

Clark, J., Heng, I. S., Pitkin, M. and Woan, G. (2008). *J. Phys. Conf. Ser.*, **122**, 012035.

Cornish, N. J. and Crowder, J. (2005). *Phys. Rev. D*, **72**, 043005.

Cornish, N. J. and Littenberg, T. B. (2007). *Phys. Rev. D*, **76**, 083006.

Cornish, N. J. and Porter, E. K. (2006). *Class. Quant. Grav.*, **23**, S761.

Cornish, N. J. and Porter, E. K. (2007). *Phys. Rev. D*, **75**, 021301.

Crowder, J. and Cornish, N. J. (2007). *Phys. Rev. D*, **75**, 043008.

Crowder, J., Cornish, N. J. and Reddinger, L. (2006). *Phys. Rev. D*, **73**, 063011.

Davis, M. H. A. (1989). In B. F. Schutz, ed., *Gravitational Wave Data Analysis*. Dordrecht: Kluwer, p. 73.

Finn, L. S. (1992). *Phys. Rev. D*, **46**, 5236.

Finn, L. S. and Chernoff, D. F. (1993). *Phys. Rev. D*, **47**, 2198.

Flanagan, E. E. and Hughes, S. A. (1998a). *Phys. Rev. D*, **57**, 4535.

Flanagan, E. E. and Hughes, S. A. (1998b). *Phys. Rev. D*, **57**, 4566.

Green, P. J., (1995), *Biometrika*, **82**, 711.

Hughes, S. A. (2003). *Ann. Phys.*, **303**, 142.

LIGO Scientific Collaboration (2007). arXiv:0711.3041 [gr-qc].

LISA International Science Team (2007). http://www.srl.caltech.edu/lisa/documents/lisa science case.pdf.

Littenberg, T. B. and Cornish, N. J. (2009). arXiv:0902.0368.

Owen, B. J. (1996). *Phys. Rev. D*, **53**, 6749.

Rover, C., Meyer, R. and Christensen, N. (2007). *Phys. Rev. D*, **75**, 062004.

Searle, A. C., Sutton, P. J., Tinto, M. and Woan, G. (2008). *Class. Quant. Grav.*, **25**, 114038.

Skilling, J. (2006). *Bayesian Analysis*, **1**, 833.

Stroeer, A., Gair, J. and Vecchio, A. (2006). *AIP Conf. Proc.*, **873**, 444.

Thorne, K. S. (1987). In S. W. Hawking and W. Israel, eds., *300 Years of Gravitation*. Cambridge: Cambridge University Press.

Veitch, J. and Vecchio, A. (2008). *Phys. Rev. D*, **78**, 022001.

10

Bayesian analysis of cosmic microwave background data

Andrew H. Jaffe

10.1 Introduction

At redshifts $z \gtrsim 1000$, with temperatures higher than those equivalent to the recombination energy of hydrogen (1 Ry = 13.6 eV), there were sufficient energetic photons in the Universe to maintain a very high ionization fraction among the atomic species (75% hydrogen with the remainder almost entirely helium). In such an ionized plasma, the mean free path of photons to Thomson scattering off of the electrons is very short compared with the Hubble scale, and the Universe is effectively opaque. As the Universe cools, the photons no longer possess enough energy to maintain the high ionization fraction, and the protons and electrons very rapidly combine into neutral hydrogen gas.[1] This neutral gas no longer scatters the photons, and hence they decouple, streaming freely through the Universe thereafter, and redshifting until they are observed today as the cosmic microwave background (CMB), with an average temperature $T_0 = 2.725 \pm 0.001$ K (Fixsen *et al.* 1996). Because the state of the plasma was not completely uniform, we observe a slightly different temperature of the gas in different directions, with an rms deviation $\langle (\Delta T/T_0)^2 \rangle^{1/2} \sim 10^{-5}$.

Observationally, as predicted by the inflationary paradigm for their generation, these temperature fluctuations appear to be well described as a statistically isotropic Gaussian field on the sky (reflecting an underlying statistically isotropic

---

[1] In fact this occurs not at $kT \simeq 13.6$ eV, but rather at the lower temperature $kT \simeq 0.3$ eV, as there are roughly 10^9 photons per baryon in the Universe and thus there are sufficient energetic photons in the tail of the Bose–Einstein distribution to maintain a high ionization fraction.

three-dimensional density field, a perturbation around the homogeneous and isotropic Friedmann–Robertson–Walker background):

$$\left\langle \frac{\Delta T}{T_0}(\hat{x}) \frac{\Delta T}{T_0}(\hat{y}) \right\rangle = C(\arccos \hat{x} \cdot \hat{y}), \tag{10.1}$$

where $C(\theta)$ is the correlation function at angular distance θ. Isotropy is embodied in the requirement that this correlation function only depends upon θ and not the angles themselves. The small fractional perturbation amplitude means that the evolution of the fluctuations is well described by linear equations, which in turn implies that an initially Gaussian distribution is subsequently maintained.

It is more convenient to work in the spherical equivalent of Fourier space, the spherical harmonic domain:

$$T(\hat{n}) = \sum_{\ell=2}^{\infty} \sum_{m=-\ell}^{+\ell} a_{\ell m} Y_{\ell m}(\hat{n}), \tag{10.2}$$

which has diagonal correlations given by

$$\langle a_{\ell m} a_{\ell' m'}^* \rangle = C_\ell \, \delta_{\ell \ell'} \delta_{mm'}, \tag{10.3}$$

where C_ℓ is the *power spectrum*, related to the correlation function by

$$C(\theta) = \sum_\ell \frac{2\ell+1}{4\pi} C_\ell P_\ell(\cos\theta), \tag{10.4}$$

or

$$C_\ell = 4\pi \int \mathrm{d}(\cos\theta) P_\ell(\cos\theta) C(\theta) \,. \tag{10.5}$$

Note that the monopole ($\ell = 0$, corresponding to the average temperature) and dipole ($\ell = 1$, dominated by the doppler shift pattern induced by our motion with respect to the rest frame of the CMB) are generated by different physics and therefore not included in the sums above, which correspond to the primordial perturbation signal.

For a specific set of cosmological parameters, the power spectrum C_ℓ – which is all we need to provide a complete statistical description of Gaussian cosmological perturbations – is completely determined, and can be calculated by solving the coupled linearized Einstein–Boltzmann equations for the distribution of matter particles, photons and dark matter in the expanding Universe. This has been implemented in the popular and publicly available CMBFAST[2] and CAMB[3] codes. Ultimately, then, we wish to invert this, and recover these cosmological parameters from noisy data of the CMB sky.

---

[2] http://www.cmbfast.org
[3] http://camb.info

We can also observe the polarization of the CMB radiation, which can confirm the cosmological measurements from the temperature field and moreover holds the promise of providing a measurement of an early background of gravitational radiation – predicted by the theory of cosmic inflation (Starobinsky 1979; Guth 1981) – and whose effect can in principle be disentangled from other cosmological perturbations. In the first part of this chapter, we concentrate on the temperature (intensity) of the CMB, and defer discussion of polarization, which is formally similar but brings added complications, to Section 10.3.

CMB data is therefore an ideal case study for Bayesian analysis (for early discussions, see Borrill 1999 and Bond *et al.* 1999). For the simplest theories there really is no information in the CMB beyond the isotropic, two-point correlation function: we can write down the appropriate prior distributions right away. (To put this in a non-Bayesian way with which many physicists will be comfortable, the CMB 'really is' described by a Gaussian distribution.)

10.2 The CMB as a hierarchical model

Raw CMB data usually come to us in the form of a set of timestreams, the readout of multiple detectors observing the sky, always smoothed with some sort of experimental beam (known as the point-spread function or response function in other areas of astronomy and physics) and combined with instrumental noise. From these data, we usually perform the following steps:

  (i) estimating a map from the timestream(s);
 (ii) estimating a power spectrum from the map; and
(iii) estimating the cosmological parameters from the power spectrum.

(In addition, there may be additional steps to deal with real-world contamination: we may have to separate out instrumental 'systematic' effects and astrophysical foreground emission which we will discuss briefly in Section 10.4.1.) This is known in statistical parlance as a 'hierarchical model', but in fact nearly any scientific data analysis involves these sorts of steps, although very often the earlier parts of the process are known as 'data processing' and the later as 'science' or 'analysis' or 'interpretation'. The relative simplicity of the data and the underlying theoretical models serve to make the procedure more transparent for the CMB. As we will see, we really can perform these steps in this order, and ignore the products of the earlier steps: the power spectrum depends on the map and not the raw timestream; the cosmological parameters depend on the power spectrum and not the map or the timestream. In the following we will make these steps more explicit and explore some of the algorithms that have been developed to perform the needed calculations.

10.2.1 CMB data

A CMB instrument observes the sky temperature $T_\nu(\hat{x})$ at frequency ν and position $\hat{x}$. But in fact any real instrument has a finite beam and bandwidth as well as noise, so a more realistic model for the observing process is

$$d(f,t) = \int d\nu \, w_f(\nu) \int d^2\hat{x} B_t(\hat{x}) \, T_\nu(\hat{x}) + n(f,t), \qquad (10.6)$$

where $w_f(\nu)$ defines an observing filter (usually labelled as centred around frequency channel f) and $B_t(\hat{x})$ defines the experimental beam (usually centred around some position labelled $p = p(t)$), and the noise in that channel at that time is $n(f,t)$. The CMB itself has a flat spectrum in thermodynamic temperature, $T_\nu = T$, so from now on we ignore this effect (although it enters into the disentanglement of astrophysical foregrounds from the cosmological signal). In practice, we assume that $B_t(\hat{x})$ is only a function of the distance between the nominal detector pointing at time t and the location $\hat{x}$, so the integral turns into an isotropic convolution, which in turn means that the spherical harmonic components of the convolved quantity can simply be written as $a_{\ell m} B_\ell$ where $a_{\ell m}$ are the components of $T(\hat{x})$ and B_ℓ of the beam function, dependent only upon ℓ for an isotropic beam.

In a more simplified form, the data can therefore be described by the model

$$d_i = \sum_p A_{ip} T_p + n_i . \qquad (10.7)$$

Here, d_i is the data gathered at a particular detector (f) and time (t), together denoted $i = (f,t)$. The label p denotes pixels on the sky (i.e., specific locations on the sky, smoothed by the beam function) and T_p gives the temperature in that pixel. The matrix A_{ip} gives the action of the detector at a particular time: for a scanning experiment, $A_{ip} = 1$ when observing pixel p with detector/time i, and 0 otherwise. For a differencing experiment such as WMAP or COBE/DMR, A_{ip} will be ± 1 for the two pixels being differenced. In these cases, we take the underlying sky to be already smoothed by the beam and ignore the effects of pixelization (we address these shortcomings briefly below in Section 10.2.2). Finally, the noise contribution is given by n_i. We will occasionally also use matrix notation:

$$\mathbf{d} = \mathbf{AT} + \mathbf{n} . \qquad (10.8)$$

For now, we will ignore any other contributions to the data, such as foregrounds or instrumental systematics (see Section 10.4.1).

To apply Bayes' theorem, we need to determine the appropriate parameterization of the quantities on the right-hand side and priors to quantify our knowledge of them. At the outset, we will take the noise contribution to be described by a

Gaussian distribution with zero mean and covariance matrix N_{ij}:

$$\langle n_i \rangle = 0, \qquad \langle n_i n_j \rangle = N_{ij} . \tag{10.9}$$

We will discuss how the noise covariance matrix is estimated below.

10.2.2 From detectors to maps

First, we wish to recover the sky signal, T_p, from the timestream(s) d_i. Given our Gaussian model for the noise, we can write the likelihood function as

$$P(d_i | T_p, N_{ij}) = \frac{1}{|2\pi\mathbf{N}|^{1/2}} \exp\left[-\frac{1}{2}(\mathbf{d} - \mathbf{AT})^{\mathrm{T}}\mathbf{N}^{-1}(\mathbf{d} - \mathbf{AT})\right]. \tag{10.10}$$

Up to a constant factor, this can be written by completing the square in the exponential as

$$P(d_i | T_p, N_{ij}) \propto \frac{1}{|2\pi\mathbf{C_N}|^{1/2}} \exp\left[-\frac{1}{2}(\mathbf{T} - \hat{\mathbf{T}})^{\mathrm{T}}\mathbf{C}_N^{-1}(\mathbf{T} - \hat{\mathbf{T}})\right], \tag{10.11}$$

with

$$\hat{\mathbf{T}} = (\mathbf{A}^{\mathrm{T}}\mathbf{N}^{-1}\mathbf{A})^{-1}\mathbf{A}^{\mathrm{T}}\mathbf{N}^{-1}\mathbf{d} \tag{10.12}$$

and

$$\mathbf{C}_N = (\mathbf{A}^{\mathrm{T}}\mathbf{N}^{-1}\mathbf{A})^{-1}. \tag{10.13}$$

If we assume a uniform (and therefore un-normalized) prior on T_p, Eq. (10.11) just gives the posterior distribution for T_p, distributed as a Gaussian with mean $\hat{T}_p$ and covariance matrix $C_{Npp'}$. (This is in fact the correct posterior in the limit of an infinitely dispersed but correctly normalized prior.) This is also exactly the generalized least-squares (GLS) solution for T_p given data d_i in the presence of correlated noise.

We will see, however, that we do not in fact need to use a prior for this step to be useful; since the data *only* enter the likelihood through $\hat{T}_p$, the latter is a *sufficient statistic* (along with its covariance matrix) and we can choose to work with it instead of the raw timestream. The map $\hat{T}_p$ can be seen as simply a form of data compression, although in practice it is a very useful intermediate step to visualize the data.

Note one special case of this formula. If the noise is white with variance σ_w^2, we have $N_{tt'} = \sigma_w^2 \delta_{tt'}$ and

$$\hat{T}_p = \frac{1}{N_p} \sum_{t \in p} d_t \qquad \text{(white noise)} \tag{10.14}$$

and

$$C_{Npp'} = \frac{\sigma_w^2}{N_p} \delta_{pp'}, \tag{10.15}$$

where the sum is over all time samples for which pixel p is observed and N_p is the number of such samples. Of course, this is just the obvious average and 'root-N' noise reduction we expect. In this trivial but unphysical case, the map can be calculated in $O(N_p)$ steps. (In fact, in this case the scaling is best expressed as $O(N_t)$, where N_t is the number of time samples.)

The cost of solving the full GLS equations, however, is naively $O(N_p^3)$, owing to the calculation of the inverse matrix, Eq. (10.13). However, to calculate the map itself we need actually only solve

$$\mathbf{C}_N^{-1}\hat{\mathbf{T}} = \mathbf{A}^{\mathsf{T}}\mathbf{N}^{-1}\mathbf{d} \tag{10.16}$$

for $\hat{\mathbf{T}}$. This can usually be done iteratively with preconditioned conjugate-gradient methods in which each iteration has cost $O(N_p^2)$ (and the number of iterations is always much less than N_p). The appropriate preconditioner (defined as something that can be used as an approximate inverse) is usually very simple: we just use the inverse of the white-noise case, Eq. (10.15) (e.g., Ashdown *et al.* 2007a,b, 2009 and references therein).

Note that, irrespective of the preconditioner, the calculation of the full noise matrix C_N does scale as (or nearly) $O(N_p^3)$: each of the $O(N_p^2)$ entries in the symmetric matrix requires $O(N_p)$ operations. For an arbitrary scanning strategy, it is very difficult to reduce these scalings, but, as we shall see next, there are cases with sufficient symmetry to simplify the needed calculations considerably.

Destripers

In practice, we often do not have to solve the full GLS equations when we take the specifics of the noise structure and the scan strategy into account. We will consider the Planck satellite as a case study (Planck Consortium 2005). Planck observes the sky by rotating at a rate of approximately one rpm, with its detectors pointed at approximately $85°$ from the rotation axis. After one hour, that axis itself is repointed by about $2.5\,\text{arcmin}$ to its normal, so the rotation axes trace out a great circle over the course of approximately seven months. The noise is well described by low-frequency correlated noise with power spectrum $1/f^\alpha$, $\alpha \simeq 1$ ('$1/f$ noise') along with white noise (constant power spectrum). The frequency at which the two have equal contributions is known as the *knee frequency*, f_{knee}. As long as $f_{\text{knee}} \lesssim (60\,\text{sec})^{-1}$, we can model the noise as a constant offset plus a high-frequency component along each one-minute circle. This enables some significant simplifications (see, for example, Delabrouille 1998; Stompor & White 2004; Ashdown *et al.* 2007a,b, 2009).

We denote a single rotation of the satellite as a *circle*, and the collection of consecutive circles between repointings as a *ring*. First, we consider the recovery of the signal along a single ring. In fact, this is just the standard mapmaking problem, restricted to the ring. This gives a map $\hat{T}_j = T(\hat{x}_j) + n_j$, where j labels the phase along the ring and n_j gives the noise contribution. However, in the simplest destriping implementations, we now make a considerable simplifying assumption: we assume that the noise has the very simple model $n_j = \alpha + w_j$, where α is a constant (for each ring) and w_j is an uncorrelated noise contribution, $\langle w_j w_{j'} \rangle = \delta_{jj'}\sigma^2/N_j$, where N_j gives the number of hits in phase-bin j. We then combine the individual maps by a least-squares solution for the α for each ring (in fact, we marginalize over them). More complicated versions of the destriping algorithm allow for more complicated low-frequency noise templates as well as correlated noise at high frequencies.

Although the description here has been in terms of a scanning strategy like that of Planck, in fact it can be applied to any strategy in which discrete areas of sky are observed for times that are long compared with the inverse of the knee frequency (Stompor *et al.* 2002).

Deconvolution mapmaking

So far, we have made a significant approximation in our formalism. In going from the more complicated model of Eq. (10.6) to the simplified Eq. (10.7) that is actually solved by the mapmakers discussed so far, we assume that the underlying signal we are trying to reconstruct is constant across each pixel, so that all measurements within a given pixel are of the same quantity. We further assume that this quantity is equal to the isotropic convolution of the beam function with the signal, represented by $B_\ell a_{\ell m}$ in harmonic space. In practice, neither of these obtain: every measurement is of a slightly different quantity. Formally, we could estimate the underlying signal smoothed with a beam of our choosing at more finely grained pixels, but this is only practical for highly symmetric scanning strategies, as discussed in Armitage and Wandelt (2004) and Armitage-Caplan and Wandelt (2008); this technique has not yet been implemented for a realistic experimental set-up.

Noise marginalization

We have thus far assumed that we know the noise correlation matrix N_{ij}, or its Fourier transform, the noise power spectrum $N(f)$, a priori. In practice, this noise power spectrum is usually estimated using a number of simplifying assumptions. First, we assume that the noise in a single detector is *piecewise stationary*. That is, the noise in different 'stationary chunks' is taken to be completely uncorrelated, and that within such a chunk the noise covariance is taken to be only a function of the interval between samples. Hence, within a chunk, $N_{ij} = N(|t_i - t_j|)$. This further enables us to consider as a more fundamental quantity the noise power spectrum $\tilde{N}(f)$, the Fourier transform of the noise covariance function, $N(t)$, which we further approximate by the discrete Fast Fourier Transform (which does use the unphysical assumption of periodicity). In practice, the noise power spectrum is determined by an iterative process (Ferreira & Jaffe 2000) in which estimates of the signal power, described in the following section, are subtracted from the data timeline, after which the smoothed periodogram of the resulting estimated noise timeline is used as the power spectrum estimate. In most cases, the noise is not marginalized over, as would be most appropriately Bayesian, but just fixed at its maximum after this iteration procedure.

Wiener filters

If instead of a uniform prior on the signal we assume a Gaussian, we still get a Gaussian posterior, now depending upon the data, the noise power spectrum, and the signal prior we have chosen. The obvious choice for a signal prior, expressing our theoretical idea that the signal is isotropic and distributed as a Gaussian, is expressed most simply in spherical harmonics as

$$P(a_{\ell m}|C_\ell) = \frac{1}{\sqrt{2\pi C_\ell}} \exp\left[-\frac{1}{2}\frac{|a_{\ell m}|^2}{C_\ell}\right]; \tag{10.17}$$

that is, as a zero-mean Gaussian with variance $\langle|a_{\ell m}|^2\rangle = C_\ell$. This translates to a zero-mean Gaussian in pixel space, $P(T|C_\ell)$, with correlation matrix

$$C_{Spp'} = \sum_\ell \frac{2\ell + 1}{4\pi} C_\ell P_\ell(\cos\theta_{pp'}) B_\ell^2, \tag{10.18}$$

where $P_\ell(\cos\theta_{pp'})$ are the Legendre polynomials evaluated at the cosine of the angular distance between pixels p and p', and B_ℓ gives the spherical transform of the beam (a more complicated expression can be written for an asymmetric beam). We combine this prior with the likelihood of Eq. (10.11) by completing the square to find

$$P(\mathbf{T}|\mathbf{d}, \mathbf{N}, C_\ell) = \frac{1}{|2\pi\mathbf{C}_W|^{1/2}} \exp\left[-\frac{1}{2}(\mathbf{T} - \mathbf{W}\hat{\mathbf{T}})^{\mathrm{T}}\mathbf{C}_W^{-1}(\mathbf{T} - \mathbf{W}\hat{\mathbf{T}})\right], \tag{10.19}$$

where the posterior maximum is at

$$\mathbf{W}\hat{\mathbf{T}} = \mathbf{C}_S(\mathbf{C}_S + \mathbf{C}_N)^{-1}\hat{\mathbf{T}} \tag{10.20}$$

with covariance

$$\mathbf{C}_W = \mathbf{C}_S(\mathbf{C}_S + \mathbf{C}_N)^{-1}\mathbf{C}_N. \tag{10.21}$$

This is the *Wiener filter*, and this calculation shows exactly when it is applicable in a Bayesian setting: if both the signal and noise contributions are assigned Gaussian distributions. Hence, we do not expect it to recover the signal correctly when the Gaussian is inappropriate (as for foreground astrophysical emission), and we understand that the usual Wiener 'oversmoothing' occurs due to the relaxation of the Gaussian posterior to the zero-mean prior in the absence of data.

10.2.3 From maps to power spectra

We have just seen the effect of assuming a Gaussian prior on the posterior distribution of the sky signal T_p, resulting in the Wiener filter estimate for the map. Instead, we can marginalize over the signal and determine the likelihood for parameters that enter into the Gaussian prior for the signal: the power spectrum values, C_ℓ. The easiest way to see this is

$$P(C_\ell|\hat{\mathbf{T}}) = \int dT \, P(C_\ell, \mathbf{T}|\hat{\mathbf{T}}) \propto \int d\mathbf{T} \, P(\mathbf{T}|C_\ell)P(C_\ell)P(\hat{\mathbf{T}}|\mathbf{T}, C_\ell), \quad (10.22)$$

where in the proportionality we use Bayes' theorem and write the joint prior as $P(\mathbf{T}, C_\ell) = P(C_\ell)P(\mathbf{T}|C_\ell)$. Hence, the likelihood is just

$$P(\hat{\mathbf{T}}|C_\ell) = \int d\mathbf{T} P(\mathbf{T}|C_\ell)P(\hat{\mathbf{T}}|\mathbf{T}, C_\ell) \quad (10.23)$$

$$\propto \frac{1}{|2\pi(\mathbf{C}_N + \mathbf{C}_S)|^{1/2}} \exp\left[-\frac{1}{2}\hat{\mathbf{T}}^{\mathrm{T}}(\mathbf{C}_N + \mathbf{C}_S)^{-1}\hat{\mathbf{T}}\right],$$

where we can do the integrals after substituting in the distributions from the previous section. That is, the map $\hat{T}_p$ can be thought of as a sum of signal and noise contributions, $\hat{T}_p = T_p + n_p$, distributed as a multivariate Gaussian with covariance given by the sum of signal ($\mathbf{C}_S$) and noise ($\mathbf{C}_N$) covariances, which is exactly what we would have written down to begin with, of course, but reminds us of the Bayesian formalism's self-consistency.

Note that in the simplest case of an all-sky experiment with uniform white noise of pixel variance σ^2, power spectrum estimation is trivial. In this case, we can perform all calculations in the spherical-harmonic basis, transforming from $\hat{T}_p$ to $\hat{a}_{\ell m} = a_{\ell m} + n_{\ell m}$, where now the noise contribution is uncorrelated with $\langle n_{\ell m} n_{\ell' m'} \rangle = \sigma^2 \delta_{\ell\ell'} \delta_{mm'}$ so

$$P(\hat{a}_{\ell m}|C_\ell) = \prod_\ell \frac{1}{|2\pi(C_\ell + \sigma^2)|^{1/2}} \exp\left[-\frac{1}{2}\frac{|\hat{a}_{\ell m}|^2}{C_\ell + \sigma^2}\right]. \quad (10.24)$$

In this case, the properties of the distribution as a function of C_ℓ are straightforward; it is maximized at the expected

$$\hat{C}_\ell = \overline{|\hat{a}_{\ell m}|^2} - \sigma^2, \qquad \text{where} \qquad \overline{|\hat{a}_{\ell m}|^2} = \frac{1}{2\ell + 1} \sum_{m=-\ell}^{\ell} |\hat{a}_{\ell m}|^2 \qquad (10.25)$$

gives the average 'observed' spectrum at each ℓ.

Whereas the likelihood function for the map, Eq. (10.11), is a Gaussian distribution with an analytic form for the mean and variance, the shape of Eq. (10.23) as a function of the C_ℓ can be quite complicated. In general it does not depend on a small number of sufficient statistics: the likelihood function should be calculated in detail. Luckily, a number of approximations have been developed to understand and explore the structure of the likelihood function.

Of course, we can simply evaluate the likelihood directly as a function of C_ℓ (as in, e.g., the COBE/DMR analysis of Gorski *et al.* 1996), but from the expressions above, each evaluation takes $O(N_p^3)$ operations (the determinant is actually the most difficult to speed up in this case), and this must be multiplied by the number of evaluations.

As a first step, we can try to find the maximum value of the likelihood and some measurement of the width of the function around the peak. This is usually done by something like Newton–Raphson iteration to find the maximum, accompanied by the evaluation of the second derivative (Hessian) matrix as a proxy for the width of the peak. [Sometimes, the Fisher matrix, which is the ensemble average Hessian for a fixed C_ℓ, is used as well, but it is not considerably less expensive to evaluate (Bond, Jaffe & Knox 1998; Tegmark 1997)]. Unfortunately, each step in the iteration naively scales as $O(N_p^3)$, with considerably difficulty of significant speed-up, except in highly symmetric cases.

As an alternative to direct manipulation of the likelihood function, we can instead find a method to draw samples from it. The most obvious Metropolis–Hastings sampling methods would themselves require an evaluation of the likelihood function for each sample, making them prohibitively expensive. Instead, *Gibbs samplers* are often used, which go back to the joint distribution $P(\mathbf{T}, C_\ell | \hat{\mathbf{T}})$ and alternately sample from the two conditional distributions $P(\mathbf{T} | C_\ell, \hat{\mathbf{T}})$ and $P(C_\ell | \mathbf{T}, \hat{\mathbf{T}})$. In this case, the former distribution is simply that of our Wiener filter, Eq. (10.19), and the latter is an inverse Gamma distribution (or an inverse Wishart distribution if we consider polarization data, to be discussed below) (Eriksen *et al.* 2004; Larson *et al.* 2007). Alternatively, a Hamiltonian Monte Carlo method can be used to sample directly from the joint distribution $P(\mathbf{T}, C_\ell | \hat{\mathbf{T}})$ (Taylor, Ashdown & Hobson 2008).

As is discussed in Part I of this volume, from such samples we can easily construct any desired mean values of our estimated C_ℓ. Similarly, although they do not characterize the full shape of the likelihood, the maximum likelihood provides a measurement of the spectrum, and the curvature a measurement of the errors about this peak. However, we emphasize that to *use* these numbers for any further calculations (as described in the following section), the Bayesian program requires the full shape of the likelihood function. Until recently, the quality of CMB data was such that various analytical ansatzen were of sufficient quality to model the functional shape (e.g., Bond, Jaffe & Knox 2000; Hamimeche & Lewis 2008). Current (and certainly future) experiments will need more precision. Various proposals have been advanced, ranging from the direct evaluation of the likelihood, to the fitting of analytical forms to Monte Carlo samples, to the use of a Blackwell–Rao estimate of the likelihood function itself (e.g., Chu *et al.* 2005). Unfortunately, all of these methods have problems, requiring considerable resources especially when extended to higher multipoles.

Indeed, at high ℓ all of these methods are computationally prohibitive to even calculate the means and/or maxima, and we are forced to rely on frequentist or hybrid methods, such as those based on the MASTER formalism (Hivon *et al.* 2002). MASTER calculates so-called 'pseudo-$a_{\ell m}$' components using a brute-force spherical apodization, and then calculates the linear relationship between the underlying power spectrum and the ensemble-average 'pseudo-C_ℓ', defined in the naive way as the average of the squared pseudo-$a_{\ell m}$ at each ℓ. FASTER (Myers *et al.* 2003) is a particularly interesting wrinkle on this method, as it uses the MASTER calculation to then perform a hybrid Bayesian calculation, assuming that the conditional likelihoods at different ℓ are independent. In a Bayesian/maximum-entropy sense, this approximation is equivalent to removing information in a controlled way, but not introducing any new assumptions.

For the simple case of an all-sky experiment with uniform white noise between pixels (i.e., white noise in the timestream with the same number of observations per pixel), the frequentist methods give answers equivalent to the Bayesian technique (Eq. 10.25): the frequentist mean gives the maximum likelihood and the variance gives the curvature. We can further reproduce the likelihood shape by using descriptions such as those discussed above (Bond, Jaffe & Knox 2000; Hamimeche & Lewis 2008). However, it must be stressed that for more complicated (i.e. realistic) experiments, there is no reason to expect this Bayesian/frequentist correspondence to hold in detail.

10.2.4 From spectra to cosmological parameters

We now wish to measure the cosmological parameters, such as the Hubble constant H_0, the matter density Ω_{m}, the primordial spectral tilt of scalar perturbations n_s, etc., which we will collectively label θ. For standard cosmological models, the parameters uniquely determine the power spectrum, $C_\ell = C_\ell(\theta)$, so the likelihood function can be written

$$P(\theta|d) = \int \mathrm{d}C_\ell P(\theta, C_\ell|d) \propto \int \mathrm{d}C_\ell P(C_\ell|\theta) P(\theta) P(d|C_\ell) \qquad (10.26)$$

$$= \int \mathrm{d}C_\ell \delta[C_\ell - C_\ell(\theta)] P(\theta) P(d|C_\ell) \qquad (10.27)$$

$$= P(\theta) P[d|C_\ell = C_\ell(\theta)] . \qquad (10.28)$$

Again, we do not need to assume a separate prior on the power spectrum. Unlike the previous step, however, we see that the likelihood function does not have a simple form as a function of C_ℓ, and cannot trivially be reduced to a function of a small number of sufficient statistics, such as the maximum and curvature of the C_ℓ likelihood. Hence, in order to measure the cosmological parameters we need to do one of the following. First, we can calculate the likelihood function directly, but as we have noted this scales as $O(N_p^3)$ and so is prohibitively expensive in many cases. Second, we can try to model the function with some small number of parameters in addition to the aforementioned maximum and curvature. Finally, we can instead draw samples from this distribution which can be used in a variety of ways. This latter method, using Markov chain Monte Carlo and related techniques such as nested sampling, is described in Part I of this volume.

10.3 Polarization

Until this point, we have assumed that our experiment measures only the temperature (equivalently, the intensity) of the CMB radiation. For much of the last decade, however, attention has been concentrated upon the measurement of the polarization of the CMB, and the majority of recent experiments are polarimeters as well as thermometers. Polarization is produced as the Universe transitions from being ionized to neutral, at which point individual electrons will emit polarized radiation aligned with any surrounding quadrupole radiation field. This polarized CMB will probe the scalar mode of the metric – the gravitational potential – which dominates the temperature, and also the tensor mode, corresponding to gravitational waves. These tensor modes are thought to be produced by (and possibly only by) metric perturbations from an early epoch of inflation. Moreover, they can in principle be separated from the scalar contribution due to their different geometrical properties. Polarization is represented by a headless two-dimensional vector field on the spherical sky. The simplest representation of polarization on the sky is in terms of its components (the pixelized values of the Stokes parameters Q_p and U_p), but these depend on the coordinate system used.

For such experiments, the data become somewhat more complicated:

$$d_i = \sum_p A_{ip} \left[T_p + Q_p \cos(2\alpha_i) + U_p \sin(2\alpha_i) \right] + n_i, \tag{10.29}$$

where now α_i gives the polarization angle of the detector, and now we wish to reconstruct not just the temperature Y_i but the Q_i and U_i components as well (referenced to the direction of the North Galactic Pole as in the WMAP analysis of Page et $al.$ (2007) and elsewhere). Formally, this equation still has the same linear form, $d = BS + n$, where now $S = (T, Q, U)$, and the operator B is considerably more complicated, containing the angle-dependent terms from Eq. (10.29). Note that data from a single time sample is used to reconstruct all of the three components of S seen at that time. This has several repercussions. First, a larger number of observations of a given pixel are required to reliably reconstruct each component. Second, this inevitably induces noise correlations between the T, Q and U components at the pixel which must be taken into account.

The Q and U components represent a headless vector field on the sphere, which itself can be decomposed into so-called 'E' and 'B' components (as they are denoted in the CMB community). By analogy with electromagnetism, the E component is curl-free, and the B component divergence-free (for suitable spherical definitions of curl and divergence). Specifically, a polarization field can be written by the extension of Eq. (10.2) to a polarization field as, in the notation of Zaldarriaga and Seljak (1997),

$$Q_p \pm iU_p = \sum_{\ell m} {}_{\pm 2}a_{\ell m} {}_{\pm 2}Y_{\ell m}(\hat{x}_p), \tag{10.30}$$

where ${}_{\pm 2}Y_{\ell m}(\hat{x}_p)$ are the spin-2 spherical harmonics (related by derivatives to the usual $Y_{\ell m}$). We can then define the scalar (pseudo-scalar) field corresponding to E (B) as

$$a_{\ell m}^E = -({}_{+2}a_{\ell m} + {}_{-2}a_{\ell m}) \qquad \text{and} \qquad a_{\ell m}^B = i({}_{+2}a_{\ell m} - {}_{-2}a_{\ell m}) \tag{10.31}$$

and then we can finally define the power spectra in analogy to Eq. (10.3):

$$\langle a_{\ell m}^A a_{\ell' m'}^{A'} \rangle = C_\ell^{AA'} \delta_{\ell\ell'} \delta_{mm'} , \tag{10.32}$$

where $(A, A') \in (T, E, B)$ (in most theories $C_\ell^{EB} = C_\ell^{TB} \equiv 0$). Equivalently, the Q and U maps themselves have correlation functions that are linear combinations of these spectra. (Kamionkowski, Kosowsky & Stebbins 1997; Zaldarriaga & Seljak 1997; Tegmark & de Oliveira-Costa 2001).

Power spectra and E/B separation

For a full-sky noise-free experiment this could in principle be inverted to recover maps of the polarization components corresponding to the E and B fields, but boundary effects in a realistic experiment make this impossible to do unambiguously. In fact, the fully Bayesian technique does not involve any E/B reconstruction or separation, but just starts from the Q and U maps and calculates the distribution of power spectra given these maps by techniques completely equivalent to those given in 10.2.3. This is straightforward but computationally expensive, and of course E and B maps are useful diagnostic tools which are not natural products of this technique (although the equivalent of a Wiener filter can also be applied to effectively separate the components).

Recently (Smith 2006; Smith & Zaldarriaga 2007) there has been an effort to directly estimate the scalar and pseudo-scalar components on the sky (rather than in harmonic space). This is a somewhat more local operation than the reconstruction of the E and B polarization fields, but at the expense of requiring a numerical derivative (always dangerous with noisy data) or significant apodization in order to use harmonic techniques.

10.4 Complications

10.4.1 Foregrounds and systematics

So far, we have assumed an idealized CMB experiment uncontaminated by astrophysical foregrounds or instrumental effects (beyond the instrumental noise whose behaviour is fully specified by the N_{ij} correlation matrix). Instrumental effects are often correlated with other pieces of 'housekeeping' data, such as the temperature of the telescope, or the rotation angle of the polarizing hardware. These instrumental effects can often be added into our mapmaking procedure as templates in a similar manner to the offsets in the destriping methods discussed above (e.g., Stompor *et al.* 2002; Johnson *et al.* 2007).

Astrophysical foregrounds are fixed on the sky, but correlated between frequency channels. There is rarely enough information to unambiguously reconstruct the foregrounds and different methods have been proposed to take into account different kinds and amounts of prior information. In some cases, template methods can be used for a fully Bayesian treatment, but other 'non-linear' methods have been developed that fit less easily into the Bayesian paradigm, which requires knowledge of the full posterior distribution of the foreground-cleaned map, and in practice requires that this distribution be a multivariate Gaussian with a simple noise correlation structure (Hobson, Ashdown & Stolyarov 2009).

10.4.2 Non-Gaussianity

However, there is one major complication that has not yet proven amenable to Bayesian analysis: non-Gaussianity. In practice, we do expect to see small departures from Gaussianity, parameterized by non-zero higher correlation functions, the multivariate equivalent of the skewness, kurtosis, etc. A Gaussian distribution is the unambiguous distribution to assign when only the two-point function is known, but no such assignment exists for the more general case. (Indeed, the Gaussian distribution arises from numerous lines of argument: it is the only distribution with zero higher-order correlations; it is the maximum-entropy distribution for a known two-point function; it arises naturally via the central limit theorem in many realistic situations.) Non-Gaussian distributions must be assigned in cases based on specific physical or mathematical models. But such models do not in general exist; instead only specific moments of the required distribution can be calculated.

All of this means that, although some non-Gaussian distributions have been proposed (e.g., Contaldi *et al.* 2000; Rocha *et al.* 2001), they have not been applied to CMB data in practice; frequentist techniques have dominated the analysis so far.

10.5 Conclusions

We have seen that the problem of CMB data analysis is an ideal case study for Bayesian techniques in cosmology. For the standard model of small perturbations and Gaussian initial conditions, we can calculate in sequence the map from a set of timestreams, the power spectrum from the map, and the cosmological parameters from the spectra. At least in some cases, these techniques can cope with contamination from instrumental and astrophysical sources. Some of these steps are computationally expensive, and a fully Bayesian solution has not been implemented for datasets as large as that expected from the Planck satellite.

References

Armitage, C. and Wandelt, B. (2004). *Phys. Rev. D*, **70**, 123007.
Armitage-Caplan, C. and Wandelt, B. (2008). arXiv:0807.4179.
Ashdown, M. *et al.* (2007a). *Astron. Astrophys.*, **467**, 761.
Ashdown, M. *et al.* (2007b). *Astron. Astrophys.*, **471**, 361.
Ashdown, M. *et al.* (2009). *Astron. Astrophys.*, **493**, 753.
Bond, J. R., Crittenden, R., Jaffe, A. H. and Knox, L. E. (1999). *Comput. Sci. Eng.*, **1**:2, 21.
Bond, J. R., Jaffe, A. H. and Knox, L. E. (1998). *Phys. Rev. D*, **57**, 2117.
Bond, J. R., Jaffe, A. H. and Knox, L. E. (2000). *Astrophys. J.*, **533**, 19.
Borrill, J. (1999). arXiv:astro-ph/9911389.
Chu, M. *et al.* (2005). *Phys. Rev. D*, **71**, 103002.
Contaldi, C. R., Ferreira, P. G., Magueijo, J. and Gorski, K. M. (2000). *Astrophys. J.*, **534**, 25.
Delabrouille, J., Gorski, K. M. and Hivon, E. (1998). *Mon. Not. Roy. Astron. Soc.*, **298**, 445.
Eriksen, E. K. *et al.* (2004). *Astrophys. J. Supp.*, **155**, 227.

Ferreira, P. G. and Jaffe, A. H. (2000). *Mon. Not. Roy. Astron. Soc.*, **312**, 89.

Fixsen, D. *et al.* (1996). *Astrophys. J.*, **473**, 576.

Gorski, K. M. *et al.* (1996). *Astrophys. J. Lett.*, **464**, L11.

Guth, A. H. (1981). *Phys. Rev. D*, **23**, 347.

Hamimeche, S. and Lewis, A. (2008). *Phys. Rev. D*, **77**, 103013.

Hivon, E. *et al.* (2002). *Astrophys. J.*, **567**, 2.

Hobson, M. P., Ashdown, M. A. J. and Stolyarov, V. (2010). Chapter 6 in this volume.

Johnson, B. *et al.* (2007). *Astrophys. J.*, **665**, 42.

Kamionkowski, M., Kosowsky, A. and Stebbins, A. (1997). *Phys. Rev. D*, **55**, 7368.

Larson, D. *et al.* (2007). *Astrophys. J.*, **656**, 653.

Myers, S. T. *et al.* (2003). *Astrophys. J.*, **591**, 575.

Page, L. *et al.* (2007). *Astrophys. J. Supp.*, **170**, 335.

Planck Consortium (2005). *Planck: The Scientific Programme*. ESA-SCI(2005)1.

Rocha, G. *et al.* (2001). *Phys. Rev. D*, **64**, 63512.

Smith, K. (2006). *New Astron. Rev.*, **50**, 1025.

Smith, K. and Zaldarriaga, M. (2007). *Phys. Rev. D*, **76**, 43001.

Starobinsky, A. (1979). *JETP Lett.*, **30**, 682.

Stompor, R. *et al.* (2002). *Phys. Rev. D*, **65**, 022003.

Stompor, R. and White, M. (2004). *Astron. Astrophys.*, **419**, 783.

Taylor, J. F., Ashdown, M. A. J. and Hobson, M. P. (2008). *Mon. Not. Roy. Astron. Soc.*, **389**, 1284.

Tegmark, M. (1997). *Phys. Rev. D*, **55**, 5895.

Tegmark, M. and de Oliveira-Costa, A. (2001). *Phys. Rev. D*, **64**, 063001.

Zaldarriaga, M. and Seljak, U. (1997). *Phys. Rev. D*, **55**, 1830.

11

Bayesian multilevel modelling of cosmological populations

Thomas J. Loredo and Martin A. Hendry

11.1 Introduction

Surveying the Universe is the ultimate remote sensing problem. Inferring the intrinsic properties of the galaxy population, via analysis of survey-generated catalogues, is a major challenge for twenty-first century cosmology, but this challenge must be met without any prospect of measuring these properties *in situ*. Thus, for example, our knowledge of the intrinsic luminosity and spatial distribution of galaxies is filtered by imperfect distance information and by observational selection effects, issues which have come to be known generically in the literature as 'Malmquist bias'.[1] Figure 11.1 shows schematically how such effects may distort our inferences about the underlying population since, in general, these must be derived from a noisy, sparse and truncated sample of galaxies.

There is a long (and mostly honourable!) tradition in the astronomical literature of attempts to cast such remote surveying problems within a rigorous statistical framework. Indeed, it is interesting to note that seminal examples from the early twentieth century (Eddington 1913, 1940; Malmquist 1920, 1922) display, at least with hindsight, hints of a Bayesian formulation long before the recent renaissance of Bayesian methods in astronomy. Unfortunately, space does not permit us to review in detail that early literature, nor many of the more recent papers which evolved from it. A more thorough discussion of the literature on statistical analysis of survey data can be found in, e.g., Hendry and Simmons (1995), Strauss and Willick (1995), Teerikorpi (1997) and Loredo (2007).

---

[1] Although 'Malmquist' is the most prevalent appellation, the literature also uses other terms – including 'Eddington–Malmquist' and 'Lutz–Kelker' – to denote biases arising in astronomical surveys from distance indicator scatter and observational selection. There is also an unfortunate history in the cosmology literature of the same term being used to mean substantially different things by different authors. For a more detailed account of the meaning, use and abuse of bias terminology see, e.g., Hendry and Simmons (1995), Strauss and Willick (1995) and Teerikorpi (1997).

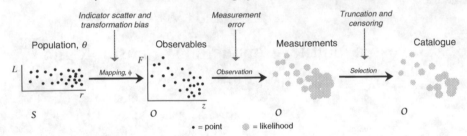

Fig. 11.1. Schematic depiction of the survey process. A population with a distribution of source parameters, S (e.g., luminosity and distance), implies, via a mapping ϕ, a distribution of observables, $\mathcal{O}$ (e.g., flux and redshift). Observation introduces measurement error; selection criteria truncate or censor the catalogued population.

The analysis of survey catalogues can have a number of scientific goals. For example, the objective may be to compare the underlying galaxy luminosity distribution with the predictions of different galaxy formation models. In this case the galaxy distances (which we must infer as an intermediate step towards estimating their luminosities) are, in Bayesian parlance, nuisance parameters. On the other hand, the goal may be to infer the distances of the surveyed galaxies, in order to test models of galaxy clustering and/or constrain parameters of the underlying cosmology. In this case it is the inferred galaxy luminosities which may be thought of as nuisance parameters. The related inference problem for galaxy fluxes of individual objects is discussed by Mortlock in Chapter 8.

In the literature of the past 20 years the second case, that of inferring galaxy distances, has proven to be fertile territory for the development and application of Bayesian methods. This is particularly true with regard to redshift-independent distance indicators, i.e., indicators whose behaviour is independent of the underlying cosmological model (and which may thus be straightforwardly used to constrain parameters of that model). A likely reason for this is that redshift-independent distance indicators suffer from large intrinsic scatter, with distance uncertainties to individual galaxies typically in the range 5% to 30%. A consequence is that care must be taken when incorporating prior distance information, since in this setting final inferences can be significantly influenced by the prior, a point we discuss further in Section 11.2.

Another hallmark of the literature on extracting information from galaxy surveys has been the recognition by some authors (e.g., Hendry and Simmons 1995; Loredo 2007) that the task of identifying 'optimal' (e.g., in the sense of unbiased and/or minimum mean square error) estimators of galaxy distance and luminosity will generally not have a unique solution, but will depend on the context in which the inferred galaxy distances or luminosities are to be used.

In Section 11.3 we discuss how Bayesian multilevel models provide a natural and powerful framework in which to formulate and implement optimal analyses of surveys, incorporating prior information and carefully accounting for selection effects and source uncertainties. To motivate this approach, and to establish context, in Section 11.2 we survey key Bayesian elements of current 'state of the art' methodology for analysis of survey data, focussing on use of redshift-independent distance indicators, and highlighting examples which have previously hinted at a multilevel approach.[2]

11.2 Galaxy distance indicators

While future astrometric space missions (e.g., GAIA and SIM) offer some prospect of applying trigonometric methods over cosmological scales, in large part the measurement of galaxy distances relies on more astrophysical, and therefore less precise, methods. Perhaps the most obvious of these methods is the cosmological redshift. To derive a distance measure from it requires the use of a cosmological model. But if one's goal is to use galaxy distance estimates to probe the parameters of the cosmological model, the safest path is to identify galaxy distance indicators whose properties are independent of redshift.

Almost all redshift-independent 'distance' indicators are really luminosity or size indicators: one uses the indicator to estimate the intrinsic luminosity or size of a galaxy (or some object therein); combined with a measurement of its apparent brightness or apparent size, one may estimate distance via the inverse-square law or the angle–distance relation (or their cosmological generalizations). For simplicity, we consider here only the case of luminosity indicators, although very similar statistical considerations apply to other types of indicator. We can estimate the luminosity either by assuming a constant, fiducial value (the so-called 'standard candle' assumption) or, better, by exploiting correlations between luminosity and some other intrinsic, but directly measurable, physical characteristic(s) of the source. (The latter approach is often referred to as a 'standardizable candle'.) Examples of these correlations include: the Tully–Fisher relation for spiral galaxies; the period–luminosity relation for Cepheid variables and the luminosity–light curve shape relation for type-Ia supernovae. The correlations may be motivated by theory, but they must be calibrated empirically, e.g., using nearby galaxies at known distances (and hence of known luminosity). Their scatter, reflecting the intrinsic spread in luminosity of the sources, renders distance indicators susceptible

---

[2] The state of the art here is a formalism developed more than a decade ago, suggesting that this has been a neglected research field for some years. In large part this has been due to limitations on the quality and quantity of indicator-based survey data available; this is a situation which is improving dramatically, with the latest galaxy surveys such as 6dF and SDSS, and should continue to do so.

to observational selection biases since one cannot observe arbitrarily faint sources to arbitrary distances.

Since the late 1980s, several authors have investigated the statistical properties of redshift-independent galaxy distance indicators, with the goal of placing their use in cosmology on a more rigorous statistical footing. A significant step forward in this regard came with the work of Willick (1994), which brought much needed clarity to the discussion by explicitly making a distinction between the tasks of calibrating a distance indicator and applying it to a galaxy survey to infer distance information. We now briefly summarize the formalism presented in Willick (1994) and adopted in subsequent papers.

11.2.1 The calibration problem

Our starting point is the joint probability distribution for a single galaxy's distance, r, apparent magnitude, m, and some third observable correlated with luminosity which, following Willick, we denote by η and refer to as the 'line width' parameter. As a concrete example, consider the Tully–Fisher relation, for which we expect the intrinsic relation between absolute magnitude and η to be linear, i.e., $M = a\eta + b$, where the coefficients a and b must be calibrated empirically. As noted earlier, an analysis may have various goals: a and b may simply be nuisance parameters, necessary to estimate galaxy distances; alternatively, they may be important target parameters in their own right.

Thus, for the joint distribution describing the properties of galaxies within a particular survey catalogue, we have

$$p(r, m, \eta) \propto r^2 n(r) \, S(m, \eta) \, \psi(m|\eta) \, \phi(\eta), \qquad (11.1)$$

where $n(r)$ and $\phi(\eta)$ denote the marginal distribution of distance and line width respectively for the galaxy population, $\psi(m|\eta)$ denotes the conditional distribution of apparent magnitude at a given line width (which depends on unknown parameters, e.g., a and b), and $S(m, \eta)$ denotes the observational selection effects (assumed here, for simplicity, not to depend on distance or direction).

Consider first the calibration of the distance indicator, which might reasonably be carried out, e.g., using a galaxy cluster, so the set of calibrators are all at the same distance.[3] In this case it is natural to work with the conditional distribution of m at given η and r (i.e., the result from using Eq. (11.1) as a prior in Bayes' theorem, with a 'likelihood' corresponding to precise measurement of η and r). Then the indicator coefficients a and b can be interpreted as the slope and zero-point of a linear regression of absolute magnitude on η; this case is known as the

---

[3] Willick also considers the calibrators at a range of true distances; this case lends itself well to a Bayesian multilevel formulation, as we discuss in Section 11.3.

'direct' indicator relation. Thus

$$P(m|\eta, r) = \frac{S(m, \eta)\, \psi(m|\eta)}{\int_{-\infty}^{\infty} S(m, \eta)\, \psi(m|\eta)\mathrm{d}m}. \tag{11.2}$$

Notice that, as expected, the marginal distributions of distance and line width drop out. The presence of the observational selection effects will bias the determination of the indicator coefficients a and b obtained via simple linear regression; however, Willick (1994) proposed an iterative scheme to overcome this problem and showed that it works well for realistic mock galaxy data.

A popular variant on the above approach is to use the so-called 'inverse' relation, i.e., (in Willick's notation) $\eta^0(M) = a'M + b'$, where the inverse coefficients a' and b' again must be determined empirically (and again may be regarded either as nuisance parameters or target parameters). This relation is most directly expressed by the conditional distribution of η at a given r and m (corresponding to given M, since we are assuming all the calibrators lie at the same distance), namely

$$p(\eta|m, r) = \frac{S(m, \eta)\, \Psi(\eta|m)}{\int_{-\infty}^{\infty} S(m, \eta)\, \psi(\eta|m)\mathrm{d}\eta}, \tag{11.3}$$

where Ψ denotes the conditional distribution of line width at given apparent magnitude. In this case, explicit dependence on the galaxy luminosity function drops out of our expression. Moreover, one can see that if the selection effects depend only on apparent magnitude and not on line width, then a straightforward linear regression of η on M will yield unbiased estimates of the indicator coefficients a' and b'. This appealing property of the inverse indicator relation had been recognized in principle much earlier by Schechter (1980) and was also placed on a rigorous statistical footing around the same time as Willick by Hendry and Simmons (1994). However, the successful calibration of a galaxy distance indicator is only the first part of the story.

11.2.2 The estimation problem

Suppose one has used the relations above to calibrate a distance indicator accurately and precisely (e.g., a and b are now precisely known). Now we seek to use the indicator in settings where there is no direct measurement of r; we must infer r from measurements of m and η. We can calculate a predicted galaxy distance, d, in the obvious way by combining the observed apparent magnitude of the galaxy with its estimated absolute magnitude inferred (via our indicator relation) from its observed line width. Moreover, since $d = d(m, \eta)$, it is straightforward to compute the joint distribution, $p(r, d)$, of true and estimated galaxy distance, and further to determine the conditional distribution of r given d. For the direct indicator, we obtain

$$p(r|d) = \frac{r^2 n(r) \, \exp\left(-\frac{[\ln r/d]^2}{2\Delta^2}\right)}{\int_0^\infty r^2 n(r) \, \exp\left(-\frac{[\ln r/d]^2}{2\Delta^2}\right) \, dr}, \tag{11.4}$$

where Δ is a constant, proportional to the (here assumed Gaussian) scatter in the direct indicator relation, i.e., the dispersion of the conditional distribution of absolute magnitude at given line width. For the inverse indicator, on the other hand, we obtain

$$p(r|d) = \frac{r^2 n(r) s(r) \, \exp\left(-\frac{[\ln r/d]^2}{2\Delta^2}\right)}{\int_0^\infty r^2 n(r) s(r) \, \exp\left(-\frac{[\ln r/d]^2}{2\Delta^2}\right) \, dr}, \tag{11.5}$$

where $s(r)$ is an integral over the galaxy luminosity function weighted by the selection effects, and expresses the probability that a galaxy at true distance r would be observable in the survey. This term is often referred to as the *selection function* for r.

The interpretation of Eqs. (11.4) and (11.5) within the framework of Bayesian inference is clear. We can think of $p(r|d)$ as representing the posterior distribution of true distance r, given some observed data d (i.e., the estimated distance, from our indicator). Moreover the *difference* between the two expressions can then be interpreted in terms of the adoption of different prior information for r: for the direct indicator, the prior information is the true distance distribution $n(r)$, while for the inverse relation the prior is the product of $n(r)$ and the selection function.

The classical Malmquist bias is manifest when we take the conditional expectation of r given d, using Eqs. (11.4) and (11.5). For both direct and inverse indicators we find that in general $E(r|d) \neq d$. However, we can correct our 'raw' distance indicator d, defining d_{corr} which satisfies

$$E(r|d_{\mathrm{corr}}) = d_{\mathrm{corr}}, \tag{11.6}$$

with the correction term referred to as a 'Malmquist correction'. Note, however, that the Malmquist correction depends explicitly on the true distance distribution $n(r)$, which in general will be unknown. Lynden-Bell *et al.* (1988) computed homogeneous Malmquist corrections, assuming that the underlying spatial distribution of galaxies is uniform, in which case

$$E(r|d) = d \, \exp\left(\frac{7}{2}\Delta^2\right) \simeq d \left(1 + \frac{7}{2}\Delta^2\right) \equiv d_{\mathrm{corr}}. \tag{11.7}$$

We should not be surprised that the correction is always positive in this case; since we are assuming homogeneity we are saying that the distance indicator scatter is more likely to scatter galaxies *downwards* from greater true distances, simply because there are more galaxies at larger r due to the rapid growth of the volume element with r. Note, however, that for the inverse indicator the assumption of a uniform prior is not appropriate: even if $n(r)$ were constant, the selection function $s(r)$ clearly will not be.

The more realistic case is, of course, where the intrinsic distribution of distance is *not* uniform. In this case the adoption of a suitable prior for $n(r)$ leads to a so-called *inhomogeneous Malmquist correction*. For the direct indicator, the source of the prior information could be, for example, the underlying density field of galaxies reconstructed from an external source, e.g., an all-sky redshift survey (Hudson 1994; Strauss & Willick 1995; Freudling *et al.* 1995; Erdogdu *et al.* 2006). The Malmquist corrections will only be valid in this case, however, provided that the external galaxy survey traces the same underlying population as the galaxies to which the distance indicator is being applied.

In an important paper, Landy and Szalay (1992) proposed an interesting alternative approach, whereby the marginal distribution of raw distances might provide a suitable estimate of the prior true distance distribution. Crucially, this method should *not* be applied using the direct indicator since the marginal distribution of raw distances provides a poor estimate of $n(r)$. On the other hand, it *does* provide a reasonable proxy for the distribution of true distances for 'observable' galaxies; i.e., the product of $n(r)$ and $s(r)$. Thus, it is probably well suited to use with the inverse indicator.

Landy and Szalay's approach, although not rigorously derived, has several attractive features. It offers a method of defining inhomogeneous Malmquist corrections that adapts to spatial inhomogeneity, without requiring external assumptions or prior information about $n(r)$ from other galaxy surveys. Indeed, it appears to be an approximation to a hierarchical Bayesian procedure, as we discuss further in Section 11.3.

11.2.3 Applications of galaxy distance indicators

Why might we regard Malmquist-corrected distance indicators, which satisfy Eq. (11.6), as optimal estimators in the first place? The answer lies largely in the uses to which they have been put. Since the late 1980s, redshift-independent distance indicators have often been used to measure galaxy *peculiar velocities*, namely the motions, over and above the Hubble expansion, induced by the net gravitational attraction of the matter distribution around them. Methods of analyzing peculiar velocities generally involve first binning and grouping galaxies together based on their *estimated* distance. By requiring that on average the true distance of each galaxy be equal to its estimated distance, one aims to ensure that on average the correct radial peculiar velocity will be ascribed to each galaxy's apparent position.

In the 1990s, astronomers developed a number of sophisticated methods to compare observed and predicted galaxy peculiar velocities, the latter the result of reconstructing the density and peculiar velocity field from position and redshift data from an all-sky redshift survey. This reconstruction requires a model for *galaxy biasing*; i.e., a description of how the distribution of luminous galaxies and dark matter are related. By comparing observed and predicted peculiar velocities one can constrain parameters of the galaxy biasing model.

From a Bayesian perspective, probably the most notable of these comparison methods was VELMOD (Willick & Strauss 1998). This assumed a simple linear relation between the galaxy and matter density fields and computed a posterior for the linear bias parameter, marginalized over the nuisance parameters of the distance indicator relation. In its explicit modelling of the galaxy distance uncertainties, en route to estimating the linear bias parameter, VELMOD shares features with the multilevel Bayesian model approach which we now describe.

11.3 Multilevel models

The issues motivating the astronomical developments just surveyed are hardly unique to astronomy. Statisticians have addressed similar issues in applications spanning many disciplines. Although none of the resulting methods is an 'exact fit' to an astronomical survey problem, the body of literature offers numerous insights that should inspire significant advances in Bayesian methodology for astronomical surveys.

A recurring theme of much of the relevant literature is the use of *multilevel models* (MLMs), a relatively recent term for a rich framework that underlies several important statistical innovations of the latter twentieth century, including empirical and hierarchical Bayes methods, random effects and latent variable models, shrinkage estimation, and ridge regression. MLMs start with a *first-level* probability model for the measurements of parameters for each of many objects (e.g., sources in a survey). The *second level* assigns a shared prior distribution to the first-level parameters (e.g., a population-level distribution for source properties); this distribution may itself have unknown parameters, dubbed *hyperparameters*. The second level leads to probabilistic dependence among the first-level parameters that implements a pooling of information that can improve the accuracy of inferences; one says the estimates 'borrow strength' from each other. Other levels may be added, e.g., to describe relationships between groups of objects.

We here focus on Bayesian treatment of MLMs, though multilevel modelling is an area where there has been significant cross-fertilization between Bayesian and frequentist approaches. We begin by describing a very simple MLM – the *normal–normal* MLM – highlighting a feature of MLM point estimates, shrinkage, that has connections to classic astronomical approaches for correcting for survey biases. We use this as a stepping stone to a more thoroughgoing Bayesian approach that moves beyond point estimates and corrections. To date, this approach has been implemented only in fairly simple astronomical settings; we end by highlighting directions for future research.

11.3.1 Adjusting source estimates: shrinkage

Suppose we have survey data for a population of sources that we will model as having a log-normal luminosity function, so the population distribution of source absolute magnitudes, M, is a normal distribution with location M_0 and scale (standard deviation) τ – these are the hyperparameters. As a simple starting point, we will suppose τ is known ($\tau = 0.5$ mag, say), and we denote the population distribution by $f(M|M_0)$. For now, we also assume there are no selection effects. At the population level, our goal is to infer M_0.

The survey produces data, D_i, for each source; we will suppose these lead to independent, Gaussian-shaped, likelihood functions for each source's unknown true absolute magnitude M_i, with maximum likelihood estimates (MLEs) $\hat{M}_i$ and uncertainties (standard deviations) σ_i. We denote these source likelihoods as $\ell_i(M_i) \equiv p(D_i|M_i) = N(M_i|\hat{M}_i, \sigma_i^2)$, with $N(\cdot|\mu, \sigma^2)$ the normal distribution with mean μ and variance σ^2. The M_i are the first-level parameters. The survey catalogue consists of a table of the $\hat{M}_i$ estimates and their uncertainties. For simplicity, we assume equal uncertainties, $\sigma_i = \sigma = 0.3$ mag.

Let $\mathbf{D} \equiv \{D_i\}$ and $\mathbf{M} \equiv \{M_i\}$ denote the collections of data and source parameters. The likelihood function is $\mathcal{L}(M_0, \mathbf{M}) \equiv p(\mathbf{D}|M_0, \mathbf{M}) = p(\mathbf{D}|\mathbf{M}) = \prod_i \ell_i(M_i)$; that is, it does not depend on M_0 because the probabilities for the source data, D_i, are fully determined (and independent) if $\mathbf{M}$ is specified. Thus the joint MLEs for the source parameters are just the independent MLEs: $\hat{\mathbf{M}} = \{\hat{M}_i\}$. But in a Bayesian calculation, estimates are determined by the posterior, not the likelihood. If M_0 is known, the joint posterior for the source parameters, conditional on M_0, is given by

$$\pi(\mathbf{M}|\mathbf{D}, M_0) \propto \prod_i f(M_i)\ell_i(M_i), \tag{11.8}$$

with the population density appearing as a prior factor for each source. Point estimates may be found from this conditional posterior, e.g., by finding the mode or posterior mean for $\mathbf{M}$. It is important to note that if we increase the amount of survey data by increasing N, such Bayesian estimates will *not* converge to the MLEs, because additional prior factors enter with each new source. That is, we are not in the common, simpler setting of a fixed number of parameters, with additional data providing likelihood factors that eventually overwhelm a single prior factor. The presence of source uncertainties implies that each new source *adds a new parameter*, so differences between Bayesian and likelihood estimates persist. This is evident in the Malmquist corrections described above. The only way for these Bayesian estimates to converge to MLEs is to add follow-up data for each source, i.e., to make all of the $\ell_i(M_i)$ functions narrower.

Considering the population parameter M_0 to be unknown, with prior distribution $\pi(M_0)$, the joint posterior for all the unknowns is given by

$$\pi(M_0, \mathbf{M}|\mathbf{D}) \propto \pi(M_0) \prod_i f(M_i|M_0)\ell_i(M_i). \qquad (11.9)$$

If our goal is to estimate the source parameters, we account for M_0 uncertainty by marginalizing over M_0, giving the source parameter marginal posterior, $\pi(\mathbf{M}|\mathbf{D}) = \int dM_0 \, \pi(M_0, \mathbf{M}|\mathbf{D})$. If instead our goal is to infer the population density, we calculate the marginal posterior for M_0, $\pi(M_0|\mathbf{D}) = \int d\mathbf{M} \, \pi(M_0, \mathbf{M}|\mathbf{D})$. In this simple normal–normal MLM, these integrals can be done analytically (we adopt a flat prior for M_0, the *hyperprior*).

The top panel of Figure 11.2 shows a population distribution with $M_0 = -21$; the circles on the line just below it indicate the true M_i values of a sample of $N = 30$ sources. Below that, diamonds indicate the MLEs for the sources, $\hat{M}_i$, for one realization of measurement error; a line segment connects each estimate with its true parent value. The MLEs are intuitively appealing estimates; they also have several appealing frequentist properties, considering an ensemble of many realizations of the measurement errors. For example, normal MLEs are unbiased (in fact, they are the best linear unbiased estimators), and they are invariant to translation in M. But considered as an ensemble, they are overdispersed with respect to the population distribution; this is visually evident in the figure. Intuitively, the MLEs here can be viewed as samples from a modified population density that is the convolution of the true population density with the error density (though we note that this convolution interpretation does not generalize to more complicated settings).

As alternatives to the MLEs, consider Bayesian estimates in two different scenarios. First, suppose we knew $M_0 = -21$ a priori. Using the (known) population distribution as the prior for each M_i produces posteriors that remain independent and normal, but with means, $\tilde{M}_i$, shifted from the MLEs toward M_0. Define $b \equiv \sigma^2/(\sigma^2 + \tau^2)$; then the posterior means (and modes) are given by

$$\tilde{M}_i = (1 - b) \cdot \hat{M}_i + b \cdot M_0, \qquad (11.10)$$

and the variance for each estimate is reduced to $(1 - b)\sigma^2$ rather than just σ^2. The squares on the third line below the panel in Figure 11.2 show these estimates.

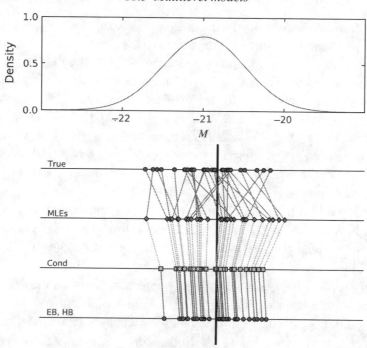

Fig. 11.2. Shrinkage in a simple normal–normal model. Top panel shows population distribution. 'True' axis shows M_i values of 30 samples. Remaining axes show estimates from measurements with $\sigma = 0.3$ normal error: MLEs, conditional (on the true mean), and empirical/hierarchical Bayes estimates.

They all move toward M_0, and thus toward each other. One says that the ensemble of estimates 'shrinks toward M_0'; this phenomenon is called *shrinkage*. They are labeled 'Cond' in the figure to indicate that we conditioned on M_0.

As an ensemble, the shrunken estimates look much more like the true values than the MLEs. These estimates are biased and are no longer invariant, but even from a frequentist perspective they may be deemed better than the MLEs: despite the bias, the shrunken estimates are, on the average (over error realizations), closer to the true values than the MLEs – i.e., they have smaller mean squared error (MSE) – as long as $N > 2$. Stein discovered this effect around 1960 and, after a decade or two of sorting out its subtleties, the use of deliberately (and carefully) biased estimators for joint estimation of related quantities is now widespread in statistics (frequentist and Bayesian), and considered one of the key innovations of late-twentieth-century statistics.

We have used strong prior information here: precise knowledge of M_0. But what if we did not know M_0 a priori? The *empirical Bayes* (EB) approach 'plugs in' an ad hoc estimate of M_0 and uses the resulting prior. The obvious estimator here is

$\bar{M}$, the average of the MLEs, whose position is indicated by the thick vertical line in the figure. Using the resulting prior produces the circle estimates on the bottom axis, still shrunken, but towards $\bar{M}$ rather than M_0. Equation (11.10) again gives the estimates, if we replace M_0 with $\bar{M}$. Note that since $\bar{M}$ depends on *all* of the MLEs, the EB estimates are no longer independent.

From a fully Bayesian point of view, plugging in $\bar{M}$ for M_0 is unjustified; M_0 is unknown, so we should consider it a parameter, assign its prior distribution, and marginalize over it, an approach called *hierarchical Bayes* (HB). The resulting estimates are identical to the EB estimates in this problem (though they need not be in general); however, the uncertainties in the HB estimates are somewhat larger than those produced in an EB calculation, reflecting uncertainty in the shrinkage point. We have motivated EB and HB shrinkage estimates via Bayesian arguments, but the frequentist advantages of shrinkage in the conditional case still hold: despite their bias, these estimates have smaller MSE than MLEs. In addition, due to their accounting for M_0 uncertainty, confidence intervals based on the HB procedure have more accurate frequentist coverage than EB intervals.

The conditional shrinkage estimates are the normal–normal model counterparts to homogeneous Malmquist and Lutz–Kelker corrected estimates (the latter do not so obviously 'shrink' because the prior in astronomical settings is much broader than a normal distribution). On the one hand, this is a source of comfort, in that some of the statistical benefits of shrinkage are presumably shared by the astronomical methods. However, the correspondence is also a source of caution and concern. Decades of study have revealed shrinkage to be a subtle phenomenon, with snares for the unwary. We highlight just a few key developments here; entries to the large literature in this area include Carlin and Louis (2000, 2009), Carlin *et al.* (2006), and Browne and Draper (2006).

Most obviously, as noted in Section 11.2.2, the homogeneous density assumption of the classic corrections is seldom justifiable; in reality, the population density is inhomogeneous and unknown. That is, conditional shrinkage is not appropriate; something along the lines of the EB or HB approaches is in order. This must be done with some care: it is known that shrinkage may not improve estimates, and may even worsen them, if done in a way that does not reflect the true distribution of first-level parameters. The EB approach – using a 'plug-in' estimate of hyperparameters to specify the population hyperprior – has appealing simplicity; the Landy–Szalay approach probably has an approximate EB justification. But it is known that EB approaches tend to underestimate final uncertainties (due to ignoring hyperparameter uncertainty). HB estimates can offer improvements, but with computational costs, and with other challenges noted below.

More subtly, shrinkage must be tuned, not only to the underlying population distribution, but also to the *inferential goal*. The shrinkage estimates just described do indeed reduce the MSE of the collection of source absolute magnitudes. But if one uses the shrunken point estimates to infer the population distribution, it turns out the point estimates are *under*dispersed and the distribution may be poorly estimated. If one instead seeks from the beginning point estimates that are optimal for estimating the population distribution (via a decision-theoretic calculation), a different shrinkage prescription is appropriate. Similarly, it is evident from the segments connecting the true M_i's with their MLEs that the ranks of the sources are shuffled, and shrinkage has not corrected it. In some settings, shrinkage estimates that improve rank estimates have been identified; they differ from those optimal for individual parameter or distribution estimates. There is a kind of 'complementarity' in relying on point estimates for subsequent inferences; estimates optimal for some questions may be misleading for other questions (Louis 1984).

A main source of these complications is the inadequacy of point estimates as summaries of a correlated, high-dimensional posterior distribution. This motivates a more thoroughly Bayesian treatment in the spirit of HB, relying on marginalization over uncertain parameters rather than use of point estimates. We pursue this approach below.

But before doing so, given the Bayesian focus of this volume, some comments at a conceptual level are appropriate here. The second level of our MLM here describes the population with a continuous density. A frequentist interpretation of this density is problematic. The volume accessible to a survey, indeed the volume within the horizon, contains a finite number of sources (galaxies, cluster, quasars, etc.). Repeating a survey will produce catalogues that largely contain the same sources (some sources near the survey detection limit may differ from one repetition to the next); the population density cannot be interpreted in terms of frequentist variability. At the first level of the MLM, measurement errors may differ among repetitions, but if the first-level uncertainties are the result of indicator scatter, itself a population-level phenomenon, first-level results will also be the same across repeated surveys.

From the Bayesian point of view, the population density describes uncertainty about population members, not variability in repeated sampling; its introduction and specification should be motivated by epistemological considerations. One way to formally motivate it is as a mechanism to introduce dependence among estimates. That is, we expect that learning the properties of many sources of a particular type should help us predict the properties of as-yet unmeasured sources of that type; this is what it means to consider the sources to comprise a population. Consistency requires that, once we obtain measurements of new sources, we cannot ignore the prior information provided by measurements of other sources that we would have used in the absence of the data. The resulting dependences in the joint posterior pools information, leading to shrinkage.

More formally still, we might justify introducing a population density by requiring the joint prior distribution for a set of source properties to be *exchangeable*, that is, invariant to permuting the labels of sources. For example, in the setting above, $p(M_1, M_2, \ldots, M_N | \theta)$ should take the same functional form if we permute the order of the $\{M_i\}$. This appears to be both a natural and a weak assumption; it is in the spirit of the common 'iid' assumption in that the marginals for each source are identical but, by allowing dependence, it sets the stage for sharing of information across the population. Surprisingly, exchangeability itself implies the MLM structure: the (continuous) de Finetti exchangeability theorem implies that any such exchangeable distribution can be written as a density-weighted mixture of identical, conditionally-independent distributions, i.e., as an MLM. In the setting here, the theorem says one may write $p(M_1, M_2, \ldots, M_N) = \int d\theta \, \mu(\theta) \prod_i f(M_i | \theta)$, where $\mu(\theta)$ defines a unit-normed measure over the form of the independent densities $f(\cdot | \theta)$. The theorem is a purely mathematical result providing a representation for symmetric functions, but in a Bayesian context it motivates introducing a continuous population density, playing the role of $f(\cdot)$, with a hyperprior playing the role of μ.[4]

11.3.2 *Poisson point process multilevel models*

While the MLM set-up above has the essential ingredients needed for us to move beyond point estimate-based population modelling, we need to generalize it in two ways to meet the needs of astronomical survey analysis. First, the analysis above took the catalogue size, N, as given. In an astronomical survey, N is instead determined by the population density and the volume surveyed; it is thus informative about the population. Second, astronomical surveys suffer from selection effects, most typically in the form of truncation in a 'blanket' survey (where sources may be missed due to detection criteria), or censoring in a targeted follow-up survey (where sources known to exist may have unmeasureable properties due to limited sensitivity). We discuss the truncation case here.

To allow catalogue size to be informative about the population density, we model the population with an inhomogeneous Poisson point process, characterized by an *intensity measure* rather than a probability distribution. Let $\mathcal{O}$ denote the observable source parameters; the Poisson point process assumption implies there is an intensity measure, $\mu(\mathcal{O})$ that, when known, allows us to write the probability for there being a source with $\mathcal{O}$ in the interval $[\mathcal{O}, \mathcal{O} + d\mathcal{O}]$ as $\mu(\mathcal{O})d\mathcal{O}$, to leading order in $\mathcal{O}$. It also presumes that this probability is independent of whether a source

---

4   Rigorously, the theorem requires that the judgement of exchangeability apply for any selection of a finite set of M_i's from an infinite set. If there is a finite limit to N, the integral representation may not be able to represent some possible exchangeable distributions, though the restriction is minor if the limit is large. See Diaconis and Freedman (1980) for details.

is found in any other (distinct) interval (provided we know the intensity $\mu(\cdot)$; i.e., this is *conditional* independence). Usually we will not know the intensity, e.g., it may depend on parameters, θ, whose values are uncertain, which we indicate by writing $\mu(\mathcal{O}; \theta)$.

To account for truncation, we introduce a survey detection efficiency, $\epsilon(\mathcal{O})$, specifying the probability that a source with parameters $\mathcal{O}$ will be detected. Although we take $\epsilon(\mathcal{O})$ as given in what follows, it is worth noting (especially for non-astronomer readers) that calculating the ϵ that characterizes a particular survey is often a very difficult task, requiring both careful measurement and calibration, and often extensive Monte Carlo simulation. Not all surveys provide an accurate detection efficiency, yet it is necessary for what follows. (In some cases it may be possible to partially infer ϵ from the available survey data; we do not cover this challenging task here.)

Finally, we highlight a point made in passing above: from the Bayesian point of view, a survey source catalogue should not be viewed as providing *estimates* of source properties, but rather as summary statistics specifying *source likelihood functions*, $\ell_i(\mathcal{O}_i) = p(D_i|\mathcal{O}_i)$, where D_i denotes the data for source i. This change in viewpoint has far-reaching implications. It can enable more accurate accounting of source uncertainties, e.g., by reporting a likelihood parameterization more complex than the traditional 'best-fit $\pm$ uncertainty', such as a parameterization describing possible likelihood skewness (as might be important near detection limits). It also opens the door to the use of marginal detections or upper limits in censored surveys, by averaging over source uncertainties with likelihoods that are far from normal, possibly peaking at zero source flux. This is discussed further in Chapter 8.

With these ingredients in hand, one can calculate the truncated Poisson point process counterpart to the MLM joint posterior density for the population and source parameters of Eq. (11.9) (see Loredo & Wasserman 1995 or Loredo 2004 for derivations):

$$\pi(\theta, \{\mathcal{O}_i\}|D) \propto \pi(\theta) \exp\left[-\int d\mathcal{O}\, \epsilon(\mathcal{O})\mu(\mathcal{O}; \theta)\right] \prod_{i=1}^{N} \ell_i(\mathcal{O}_i)\mu(\mathcal{O}_i; \theta). \quad (11.11)$$

Marginal posteriors for θ or $\{\mathcal{O}_i\}$ may be calculated as was done above, though obviously the calculations can be challenging for astrophysically interesting models. Our own applications to date have been to situations with parametric population models, where $\mathcal{O}$ was one dimensional (magnitudes of trans-Neptunian objects (TNOs); Petit *et al.* 2007 and references therein), or three dimensional (fluxes and directions of gamma-ray bursts; Loredo & Wasserman 1998a,b). In these cases, the N integrals over $\mathcal{O}_i$ were done by quadrature.

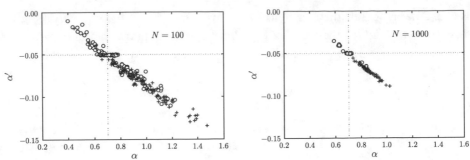

Fig. 11.3. Scatter plots of posterior modes for population parameters, using data simulated from a rolling power law, from a Bayesian analysis marginalizing over source parameters (circles), and a maximum likelihood analysis using best-fit source estimates (+ symbols). Left panel is for simulated surveys of $N = 100$ sources; right panel is for $N = 1000$.

As a simple example focusing on one aspect of the MLM approach, the value of marginalizing over source uncertainty, consider a magnitude survey (i.e., the 'number counts' setting) where $\mathcal{O} = m$, the apparent magnitude of a source. Suppose the source population has a rolling power-law distribution of fluxes, so we may write the magnitude distribution as $\mu(m) = A \times 10^{[\alpha(m-23)+\alpha'(m-23)^2]}$, where A is the density per unit magnitude at $m = 23$, and α and α' give the slope of the number–magnitude distribution, and its rate of change with m, at $m = 23$. We simulated sources from this distribution, and simulated detections and measurements for a very simple survey performing source detection and measurement via photon counting. The survey parameters were chosen so that the dimmest detected sources have magnitude uncertainties ~ 0.15 magnitudes. We analyzed data from many simulated surveys, all of a population with $\alpha = 0.7$ and $\alpha' = -0.05$ (values that describe some TNO data), and estimated α and α' by finding the mode of the marginal posterior for these parameters, marginalizing Eq. (11.11) over all the m_i and the amplitude, A. The resulting estimates, for surveys of size $N = 100$, are plotted as the open circles in the left panel of Figure 11.3; the cross-hair shows the true parameter values. We also calculated maximum likelihood estimates, i.e., using only the best-fit m_i estimates (not marginalizing). These are plotted as '+' symbols in the figure. The Bayesian estimates are distributed roughly symmetrically about the truth; the MLEs are clearly biased toward large α and small α', though the uncertainties are large enough that the estimates are sometimes accurate.

The right panel shows results from the same calculation, but with $N = 1000$. The Bayesian estimates have converged closer to the truth. In contrast, the MLEs have converged *away* from the truth. This example highlights the value of marginalization, particularly in settings with measurement error. In such settings, each new source adds parameters to the problem; one is not in the fixed parameter dimension setting in which our statistical intuitions are trained. As a result, the effects of source uncertainties do not 'average out'; instead, it becomes *more* rather than less important to carefully account for them as survey size grows. This aspect of population modelling has been repeatedly rediscovered by astronomers. The earliest discovery we know of is Eddington's treatment of what has become known as Eddington bias (Eddington 1940). A recent rediscovery in a cosmological context is contemporary as we write: Sheth's work on the effects of photometric redshift errors on modelling galaxy and quasar populations (Sheth 2007). Sheth advocates an ad hoc deconvolution algorithm; we think Bayesian multilevel modelling offers a vastly more flexible and accurate framework for addressing such problems.

11.4 Future directions

The fully Bayesian MLM approach just described has so far been implemented only for fairly simple models of modest-sized surveys (though the challenging VELMOD analysis of Tully–Fisher data by Willick and Strauss (1998), mentioned in subsection 11.2.3, includes many of the key elements). Future research must explore applications of increased complexity in three different dimensions: survey size, source parameter dimension, and population model complexity.

An obvious need is development of numerical algorithms appropriate for multivariate observables and large surveys, perhaps invoking approximations or using Monte Carlo methods for source parameter marginalization. With respect to multivariate observables, it will be insightful to work out the detailed connections between the MLM approach and other methods, such as those surveyed in Section 11.2. For example, when the observables are flux and a distance indicator, an analogue to the direct indicator method should 'fall out' of an MLM calculation when the source likelihoods provide precise estimates of the indicators.

Implementing Bayesian MLM with more complex population models will also demand development of clever algorithms. But more subtle and interesting challenges arise as model complexity increases. These challenges arise because of the 'softening' of the impact of source measurements on inferences, due to uncertainties. We saw above that this 'changes the rules' in the sense of causing violations of naive intuition about uncertainties averaging out as N grows. But this is only one of several issues complicating life with multilevel models.

Some of these issues mimic problems associated with non-parametric modelling, and this is no accident. Though a common informal 'definition' of a non-parametric model is a model with an infinite number of parameters, a more insightful definition is a model in which the *effective* dimension of the parameter space can grow with sample size. In fact, the 'many normals' problem – essentially the normal–normal MLM, with the actual dimension growing linearly with sample size – is sometimes used as a surrogate for more complex non-parametric models in theoretical analyses (Wasserman 2005).

A prime issue which non-parametric modellers must face is assessing how priors over large-dimensional spaces may influence inferences. Similar concerns arise for MLMs. For example, for estimating the mean and standard deviation of a normal distribution using *precise* measurements, common default priors are flat for the mean and log-flat for the standard deviation. We saw above that a flat prior for M_0 in the normal–normal MLM produced sensible inferences. But had we considered the population standard deviation τ to be unknown, we would have discovered that a log-flat τ prior leads to an improper (un-normalizable) posterior. Instead, priors flat in τ or τ^2 (among others) are advocated, with various justifications (sometimes including the good frequentist performance of the resulting estimates; see, e.g., Berger *et al.* 2005; Gelman 2006). This indicates that as astronomers increase the complexity of MLMs for surveys, care must be taken with priors; statisticians have helpful insights to offer here. It also suggests that model checking, in the spirit of goodness-of-fit tests, is important for MLMs. Their rich structure makes conventional model checking methods inappropriate, but there is recent research on model checking methods tailored to MLMs (e.g., Sinharay & Stern 2003; Bayarri & Castellanos 2007).

An alluring direction for increasing model complexity is to make the population model itself truly non-parametric. One motivation comes from existing non-parametric methods designed to flexibly account for selection effects, such as the C^- method of Lynden-Bell (1971) or the stepwise maximum likelihood method (Efstathiou *et al.* 1988). These methods ignore source uncertainties; finding counterparts in the MLM framework promises to broaden applicability of such approaches, and unify them with approaches relying on Malmquist-style corrections (insofar as MLMs have shrinkage 'built in'). But the fact that *parametric* MLMs already have some of the issues of non-parametric modelling suggests that non-parametric multilevel modelling will be tricky, requiring even more care with assessing robustness to priors. Fortunately, there are successful examples in the statistics literature to build upon (e.g., Müller & Quintana 2004).

Sensitivity to priors can make newcomers to Bayesian methods consider retreating to frequentist territory. But there is little solace there; the subtleties of complex multilevel Bayesian modelling reflect genuine complexity in the task of modelling surveys, complexity that has frequentist implications. For example, the best-studied methods for non-parametric analysis of survey data with uncertainties (yet to be extended to include truncation) rely on deconvolution. Though the methods are straightforward, it is known that the resulting population estimates have discouragingly slow rates of convergence (often only logarithmic in N, as opposed to the $\sqrt{N}$ we are accustomed to for parametric inference without measurement errors). In fact, leading developers of such methods have recently turned to Bayesian methods, where careful attention to structure in the prior can lead to methods with improved performance (e.g., Berry *et al.* 2002).

We think development of *semi-parametric* population models for use in MLM analyses of survey data may be an especially fruitful research direction. For example, we envision non-parametric modelling of the density distribution of galaxies, combined with parametric modelling of the luminosity function (e.g., by a Schechter function or a mixture of a few Schechter functions), as a promising approach, allowing adaptivity to complex spatial structure, but hopefully providing good convergence rates for learning the luminosity function. But there are several challenges to conquer on the path from the current state of the art to such a goal.

Acknowledgments

We are grateful to many collaborators and colleagues who have contributed to our understanding of survey biases and population modelling, especially David Chernoff, Woncheol Jang, John Simmons and Ira Wasserman. We also gratefully acknowledge the Statistical and Applied Mathematical Sciences Institute (SAMSI), whose 2006 Astrostatistics Program assembled astronomers and statisticians to discuss these issues. Loredo was partly supported by NSF grant AST-0507589 and by NASA grants NAG5-12082 and NNG06GH84G for work reported here.

References

Bayarri, M. J. and Castellanos, M. E. (2007). *Stat. Sci.*, **22**, 322.

Berger, J. O., Strawderman, W. and Tang, D. (2005). *Ann. Stat.*, **33**, 606.

Berry, S. M., Carroll, R. J. and Ruppert, D. (2002). *JASA*, **97**, 160.

Browne, W. J. and Draper, D. (2006). *Bayesian Analysis*, **1**, 473.

Carlin, B. P. and Louis, T. A. (2000). *Bayes and Empirical Bayes Methods for Data Analysis*. London: Chapman & Hall/CRC.

Carlin, B. P. and Louis, T. A. (2009). *Bayesian Methods for Data Analysis*. London: Chapman & Hall/CRC.

Carlin, B. P., Clark, J. S. and Gelfand, A. E. (2006). In J. Clark and A. E. Gelfand, eds., *Hierarchical Modelling for the Environmental Sciences*. Oxford: Oxford University Press, p. 3.

Diaconis, P. and Freedman, D. (1980). *Ann. Prob.*, **8**, 745.

Eddington, A. S. (1913). *Mon. Not. Roy. Astron. Soc.*, **73**, 359.

Eddington, A. S. (1940). *Mon. Not. Roy. Astron. Soc.*, **100**, 354.

Efstathiou, G., Ellis, R. S. and Peterson, B. A. (1988). *Mon. Not. Roy. Astron. Soc.*, **232**, 431.

Erdogdu, P. *et al.* (2006). *Mon. Not. Roy. Astron. Soc.*, **373**, 45.

Freudling, W. *et al.* (1995). *Astron. J.*, **110**, 920.

Gelman, A. (2006). *Bayesian Analysis*, **1**, 515.

Hendry, M. A. and Simmons, J. F. L. (1994). *Astrophys. J.*, **435**, 515.

Hendry, M. A. and Simmons, J. F. L. (1995). *Vistas Astron.*, **39**, 297.

Hudson, M. J. (1994). *Mon. Not. Roy. Astron. Soc.*, **266**, 468.

Landy, S. D. and Szalay, A. S. (1992). *Astrophys. J. Lett.*, **391**, 494.

Loredo, T. J. (2004). In R. Fischer, R. Preuss and U. von Toussaint, eds., *AIP Conf. Proc.*, **735**, 195.

Loredo, T. J. (2007). *ASP Conf. Ser.*, **371**, 121.

Loredo, T. J. and Wasserman, I. M. (1995). *Astrophys. J. Supp.*, **96**, 261.

Loredo, T. J. and Wasserman, I. M. (1998a). *Astrophys. J.*, **502**, 75.

Loredo, T. J. and Wasserman, I. M. (1998b). *Astrophys. J.*, **502**, 108.

Louis, T. A. (1984). *JASA*, **79**, 393.

Lynden-Bell, D. (1971). *Mon. Not. Roy. Astron. Soc.*, **155**, 95.

Lynden-Bell, D. *et al.* (1988). *Astrophys. J.*, **326**, 19.

Malmquist, K. G. (1920). *Medd. Lund Astron. Obs.*, Ser. 2, No. 22, 359.

Malmquist, K. G. (1922). *Medd. Lund*, Ser. I, **100**, 1.

Müller, P. and Quintana, F. A. (2004). *Stat. Sci.*, **19**, 95.

Petit, J.-M., Kavelaars, J. J., Gladman, B. J. and Loredo, T. J. (2007). In A. Barucci, H. Boehnhardt, D. Cruikshank and A. Morbidelli, eds., *Kuiper Belt*. Tucson, AZ: University of Arizona Press and Lunar and Planetary Institute (in press).

Schechter, P. L. (1980). *Astron. J.*, **85**, 801.

Sheth, R. K. (2007). *Mon. Not. Roy. Astron. Soc.*, **378**, 709.

Sinharay, S. and Stern, H. S. (2003). *J. Stat. Plan. Infer.*, **111**, 209.

Strauss, M. A. and Willick, J. A. (1995). *Phys. Rep.*, **261**, 271.

Teerikorpi, P. (1997). *Annu. Rev. Astron. Astrophys.*, **35**, 101.

Wasserman, L. (2005). *All of Nonparametric Statistics*. New York: Springer-Verlag.

Willick, J. A. (1994). *Astrophys. J. Supp.*, **92**, 1.

Willick, J. A. and Strauss, M. A. (1998). *Astrophys. J.*, **507**, 64.

12

A Bayesian approach to galaxy evolution studies

Stefano Andreon

12.1 Discovery space

We astronomers mostly work in the 'discovery space', the region where effects are statistically significant at less than three-sigma or near boundaries in data or parameter space. Working in the discovery space is a normal astronomical activity; few published results are initially found at large confidence. This can lead to anomalous results, e.g., positive definite quantities (such as mass, fractions, star formation rates, dispersions, etc.) are sometimes found to be negative or, more generally, quantities are sometimes found at unphysical values (completeness larger than 100%, $V/V_{max} > 1$ or fractions larger than 1, for example). Working in the discovery space is typical of frontier-line research because almost every significant result reaches this status after having appeared first in the discovery space, and because a good determination of known effects or trends usually triggers searches for finer, harder to detect, effects, mostly falling once more in the discovery space.

Many of us are very confident that commonly used statistical tools work properly in the situations in which we use them. Unfortunately, in the discovery space, and sometimes outside it, we should not take this for granted, as shown below with a few examples. We cannot avoid working in this grey region, because to move our results into the statistically significant area we often need a larger or better sample. In order to obtain this, we first need to convince the community (and the telescope time allocation committees) that an effect is probably there, by working in the discovery space. Furthermore, awaiting a larger sample leaves the unappealing possibility that someone else may publish our result and they, not us, will be credited for the discovery. Of course, this assumes that a larger sample exists and is accessible, which is not always the case. There is just one Universe (and sky), already fully observed in the microwave. A group formed by 10 galaxies has no more

265

than 10 galaxies available to obtain its velocity dispersion. Gamma-ray bursts (and any transient event) may not be long-lived enough to allow us to collect enough photons to place our target measurement, say polarization, in the statistically significant area. Working in the discovery space region is therefore an essential part of astronomers' work.

Standard tools may fail (especially if misused) in many ways. In the next two sections we will show some examples of failure in two idealized experiments, and we will show that the Bayesian approach does not suffer from these failures. In the first example, we show that the maximum of the likelihood (the best fit) may not be a good estimate of the true value: averaging the likelihood is preferable to maximizing it. The second example shows that sometimes the observed value is biased, and highlights the bad things that may occur when the prior is ignored. These two examples are very simple as compared to true problems, and they have been chosen so as to make obvious that the best-fit value or observed value may be bad or biased. The third example shows that when the sample size is small, even simple operations on data, such as performing an average or fitting it with a function, is a potentially risky operation (and something that fails even in a simple case is unlikely to work correctly in a complex one). In Section 12.5 we consider two realistic examples, showing failures of standard methods that are more difficult to spot, but are of the same nature as the easily spotted failures – all of them come from contradicting axioms of probabilities. In the first example, we want to measure the width of a distribution in the presence of a contaminating population. A mixture modelling of inhomogeneous Poisson processes easily solves this problem. In the second example, we wish to fit a trend in the presence of a contaminating population and we will use a mixture of regressions. We finally conclude the chapter by showing that the Bayesian approach allows us to properly understand the number returned by tests like Kolmogorov–Smirnov, χ^2, etc., named in the statistical jargon as the p-value.

12.2 Average versus maximum likelihood

Maximum likelihood estimates (called 'best fit' by astronomers) are one of the most widely used tools in astronomy and it is usually taken for granted that maximizing the likelihood (or minimizing $\chi^2 = -2\ln\mathcal{L}$) will *always* give the correct result.

Mixture distributions naturally arise in astronomy when data come from two populations, in such cases where

(a) a signal is superposed on a background,
(b) there are interlopers in the sample,

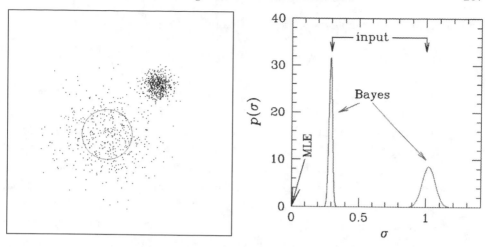

Fig. 12.1. *Left panel*: An example of a mixture distribution in two dimensions. Readers may consider that we displayed the spatial distribution of photons coming from two astronomical extended sources as observed by a perfect instrument (no background, perfect angular resolution, etc.), or the distribution of two galaxy populations in a two-dimensional parameter space, or whatever else. *Right panel*: true (input) values, maximum likelihood estimate (MLE) and posterior (Bayes) probability distributions for the example in the left panel.

(c) there are two distinct galaxy populations (by colour, morphology, dynamical properties, etc.),

(d) an image of the sky is taken using an instrument with a field of view large enough to accommodate more than one source, or

(e) a galaxy spectrum is observed in which we note the presence of two stellar populations.

Let us consider the simple case of a mixture (sum) of two Gaussians:

$$p(y_i|\mu_1,\sigma_1,\mu_2,\sigma_2,\lambda) = \lambda\mathcal{N}(y_i|\mu_1,\sigma_1^2) + (1-\lambda)\mathcal{N}(y_i|\mu_2,\sigma_2^2), \qquad (12.1)$$

where (μ_j,σ_j) $j = 1, 2$ are the location (or centre) and scale (or width) of the two Gaussians

$$\mathcal{N}(y_i|\mu_j,\sigma_j^2) = \frac{1}{\sqrt{2\pi}\sigma_j} e^{-\frac{(y_i-\mu_j)^2}{2\sigma_j^2}}, \qquad (12.2)$$

λ and $1 - \lambda$ are the proportions of the two components, and y_i is the ith datum. Figure 12.1 shows a two-dimensional example.

We want to determine the locations and scales of the two Gaussians with the data at hand. The likelihood of independently and identically distributed data is given by the product, over the data y_i, of the terms in Eq. (12.1):

$$p(y|\mu_1, \sigma_1, \mu_2, \sigma_2, \lambda) = \prod_i p(y_i|\mu_1, \sigma_1, \mu_2, \sigma_2, \lambda). \tag{12.3}$$

We usually maximize the likelihood in current problems, blindly assuming that the maximum likelihood values are good estimates of the true value. Here, though, the parameters that maximize the likelihood above (the best fit) are not near their true value (e.g., those used to draw the points in Figure 12.1), but occur when $\mu_j = y_i$ and $\sigma_j \to 0$. In fact, the likelihood goes to infinity as $\sigma_j \to 0$, because the ith term of Eq. (12.1) diverges and the remaining ones take a finite, non-zero, value.

The problem is a general characteristic of mixtures of probability distributions of non-fixed variance, not only of Gaussians. The problem does not disappear 'in the long run', i.e., by obtaining a sufficiently large sample (that we don't have, or which time allocation committees are reluctant to allocate, or perhaps does not exist). On the contrary, our chances of failure increase with sample size, because there is an increasing number of values for which the likelihood goes to infinity, one per datum.

Therefore, maximizing the likelihood, even for unlimited data, does not return the size of two astronomical sources, or the velocity dispersion of a cluster in the presence of interlopers, or many other quantities in the presence of two populations or signals (or an interesting and an uninteresting population or signal). Even worse, there is no 'warning bell', i.e., something that signals that something is going wrong, until an infinity is found when maximizing the likelihood. In real applications, as those described in Sections 12.4 and 12.5, nothing as bad as an infinity appears, and thus there is no 'warning bell' signalling that something is going wrong.

The Bayesian approach is not affected by such problematics: it never instructs us to maximize any unknown parameter, because the sum axiom of probability tells us to sum (or integrate) over unknown quantities, so that their effect is averaged over all plausible values.

The right panel of Figure 12.1 shows the posterior distribution for the data shown in the left panel, adopting a constant prior up to very large values (the precise values are irrelevant for this parameter estimation problem). The posterior is well behaved and it is centred on the input value. The likelihood, instead, has hundreds of infinities, one per datum, all at $\sigma = 0$.

12.3 Priors and Malmquist/Eddington bias

Number counts are steep. It is well known to astronomers that the true value, μ, of the source intensity differs from the measured counts, n, of the source when n

is small: even in the presence of symmetric errors, an object with n counts more probably comes from the numerous population of objects having $\mu < n$ than the rare population having $\mu > n$. Therefore, objects with n counts have, likely, $\mu < n$ (e.g., Eddington 1913). A similar effect arises for parallaxes, star counts, velocity dispersions and any noisy determination of a quantity concerning an object drawn from a population that shows an important numerical change over the range included by the error on the measurand[1] (on μ, in our example). Restated in the statistical jargon, in parameter estimation problems where the prior, $p(\mu)$, has a large change in the μ range where the likelihood is slowly varying, the prior cannot be neglected.

As a quantitative example, tailored around the X-Bootes survey (Kenter *et al.* 2005), we consider a Poisson process $p(n|\mu) = \mu^n e^{-\mu}/n!$, with rate μ and a power-law prior of logarithmic slope α, $p(\mu) = \mu^{-\alpha}$ (the latter is called number counts by astronomers). Having observed four photons (i.e., $n = 4$), the maximum likelihood estimate of the source rate is $\hat{\mu} = 4$, but astronomers know by experience that this value is wrong: with better data we find, most of the time, that actually $\mu < 4$. Assuming a Euclidean slope $\alpha = 2.5$ (the observed value of the slope of number counts at the rate of interest), the posterior $p(n|\mu)p(\mu)$ is $\propto \mu^n e^{-\mu} \mu^{-2.5}/n!$. The posterior mean (or, in astronomical terms, the Eddington-corrected value) of the source rate is 2.5 photons, 40% less than the originally observed value. See Figure 12.2 for detail.

The same holds true, as mentioned, for many noisy quantities, such as the determination of a velocity dispersion with just a few velocities or the estimate of the cluster richness N_{200} (see Andreon & Hurn 2009 for details on the latter).

What is known in astronomy as Malmquist or Eddington bias is the manifestation of the important role of the prior when measurements are imprecise, i.e., the fact that a correct inference proceeds along the Bayes' theorem. The prior (the correction for Malmquist bias) moves the result away from the observed value, or the maximum likelihood estimate, and brings it near the true value. The example shows that priors (the fact that there are many more fainter systems than bright ones) cannot be ignored in inferences, if one does not wish to be wrong most of the time. Priors are specific to the Bayesian approach and, usually, non-Bayesian methods consider them something to be avoided.

It is therefore apparent that prior-free and maximum likelihood methods are in trouble.

---

[1] The measurand is the parameter being quantified. It usually differs from the outcome of the measurement because of the noise.

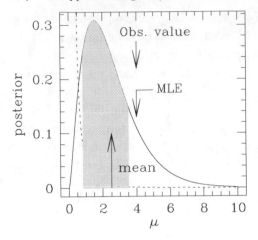

Fig. 12.2. Posterior distribution (solid curve) of the source flux, having observed four photons and knowing that source number counts have a Euclidean distribution (dashed curve). The maximum likelihood estimate (MLE) is also reported.

12.4 Small samples

Astronomers are often faced with computing an average by combining a small number of estimates, or fitting a trend or a function from just a few data points. We almost always start with a few measurements of an interesting quantity, say the rate at which a galaxy mass increases. In most of the cases, the measurand may be parameterized in several ways. For example, if the aim is to measure the relative evolution of luminous (L) and faint (F) red galaxies, a central topic in galaxy evolutionary studies, should we study L/F, F/L, or does the chosen parameterization not matter? Both parameterizations have been adopted in recent astronomical papers (and the author, in Andreon (2008), took a third, different, parameterization!). Specific star formation rates (sSFR) and e-folding times, τ, are approximatively reciprocal measures (long e-folding times correspond to small sSFR). To the author's knowledge, none of the proposed parameterizations (e.g., L/F vs. F/L or sSFR vs. τ) has a special status. Unfortunately, when the sample size is small, results obtained via commonly used formulae (e.g., weighted average, best fit, etc.) depend on the adopted parametric form. For example, an average value, computed by a weighted sum, or a fit performed by minimizing the χ^2, has a special meaning, because the result depends on which parameterization is being adopted.

As an example, let us consider just two data points, $(f/l)_1 = 3 \pm 0.9$ and $(f/l)_2 = 0.3333 \pm 0.1$. The error weighted average $\langle f/l \rangle$ is 0.37. The reciprocal values $((l/f)_i = 1/(f/l)_i; 0.3333 \pm 0.1$ and $3 \pm 0.9)$ have an error weighted average equal again to 0.37, quite different from the reciprocal of $\langle f/l \rangle$, $1/0.37 = 2.7$. Therefore $\langle f/l \rangle \neq 1/\langle l/f \rangle$, and they differ by much more than their error (obviously, the error on the mean is smaller than the error on any data point). At first sight, by choosing the appropriate parameterization, the astronomer may select the number he wants, a situation surely not recommended by the scientific method. Similar problems are present with two data points differing by just one-sigma, or in general with small samples.

One may argue that in the above case the number of points is so small that no one will likely make an average of them. However, one almost always starts by averaging two or three values, or looks for trends, or fits a function using a number of data points only slightly exceeding the number of parameters. Often, the few points are the result of a large observational effort (e.g., obtained through analyzing thousands of galaxies, as in the example above) and it is very hard to assemble a larger sample. Thus the average of a few numbers is almost all we can do. For example, how many estimations of the SFR at $z \sim 6$ exist? Should we not combine the very few available in some way to take profit of all them? Small sample problems are often 'hidden' in large samples: even large surveys, such as the SDSS, 2dF, VVDS and CNOC2 surveys including tens or hundreds of thousand of galaxies, estimate galaxy densities using sub-samples equivalent to just one to ten galaxies. Finally, how many of us have checked, before performing a fit or an average, whether the sample size is large enough to be insensitive to the chosen parameterization?

The described problem originates from the freedom, in the frequentist paradigm, of choosing an estimator of the measurand ($\langle f/l \rangle$ or $1/\langle l/f \rangle$ for example). All estimators (satisfying certain conditions) will converge on the true value of the estimand in the long run, but without any assurance that such a regime is reached with the sample size in hand. Until this regime is reached, different estimators will return different numbers. Bayesian methods do not present this shortcoming, because they already hold with $n = 2$ and do not pass through the intermediate and non-unique step of building an estimator of the measurand.

This example shows that frequentist methods return the value of an estimator of the measurand, not the measurand itself or its probability distribution. While these differences are easy to appreciate in our example, it is not always so with real cases dealing with dispersion, slope or intrinsic scatter, discussed in the next sections.

12.5 Measuring a width in the presence of a contaminating population

We now focus on how to measure the scale (dispersion) of a distribution (say, of velocities v), knowing that the sample is contaminated by the presence of interlopers, but without the knowledge of which object is an interloper. The main idea is not to identify or de-weight interlopers in the scale estimate, but to account for them statistically, precisely as astronomers do with photons when estimating the flux of a source in the presence of a background.

We assume that data come from two populations: background galaxies, whose distribution is assumed to be a homogeneous (i.e., the intensity is independent of v) Poisson random process, and cluster galaxies, whose distribution is assumed to be a Poisson process whose intensity is Gaussian-distributed in v, i.e.,

$$I(v_i|...) = N_{\text{clus}}\mathcal{N}(v_i|v_{\text{clus}}, \sigma_{v_i}^2 + \sigma_{\text{clus}}^2) + \frac{N_{\text{bkg}}}{\Delta v}, \qquad (12.4)$$

where Δv is the velocity range over which velocities are considered (say, ± 5000 km s^{-1} from the cluster preliminary velocity centre), σ_{v_i} is the velocity error, N_{clus} and N_{bkg} are the number of cluster and background galaxies, and v_{clus} and σ_{clus} are our most interesting quantities: the cluster redshift and velocity dispersion.

Simple algebra shows that the likelihood of independently and identically distributed data, $p(v|I(v))$, is

$$p(v|I(v)) \propto \prod_i I(v_i|...)\; e^{-\int_v I(v|...)}. \qquad (12.5)$$

Combined with prior probability distributions for the parameters, this likelihood function yields, via Bayes' theorem, the posterior distribution for the function parameters, given the data. Uniform priors, zeroed at unphysical values of the parameters, are often adequate for the samples available. Markov chain Monte Carlo with a Metropolis sampler (Metropolis *et al.* 1953) may be used to sample the posterior. The chain provides a sampling of the posterior that directly gives credible intervals for whatever quantity, either for the parameters or any derived quantity: for an interval at the desired credible level it is simply a matter of taking the interval that includes the relevant percentage of the samplings.

Most literature estimates of cluster velocity dispersions are, instead, based on the family of estimators presented by Beers, Flynn and Gebhardt (1990), often called 'robust'. We now compare the performances of the 'robust' method and the Bayesian approach.

Let us consider a simulated 'cluster' having $\sigma_v = 1000$ km s^{-1} composed of 500 galaxies with Gaussian-distributed velocities, superposed over a background of 500 uniformly distributed (in velocity) interlopers, within ± 3000 km s^{-1}. Note that within 1000 km s^{-1} from the cluster centre there are on average $500 \times 0.68 = 340$

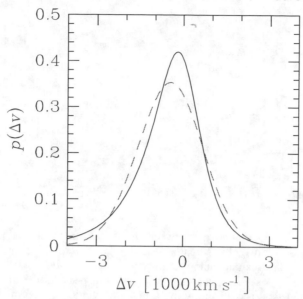

Fig. 12.3. Perturbed velocity distribution (solid line), given by Eq. (12.6), and a Gaussian with identical first two moments (dashed line). The former is used to generate hypothetic data, the latter is assumed to estimate σ_v.

members and $500/3 = 83$ background galaxies, i.e., the contamination is here just 20%. The large sample size has been adopted to let the data speak for themselves. Applying the methods of Beers *et al.* (1990) yields $\hat{\sigma}_v = 1400\,\mathrm{km\,s}^{-1}$, which is an excessively large estimate of σ_v (and hence of mass). The posterior mean is $940 \pm 85\,\mathrm{km\,s}^{-1}$ which is closer to the 'true' (input) value. This simulation shows the presence of a systematic bias in the Beers *et al.* estimator, even for a large sample. Actually, the bias is independent of the sample size, provided the relative fraction of cluster and interlopers is maintained.

We now acknowledge that in real-world experiments, we do not precisely know the model from which data are drawn (e.g., Is the velocity distribution perfectly Gaussian? Does it have more power in the wings or is it slightly tilted?). Let us therefore suppose that cluster substructure perturbs the velocity distribution, which we now assume to be described by

$$p(v) \propto e^{v/1000} \left(1 + e^{2.75v/1000}\right)^{-1}, \qquad (12.6)$$

depicted in Figure 12.3 (solid line). The function has first and second moments (mean and dispersion) equal to -460 and $1130\,\mathrm{km\,s}^{-1}$, respectively, excess kurtosis and a non-zero skewness. We simulate 1000 (virtual) clusters of 25 members each (and no interlopers) drawn from the distribution above (Eq. 12.6), but we compute the velocity dispersion using Eq. (12.4), i.e., with a likelihood function

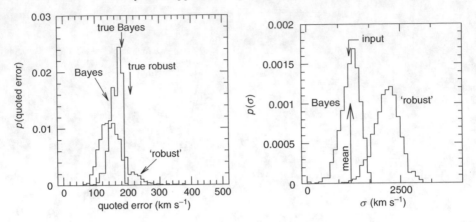

Fig. 12.4. *Left panel:* Comparison between the distributions of the quoted error (histograms) by 'robust' (biweight estimator of scale) and by our Bayesian method for 1000 simulations of 25 galaxies uncontaminated by background. The 'robust' error estimate is noiser (the histogram is wider) and somewhat biased, because the histogram is mostly on the left of the true error (given by the standard deviation of returned velocity dispersions). Bayesian errors (posterior standard deviation) are less biased and show a lower scatter. *Right panel:* The true (input) value of the velocity dispersion, and the histogram of recovered values by the biweight estimator of scale (right histogram) and by our Bayesian method (left histogram) for 1000 simulations of a sample of 25 galaxies, 50% contaminated by background. In the presence of a background, the 'robust' estimate of the velocity dispersion is biased.

appropriate for members drawn from a Gaussian, to make our study more realistic. The average of the determined posterior means is $1140\,\mathrm{km\,s^{-1}}$ (vs. the $1130\,\mathrm{km\,s^{-1}}$ input value) with a standard deviation of $185\,\mathrm{km\,s^{-1}}$. The uncertainty (posterior standard deviation), averaged over simulations, is $163\,\mathrm{km\,s^{-1}}$, close (as it should be) to the scatter of the posterior means. The uncertainty has a negligible scatter, $18\,\mathrm{km\,s^{-1}}$, indicating the low noise level of each individual uncertainty determination. The uncertainty of the dispersion error is four times better with a Bayesian estimation than using the methods of Beers *et al.*, displaying a scatter of $70\,\mathrm{km\,s^{-1}}$, and returning uncertainties as small as $73\,\mathrm{km\,s^{-1}}$ and as large as $865\,\mathrm{km\,s^{-1}}$ for data that are supposed to give a unique, fixed, value of uncertainty (see the left panel of Figure 12.4).

As a more difficult situation, we now consider a sample drawn, as before, from a distribution different from the one used for the analysis, but furthermore ~50% contaminated by interlopers and consisting of half as many members: 13 galaxies are drawn from the distribution above (Eq. 12.6), superposed with a background of 12 galaxies uniformly drawn from $\pm 5000\,\mathrm{km\,s^{-1}}$. Within $1130\,\mathrm{km\,s^{-1}}$ (i.e., $1\sigma_v$) from the centre the average contamination is about 20%. The average of the obtained posterior means is $1160\,\mathrm{km\,s^{-1}}$ (vs. the $1130\,\mathrm{km\,s^{-1}}$ input value).

The average uncertainty is $390\,\mathrm{km\,s^{-1}}$, with a low ($80\,\mathrm{km\,s^{-1}}$) scatter. The biweight estimator returns, on average, a strongly biased estimate, $2135\,\mathrm{km\,s^{-1}}$, see Figure 12.4.

As mentioned, mixtures often arise in astronomy and our Eqs. (12.4) and (12.5) hold equally for any Poisson signal superposed on a background, such as the distribution of galaxies in colour or the spatial distribution of X-ray photons, or galaxies, or whatever. We just need to re-name variables with names appropriate to the measurand and, eventually, consider a more complex model, for example if the background distribution is not uniform. In fact, the solution illustrated in this section has been developed to measure the X-ray core radius of a cluster of galaxies barely detected (Andreon *et al.* 2008) and later used to measure the cluster velocity dispersion.

12.6 Fitting a trend in the presence of outliers

Now we consider an apparently different problem: we observed some quantities x and y and we want to estimate some parameters describing how these two quantities vary as a function of each other. In astronomy, examples of these regressions are the Tully–Fisher, Faber–Jackson, and colour–magnitude relations, the fundamental plane, cluster scaling relations, and the Ghirlanda relation (for gamma-ray bursts). Many articles present their own method for the determination of these parameters (direct, inverse, orthogonal, bivariate correlated error and intrinsic scatter, measurement errors and intrinsic scatter fit). The Bayesian approach allows a simple solution, even in the difficult case of a linear fit in the presence of heteroscedastic (i.e., of different magnitude) errors on both variables and an intrinsic scatter (i.e., not accounted for by experimental errors), and censored or truncated data. In such a case, and for an ignorable data collection process (see below) and for variables having names appropriate for the colour–magnitude relation, slope a, intercept c, intrinsic scatter σ_{intr} of the colour–magnitude relation, and Gaussian photometric errors, the likelihood is a Gaussian (e.g., D'Agostini 2003, 2005; Gelman *et al.* 2004):

$$p(m_i, \mathrm{col}_i | a, c, \sigma_{\mathrm{intr}}, \text{no bkg}) \propto \mathcal{N}(\mathrm{col}_i | a\, m_i + c, \sigma_{\mathrm{intr}}^2 + \sigma_{\mathrm{col}_i}^2 + a^2 \sigma_{m_i}^2), \quad (12.7)$$

where σ_{m_i} and σ_{col_i} are the errors on magnitude and colour of the ith galaxy. The solution is quite intuitive: the colour–magnitude relation has a width given by the sum in quadrature of the intrinsic scatter, colour errors and magnitude errors propagated on the colour (via the slope a). In spite of the solution's simplicity, many pages are spent in journals trying to decide which approximate procedure (usually far more complicated than the equation above) should be used in which cases, all of which can be shown to be approximations of Eq. (12.7). Of course, a change of variable names makes the result useful for whatever scaling relation.

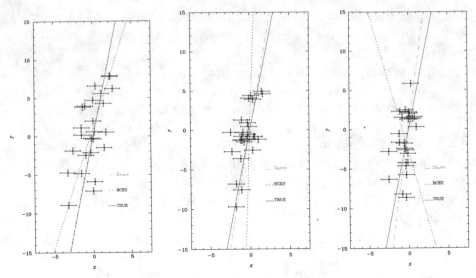

Fig. 12.5. Three simulated datasets of 25 objects (points), true trends from which data are generated (solid line), recovered trend by BCES (dotted line) and mean (Bayesian) model (dashed line). In our 1000 simulations, BCES results worse than those shown in the central and right panels occur in about ten per cent of the cases (see text for details).

To avoid recourse to maths, let us perform numerical simulations and compare the Bayesian approach and the state-of-the-art non-Bayesian astronomical method, BCES (Akritas & Bershady 1996). BCES accounts for intrinsic scatter and for heteroscedastic errors. We considered a sample of 25 objects obeying a linear trend of slope $a = 5$ with an intrinsic dispersion $\sigma_{intr} = 1$. The data have Gaussian errors, $\sigma_x = 1$ and $\sigma_y = 0.4$. In detail, the true x values have been drawn from a Gaussian having $\sigma_\tau = 1$ centred on $\mu_\tau = 0$. The true y values are given by $y = 5x$ (i.e., $c = 0$ in Eq. (12.7)). Observed x values and y values are computed by adding to each true x and y some noise (a Gaussian variate with $\sigma_x = 1$ and $\sigma_y = 0.4$, respectively). Because of the intrinsic scatter, y is perturbed by adding a Gaussian variate with $\sigma_{intr} = 1$. Figure 12.5 shows three simulated datasets. Qualitatively, these plots look similar to, or better than, many $L_X - \sigma_v$ relations seen in astronomical journals. We produced 1000 simulations of 25 data points. For each simulation we compute the slope and slope error as determined by BCES. We also compute the slope posterior mean and standard deviation, assuming uniform priors for all parameters but for the slope a, for which we take a uniform prior on the angle α ($a = \tan \alpha$). Since in our problem σ_x is comparable to the x range and the x distribution is far from being uniform (in the statistical jargon 'the data collection process is not ignorable'), the likelihood continues to be described by a Gaussian, as Eq. (12.7), but in a two-dimensional (y, x) space. Performing the

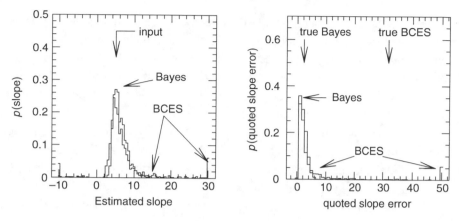

Fig. 12.6. Comparison between BCES and the Bayesian approach for our linear regression problem. We considered 1000 simulations of a sample of 25 objects uncontaminated by interlopers. *Left panel:* BCES sometimes returns badly wrong slope estimates: the BCES histogram distribution shows outliers, cumulated at −10 and 30. *Right panel:* True value of the slope error (vertical arrows), as measured by the scatter of returned slope minus input slope, and distribution of quoted errors (histograms). BCES is overly optimistic about the quality of its error; the large majority of the error estimates are small when the scatter between input and output slope is large. Furthermore, BCES displays a large scatter in the returned error, for data samples supposed to give identical values of uncertainty.

algebra associated with the matrix product gives a Gaussian $\mathcal{N}(\mu, \sigma^2)$ with parameters μ and σ^2:

$$\sigma^2 = (a^2\sigma_\tau^2 + \sigma_{\mathrm{intr}}^2 + \sigma_y^2)(\sigma_\tau^2 + \sigma_x^2) - a^2\sigma_\tau^4, \tag{12.8}$$

$$\mu = (\sigma_\tau^2 + \sigma_x^2)(y_i - c - a\mu_\tau)^2 - 2a\sigma_\tau^2(y_i - c - a\mu_\tau)(x_i - \mu_\tau)$$
$$+ (a^2\sigma_\tau^2 + \sigma_{\mathrm{intr}}^2 + \sigma_y^2)(x_i - \mu_\tau)^2. \tag{12.9}$$

As $\sigma_\tau \to \infty$, the likelihood converges to Eq. (12.7), i.e., Eq. (12.7) is an approximation of the present equation. Of course, the parameters used to produce the data (σ_τ, μ_τ, σ_{intr}, a and c) are assumed to be unknown in both analyses.

The left panel of Figure 12.6 shows that both the BCES method and the Bayesian approach return slopes whose distribution is centred on the input value, at least for our setting. However, BCES sometimes returns slopes very different from the input one (study the histogram wings, in particular around −10, 30 where we have cumulated more extreme values). The Bayesian approach does not show such catastrophic failures. The right panel of Figure 12.6 shows the distribution of the quoted errors. The important thing here is not how large (or small) the error claimed by a method is, i.e., the location of the plotted histogram, but the veracity of the claimed error, i.e., whether the quoted error distribution is located near or far from the true error (vertical arrow). The true error is computed as the scatter between the returned slope and the input slope. On average, BCES optimistically estimates errors by a large factor, mainly because in ten per cent of the cases it presents a catastrophical failure. The Bayesian method performs better in this respect: the quoted slope uncertainty is equal to the scatter between the input and output slopes, as it should be. Second, BCES displays a large scatter in the quoted slope error, for data samples supposed to give similar values of uncertainty.

To summarize, although BCES is not systematically in error, in ten per cent of our simulations, BCES returns badly wrong slopes with badly underestimated errors. In a real application, true values are unknown, and in such a case there is no way of knowing whether the BCES result is one of the frequently good values or a bad one. The Bayesian method better performs because it is better behaved, and there are no such catastrophic failures.

As formulated above, the problem does not account for our everyday experience: real samples are contaminated by interlopers, i.e., objects unrelated to the ones we are interested in. Now our model will be a *mixture of two regressions*, one carrying the signal (the cluster colour–magnitude relation) and the other describing the background (galaxies, objects on the line of sight), with the usual difficulty that we do not know which galaxy belongs to the cluster and which one is simply projected along the cluster line of sight. A real case (the cluster Abell 1185, from Andreon *et al.* 2006a) is shown in Figure 12.7. The distribution of background galaxies in the m, col space is not uniform, and therefore the background is modelled by an inhomogeneous process, $B(m_i, \text{col}_i | m, \text{col})$. Therefore, the likelihood of the ith galaxy, $p(m_i, \text{col}_i | a, c, \sigma_{\text{intr}})$, is given by the mixture of two distributions:

$$p(m_i, \text{col}_i | a, c, \sigma_{\text{intr}}, \alpha, M^*\phi^*) = \Omega_j B(m_i, \text{col}_i | m, \text{col}) \qquad (12.10)$$
$$+ \delta_c \Omega_j \mathcal{N}(\text{col}_i | a\, m_i + c, \sigma_{\text{intr}}^2 + \sigma_{\text{col}_i}^2 + a^2 \sigma_{m_i}^2) S(m_i | \alpha, M^*\phi^*).$$

In this equation we considered the usual case, where we have a control field in the form of data from a sky region uncontaminated by the cluster contribution. In such a case, $\delta_c = 1$ for cluster datasets, $\delta_c = 0$ for the other datasets, Ω_j is the studied solid angle. Otherwise, it is just a matter of replacing δ_c with a radial profile. S is the usual Schechter (1976) function, with α, M^* and ϕ^* parameters, that describes the luminosity function of galaxies. We have also assumed that the data collection model is ignorable, for mathematical convenience.

As in Eq. (12.5), the likelihood of independently and identically distributed data is given by the product, over the data m_i, col$_i$ of the individual likelihood terms. As shown there, the likelihood includes an integral term, given by the integral of the model over the values ranges. The integral should be performed on the appropriate colour and magnitude ranges (those accessible to the data) and it is equal to the *expected* number of galaxies. This term disfavours models that predict a

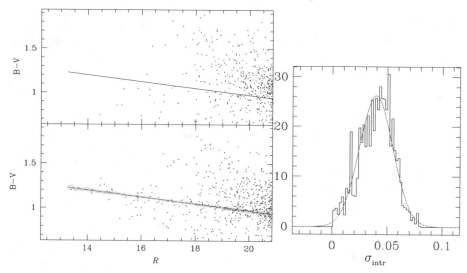

Fig. 12.7. *Left panels:* Colour–magnitude diagrams for background galaxies (upper panel) and cluster+background galaxies (lower panel). These are true data for the cluster Abell 1185, presented in Andreon *et al.* (2006a). The solid line is the mean colour–magnitude relation of cluster galaxies computed as described in the text. The shaded region marks the 68% highest posterior interval. *Right panel:* Posterior probability distribution of the colour–magnitude intrinsic scatter. The jagged nature of the distribution is due to the finite lengh of the MCMC chain. A Gaussian with first two moments matching the distribution is overplotted to guide the eye.

number of galaxies very different from the observed one. If errors on m (mag) are not negligible, S in Eq. (12.10) should be replaced by the convolution between the Schechter function and the error function. The inference proceeds as usual, by choosing a prior, computing the posterior, and summarizing the result of the computation above with a few numbers, those of scientific interest.

The problem of determining an intrinsic scatter around a linear trend in the presence of outliers or a background population is so difficult that, to our best knowledge, there are no non-Bayesian solutions to be compared with our Bayesian approach. We cannot, therefore, simulate some data and compare the performances of different methods because of the lack of a contender.

Had we (mis)used BCES, then the obtained slope of the colour–magnitude relation shown in Figure 12.7 would be completely wrong, and equal to the one of the background population, outnumbering, by a factor of four, the cluster population. This occurs because BCES is not built to be robust against a contaminating population. An ad hoc solution often used in astronomical papers is to remove the slope dependency by subtracting off an expected one (e.g., one observed at $z = 0$, assuming no slope evolution), measuring the scatter using 'robust' methods, as described in the previous section, and finally quadratically subtracting colour errors from the measured scatter, following Stanford *et al.* (1998). This procedure often leads to intrinsic scatter with 68% error bars extending to negative values, including some examples in Stanford *et al.* (1998). We have already discussed the shortcomings of using the 'robust' estimate of the scatter. The Bayesian method does not require ad hoc methods, does not make assumptions on the trend slope, and always returns positive intrinsic dispersions, as in the case of Abell 1185 shown in Figure 12.7.

Readers interested in fractions f_b bounded in the $[0, 1]$ range, or hardness ratios, $(H - S)/(H + S)$ bounded in the physical range for data contaminated by a background, may consult Andreon *et al.* (2006b) and D'Agostini (2004). The hardness ratio has the same mathematical properties of $1 - 2f_b$, i.e., one minus twice the (blue) fraction discussed in these two works.

12.7 What is the number returned by tests such as χ^2, KS, etc.?

Many articles measure the 'probability of rejecting the null hypothesis' using some statistical tests, for example, Kolmogorov–Smirnov's, Kendall's, Spearman's rank correlation, F-, Student's t-, Wilcoxon rank, and χ^2 tests. Many of us have noted oddities with the numbers (called p-values) returned by them: by taking two statistical tests we sometimes found widely different 'probabilities', e.g., 0.001 and 0.861. How can this be the case, since the desired result is a single unique value? In our example, which one is the good probability, the one rejecting the null or the other one? The mere existence of a variety of tests, as opposed to a single one, is an indication that no test always gives the desired number. Actually, p-values are not the probability of the hypothesis, which is the desired probability. They are the probability of observing more discrepant values of the chosen statistic for hypothetical data drawn from the null hypothesis, that is, the probability of rejecting the null hypothesis when it is true. There is nothing strange that two different statistics (measures, say, of height and width) of data drawn under the null hypothesis takes different values.

The difference between the p-values and probability of the hypothesis can be better understood with an astronomical example: the detection of faint sources. In such a case, the null hypothesis to reject is 'no source is there'. Let I_0 be the flux measured at the target position. A usual way to compute the detection confidence is by measuring how frequently one observes larger values, $>I_0$, under the null, i.e., in areas free from sources: $p(>I_0|\text{background})$. The p-value is, precisely, the measured frequency. For many famous tests, like those mentioned at the start of the section, the probability distribution of the test statistic is analytically known and there is no need of further data (the background) in order to compute the distribution of the test statistics.

Let us suppose we have found a p-value of 0.003, i.e., that measurements free of sources gives $p(>I_0|\text{background}) = 0.003$. Does this mean that the target is real at one minus the p-value confidence, $p(\text{source}|I_0) = 1 - 0.003 = 0.997$, i.e., is real at 99.7% confidence? Certainly not. Qualitatively, if sources fill a small portion of the sky, there is a lot of sky left to the background. Then, statistical fluctuations of the background, even rare ones, may overwhelm the number of true sources. In such a case, only a very tiny fraction of detections are true, not 99.7% as the p-value leads us to believe. More quantitatively: let x be the portion of sky occupied by sources (when observed in the same observational set-up that gives a p-value of 0.003), and N the number of independent beams in the sky. Note that x is the probability a priori that there is a source in a beam. Then, xN beams are occupied by sources and $(1 - x)N$ are not. Assuming a 100% detection efficiency, xN are true sources detected, and $0.003(1 - x)N$ will be instead false positive detections. Thus, there will be $xN + 0.003(1 - x)N$ detections, but among them only xN are true sources. The probability of being a true source, $p(\text{source}|I_0)$, is given by the fraction of true detected sources over total number of detections: $x/(x + 0.003(1 - x))$. If $x = 0.07\%$, a value appropriate for typical Chandra exposures, then sources believed to be detected at 99.7% confidence (or better, with a p-value of 0.003) have only a 19% probability of being real. Adopting instead a 5% p-value, we end up with a catalogue composed of entries that are junk 99 times out of 100, instead of being true sources 95% of the time, as the p-value suggests. Only in fortunate cases (appropriate values of x) will one have similar numbers for the p-value and the probability of rejecting the null hypothesis. Therefore, these two probabilities are conceptually different and take different values.

As shown in the example, the desired probability does depend on the a priori probability of the (null) hypothesis (x in the example). However, virtually all non-Bayesian astronomical papers compute p-values but call them 'the probability of rejecting the null hypothesis'. For example, in testing the reality of a trend, the Spearman rank correlation test is often used, and the one minus the p-value is quoted as the probability of rejecting the null ('no trend') hypothesis. Such a practice ignores the essential role played by the a priori probability of the competing hypothesis, which, in principle, may convert a '95%' confident result into an inconclusive result, as in our example. The Bayesian approach is based on probabilities for the hypothesis – it cannot ignore them – and in our example Bayes' theorem takes a form very similar to the one we have used to evaluate the desired probability.

12.8 Summary

The Bayesian approach solves some difficulties encountered with other procedures. It works in the regime of typical researcher activity: when looked-for effects are marginally significant, or near boundaries, such as when the small intrinsic dispersion of the colour–magnitude relation is to be determined, or when there is no agreement among astronomers how to set up the right procedure, as for the regression problem in the absence of a background. It also offers a solution when no other is there, as in the case of the fit of a trend in the presence of a contaminating population. It works when otherwise obtained results are unsatisfactory, as for velocity dispersions or for cases when other procedures return unphysical values. The Bayesian approach already includes corrections for biases, as for the Eddington bias. The ultimate reason for its good performance is highlighted in two idealized cases at the start of this chapter: (a) it obeys the sum axiom of probability and thus averages (marginalizes) over unknown quantities, instead of maximizing the value of some ad hoc estimators; and (b) it performs inferences following the Bayes' theorem instead of considering priors as something to be avoided. Finally, the Bayesian approach clarifies what other methods are actually computing, for example, the meaning of the number returned by Kolmogorov–Smirnov, χ^2, and Wilcoxon rank tests.

Let us conclude this chapter by remembering that the scientific method suggests preference for a procedure known to work over one whose reliability is uncertain.

References

Akritas, M. G. and Bershady, M. A. (1996). *Astrophys. J.*, **470**, 706.

Andreon, S. (2008). *Mon. Not. Roy. Astron. Soc.*, **386**, 1045.

Andreon, S. and Harn, M. (2009). *Mon. Not. Roy. Astron. Soc.*, in press.

Andreon, S., Cuillandre, J.-C., Puddu, E. and Mellier, Y. (2006a). *Mon. Not. Roy. Astron. Soc.*, **372**, 60.

Andreon, S., Quintana, H., Tajer, M., Galaz, G. and Surdej, J. (2006b). *Mon. Not. Roy. Astron. Soc.*, **365**, 915.

Andreon, S., De Propris, R., Puddu, E., Giordano, L. and Quintana, H. (2008). *Mon. Not. Roy. Astron. Soc.*, **383**, 102.

Beers, T., Flynn, K. and Gebhardt, K. (1990). *Astron. J.*, **100**, 32.

D'Agostini, G. (2003). *Bayesian Reasoning in Data Analysi: A Critical Introduction*. Singapore: World Scientific Publishing.

D'Agostini, G. (2004). arXiv:physics/0412069.

D'Agostini, G. (2005). arXiv:physics/0511182.

Eddington, A. S. (1913). *Mon. Not. Roy. Astron. Soc.*, **73**, 359.

Gelman, A., Carlin, J., Stern, H. and Rubin, D. (2004). *Bayesian Data Analysis*. London: Chapman & Hall/CRC.

Kenter, A. *et al.* (2005). *Astrophys. J. Supp.*, **161**, 9.

Metropolis, N., Rosenbluth, A., Rosenbluth, M., Teller, A. and Teller, E. (1953). *J. Chem. Phys.*, **21**, 1087.

Schechter, P. (1976). *Astrophys. J.*, **203**, 297.

Stanford, S. A., Eisenhardt, P. R. and Dickinson, M. (1998). *Astrophys. J.*, **492**, 461.

13

Photometric redshift estimation: methods and applications

Ofer Lahav, Filipe B. Abdalla and Manda Banerji

13.1 Introduction

Estimating the distance to an astronomical object is a fundamental problem in astronomy. Distances to galaxies are usually estimated from their redshifts:

$$z = \frac{\lambda_{\mathrm{obs}}}{\lambda_{\mathrm{em}}} - 1 \,, \tag{13.1}$$

where λ_{obs} and λ_{em} are the observed and emitted wavelengths. Ideally the redshift is derived from a high-resolution spectrum. The alternative is to use the photometric redshift technique. Using the colours of a galaxy observed in a selection of medium- or broad-band filters, we can gain a crude approximation of the galaxy's spectral energy distribution (SED), from which its redshift and spectral type may be found. In Figure 13.1 we show how the colours of a galaxy change with redshift, and the major features that can be picked up in each of the bands.

The technique is very efficient compared with spectroscopic redshifts, since the signal-to-noise in broad-band filters is much greater than the signal-to-noise in a dispersed spectrum. Furthermore, a whole field of galaxies may be imaged at once, while spectroscopy is limited to individual galaxies or those that can be positioned on slits or fibres. However, photometric redshifts are only approximate at best and are sometimes subject to large errors. For many applications though, large sample sizes are more important than precise redshifts and photometric redshifts may be used to good effect.

Photometric redshifts date back to Baum (1962). For reviews see Weymann *et al.* (1999) and Koo (1999). The methods have been used extensively in recent years on the ultra-deep and well-calibrated Hubble Deep Field observations (e.g., Connolly, Szalay & Brunner 1998).

283

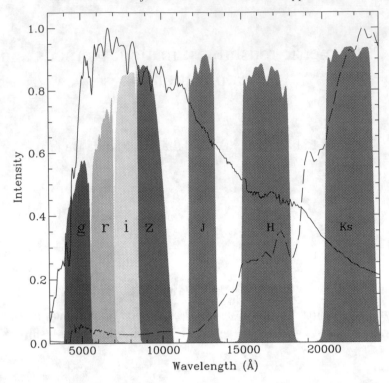

Fig. 13.1. The SED of an elliptical galaxy at redshift $z = 0$ (solid line) and the same galaxy at $z = 3$ (dashed line) as seen through a set of broad-band optical ($griz$) and near-infrared (JHK_s) filters. The Balmer break at 4000 Å moves from the g-band at $z = 0$ to the near-infrared bands as the galaxy is redshifted to $z = 3$. (Figure courtesy of Eduardo Cypriano.)

The photo-z approach has regained popularity with the planning of wide-field deep imaging surveys, e.g., SDSS[1], CFHTLS[2], KIDS, Subaru Hyper-Suprime Cam, Pan-STARRS[3], the Dark Energy Survey[4] (combined with the VISTA Hemisphere Survey[5]), LSST[6], DUNE/Euclid[7], SNAP[8], PAU[9] and Skymapper[10]. The accuracy with which photometric redshifts can be calculated for these surveys is crucial for the determination of cosmological parameters such as the dark energy equation of state.

[1] http://www.sdss.org
[2] http://www.cfht.hawaii.edu/Science/CFHLS
[3] http://pan-starrs.ifa.hawaii.edu/public/
[4] https://www.darkenergysurvey.org
[5] http://www.ast.cam.ac.uk/~rgm/vhs
[6] http://www.lsst.org
[7] http://www.dune-mission.net
[8] http://snap.lbl.gov/
[9] http://www/ice.csic.es/pau/
[10] http://www/mso.anu.edu.au/skymapper/

For a statistician, photo-z estimation is an inverse problem: going from the colour space to the redshift, type and other galaxy properties. There are two common approaches to the problem: (1) template methods (e.g., HYPERZ) and (2) training methods (e.g., ANNZ) as described below. We note that training methods can also be used on synthetic data, so they are not necessarily 'empirical'.

We also note that there could be different deliverables required from the photo-z estimation technique, e.g., a catalogue where a photo-z and associated error is given per galaxy, or, as required for certain applications, just the redshift distribution $N(z)$. A final comment is that different techniques have been developed for the photo-z for field galaxies and cluster galaxies. The focus of this review is on field galaxies.

13.2 Template methods

The most basic approach is the template-fitting technique. A commonly used package is HYPERZ (Bolzonella *et al.* 2000). This involves compiling a library of template spectra, either theoretical SEDs from population synthesis models (e.g., GISSEL, Bruzual & Charlot 1993, BC hereafter) or empirical SEDs (e.g., Coleman, Wu & Weedman 1980, CWW hereafter). The expected flux through each survey filter is calculated for each template SED on a grid of redshifts, with corrections for the interstellar medium, intergalactic medium and Galactic extinction where necessary. A redshift and spectral type are estimated for each observed galaxy by minimizing χ^2 with respect to redshift, z, and spectral type, SED, where

$$\chi^2(z, \text{SED}) = \sum_i \left(\frac{f_i - \alpha(z, \text{SED}) t_i(z, \text{SED})}{\sigma_i} \right)^2, \qquad (13.2)$$

f_i is the observed flux in filter i, σ_i is the error in f_i, $t_i(z, \text{SED})$ is the flux in filter i for the template SED at redshift z. $\alpha(z, \text{SED})$, the scaling factor normalizing the template to the observed flux, is determined by minimizing Eq. (13.2) with respect to α, giving

$$\alpha(z, \text{SED}) = \frac{\sum_i \frac{f_i t_i(z, \text{SED})}{\sigma_i^2}}{\sum_i \frac{t_i(z, \text{SED})^2}{\sigma_i^2}}. \qquad (13.3)$$

The template-fitting photometric redshift technique makes use of the available and reasonably detailed knowledge of galaxy SEDs, and in principle may be used reliably even for populations of galaxies for which there are few or no spectroscopically confirmed redshifts. However, crucial to its success is the compilation of a library of accurate and representative template SEDs. Empirical templates are typically derived from nearby bright galaxies, which may not be truly representative of high-redshift galaxies. Conversely, while theoretical SEDs can cover a large range of star-formation histories, metallicities, dust extinction models, etc., not all combinations of these parameters (at any particular redshift) are realistic, and the ad hoc inclusion of superfluous templates increases the potential for errors when using observations with noisy photometry.

13.3 Bayesian methods and non-colour priors

An extension of the above likelihood (χ^2) approach is to incorporate priors, using the Bayesian framework. Benitez (2000) formulated the problem as follows. The probability of a galaxy with colour C and magnitude m having a redshift z is

$$p(z|C, m) = \frac{p(z|m)p(C|z)}{p(C)} \propto p(z|m)p(C|z), \tag{13.4}$$

where the term $p(C|z)$ is the conventional redshift likelihood employed, e.g., by HYPERZ, and $p(C)$ is just a normalization. The new important ingredient is $p(z|m)$, which brings in the prior knowledge of the magnitude–redshift distribution. With the aid of the extra information (prior), this approach is effective in avoiding catastrophic errors of placing a galaxy at a wrong and unrealistic redshift.

As already mentioned, photo-z estimation can be viewed as mapping from colour space to a galaxy's redshift. In fact, Jain, Connolly and Takada (2007) have proposed a method called 'colour tomography'. The idea is to use limited colour information from two or three bands to derive a coarse redshift distribution $N(z)$, thus bypassing the need for a photo-z per galaxy.

Colour information is not the only input that can be used by a photometric redshift method. Structural properties of a galaxy such as its size (e.g., Firth, Lahav & Somerville 2003), concentration indices (e.g., Oyaizu *et al.* 2008) and radial light profiles (e.g., Wray & Gunn 2008) have previously been used as inputs. These are useful as the size, for example, is a function of distance and therefore can be used to get redshift information. We also note that galaxy magnitudes can be measured using different aperture sizes. This will result in different noise levels in the galaxy photometry and therefore colour gradients.

Recently, a new idea has been proposed – to use the galaxy surface brightness as a prior (e.g., Kurtz *et al.* 2007; Stabenau, Connolly & Jain 2008). As the surface brightness dimming is proportional to $(1 + z)^{-4}$ in any conventional cosmology, it can serve as a natural prior. Indeed, using this prior eliminates a large fraction of outliers.

13.4 Training methods and neural networks

Another approach can be used when one has a sufficiently large (e.g., $\sim$100–1000, depending on the redshift range) and representative sub-sample with spectroscopic redshifts. Then one can fit a polynomial or other function mapping the photometric data to the known redshifts and use this to estimate redshifts for the remainder of the sample with unknown redshifts (e.g., Connolly *et al.* 1995; Brunner, Szalay & Connolly 2000; Sowards-Emmerd *et al.* 2000). With this approach, errors in the estimated redshifts may also be calculated analytically or via Monte Carlo simulations.

An extension of the latter approach is to use artificial neural networks (ANNs). ANNs have been used before in astronomy for, amongst other applications, galaxy morphological classification (e.g., Naim *et al.* 1995; Lahav *et al.* 1996), morphological star/galaxy separation (e.g., Bertin & Arnouts 1996; Andreon *et al.* 2000) and stellar spectral classification (e.g., Bailer-Jones, Irwin & von Hippel 1998). Essentially, an ANN takes a set of inputs (e.g., logarithms of fluxes – i.e., magnitudes – in different filters) for each object, applies some non-linear function, and outputs a value (e.g., the estimated redshift). The ANN is first trained, i.e., the coefficients (weights) of the function are optimized, by using a training set where the desired output is known. The ANN may then be used on any number of other objects with similar inputs (i.e., magnitudes in the same filter set) but unknown outputs (i.e., redshifts).

An ANN comprises a set of input nodes, one or more output nodes, and one or more hidden layers each containing a number of nodes (Figure 13.2); see, e.g., Bishop (1995) for more details. For example, 5:2:1 takes 5 inputs (colours), has 2 nodes in a single hidden layer and gives a single output (redshift). The nodes are connected and each connection carries a weight, which together comprise the vector of coefficients $\mathbf{w}$ that are to be optimized. The input parameters for each object are represented by the vector $\mathbf{x}$ (e.g., the magnitudes in a set of filters). Given a training set of inputs $\mathbf{x}_k$ and desired outputs z_k (e.g., the redshift), the ANN is optimized by minimizing the cost function

$$E = \frac{1}{2} \sum_k [z_k - F(\mathbf{w}, \mathbf{x}_k)]^2. \tag{13.5}$$

The function $F(\mathbf{w}, \mathbf{x}_k)$ is given by the network. A function g_p is defined at each node p, taking as its argument,

$$u_p = \sum_j w_j x_j, \tag{13.6}$$

where the sum is over the input nodes to p. These functions are typically taken (in analogy to biological neurons) to be sigmoid functions such as $g_p(u_p) = 1/[1 +$

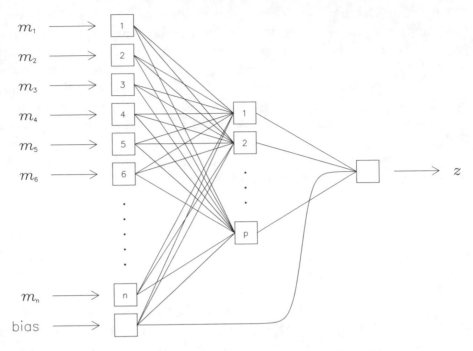

Fig. 13.2. A schematic diagram of an ANN with input nodes taking, for example, magnitudes $m_i = -2.5 \log_{10} f_i$ in various filters, a single hidden layer, and a single output node giving, for example, redshift z. The architecture is $n{:}p{:}1$ in the notation used in this chapter. Each connecting line carries a weight w_j. The bias node allows for an additive constant in the network function defined at each node. More complex nets can have additional hidden layers. (Illustration from Collister & Lahav 2004.)

$\exp(-u_p)]$, which is used here. An extra input node – the bias node – is automatically included to allow for additive constants in these functions. The combination of these functions over all the network nodes makes up the function F. The algorithm takes as its input a network architecture, a training set and a random seed to initiate the weight vector $\mathbf{w}$, and uses an iterative quasi-Newton method (see, e.g., Bishop 1995) to minimize the cost function. To ensure that the weights are regularized (i.e., they do not become too large), an extra quadratic cost term is commonly added to Eq. (13.5). After each training iteration, the cost function is evaluated on a separate validation set. After a chosen number of training iterations, training terminates and the final weights chosen for the ANN are those from the iteration at which the cost function is minimal on the validation set. This is useful to avoid over-fitting to the training set if the training set is small. The Neural Network approach ('ANNz') was implemented by Firth, Lahav and Somerville (2003), and by Collister and Lahav (2004).

While choosing a template library that is representative is a source of concern for the template-fitting method, ANNs automatically fit the true range of galaxy SEDs. Another potential advantage of ANNs relative to the template-fitting method is that the weights applied to each filter may be more optimal than simple χ^2 weighting. In addition, one can also feed in other observational inputs, such as image size or surface brightness, morphology and concentration parameters, where such data are available. We note that a single input node with a magnitude input can give redshift information, simply because the flux of a galaxy is inversely proportional to the distance squared.

Can ANNs be viewed as Bayesian? If the training set is 'faithful' then the ANN learns from a dataset very similar to the target distribution. If the training set is not representative then the method may result in biases (see below). See Bishop (1995) and Lahav *et al.* (1996) and references therein for the interpretation of neural networks in the Bayesian framework.

13.5 Errors on photo-z

Each method described above has a way of obtaining error estimates which reflects the degree of confidence with which one can rely on the photometric redshift obtained.

In template methods the χ^2 can be computed for each potential redshift value. This produces a probability distribution function and the errors are obtained from the points where the probability falls one-sigma away from the best-fit value. As other techniques use different methods for photometric redshift estimation they have to obtain error estimates in a different way. For example, in a Neural Network method, the error estimate is obtained through a chain rule that relates errors in the original magnitudes to errors on the final redshift (Collister & Lahav 2004). The estimated magnitude errors can then be used to obtain a redshift error estimate. Recently a more general method based on nearest neighbours has been devised and can be applied to any sample with a training set independently of the photo-z estimator used. This method, described in Oyaizu *et al.* (2007), uses the differences between the spectroscopic redshift and the photometric redshift of the galaxies in a training set as a function of the magnitudes. An error estimate is obtained by a nearest neighbour interpolation of this training data in magnitude space and produces very good results.

The error, or more generally the probability distribution function attached to each redshift, is a very important quantity. For example, galaxies with large errors in the photo-z can be removed ('clipped') from the sample before using them for any kind of cosmological analysis. Depending on the science question being addressed, there will be a trade-off between the amount of clipping applied to remove undesired outliers, and the number of galaxies required to keep the shot-noise levels low. For an illustration of the clipping procedure as applied to weak lensing and baryon acoustic oscillation studies, see Abdalla *et al.* (2008a) and Banerji *et al.* (2008), respectively.

13.6 Optimal filters

The performance of any photo-z method depends directly on the input colours and therefore the number of filters, their wavelength range (response function) and the exposure time in each. The choice of optimal filters is very much dependent on the science one wants to achieve. Naively speaking, dividing a single filter into two with equal light falling in each will always seem more attractive from a photometric redshift perspective due to the additional colour information obtained compared with the single-filter case. However, if one considers the noise associated with reading more CCDs, as well as the extra telescope time needed to collect photons with two different filters, then there is an optimal number of filters which would deliver the science required. This optimization has been done internally by the DES team and it was found that for galaxies between redshift 0 and 1.5, an optimal number of filters is 4 to 5, with a larger fraction of time spent on the redder filters to take account of the extra sky noise present here. Furthermore, if any near-infrared (NIR) bands are included, the number of NIR bands makes little difference to the final redshift accuracy provided the total integration time is conserved.

Similar conclusions were reached using simulations for deeper surveys in the NIR, such as the proposed Euclid satellite (Abdalla *et al.* 2008a). However, these general arguments break down when the required science becomes more specific. For example, if one is interested in high-redshift quasar studies, the Y band becomes much more important, as the Lyman break is bracketed by that band at a redshift of 6. Alternatively, if one is interested in features along the line of sight, such as baryonic acoustic oscillations (BAOs), then a much larger number of filters would be needed, as filters as wide as 1000 Å cannot obtain the resolution to get the accuracy in redshift needed to see the BAOs along the line of sight. On the other hand, if low-redshift objects are of interest, then the 4000 Å break has to be bracketed by a u-band filter and bluer bands become important.

13.7 Comparison of photo-z codes

It is interesting to see how different photometric redshift estimation methods compare in evaluating photo-z's for the same dataset. There are various publicly available software packages currently available for photo-z estimation. These are listed in Table 13.1. In addition to these, many other methods have been used extensively in the literature, but have not as yet been made public.

Table 13.1. *Publicly available software packages for photo-z estimation.*

Code	Authors/Web link	Method
HYPERZ	Bolzonella *et al.* 2000 http://webast.ast.obs-mip.fr/hyperz/	Likelihood
BPZ	Benitez 2000 http://acs.pha.jhu.edu/~txitxo/bpzdoc.html	Bayesian priors
ANNZ	Collister & Lahav 2004 http://zuserver2.star.ucl.ac.uk/~lahav/annz.html	Neural networks
IMPZ	Babbedge *et al.* 2004 http://astro.ic.ac.uk/~tsb1/Impzlite/ImpZlite.html	Template
ZEBRA	Feldmann *et al.* 2006 www.exp-astro.phys.ethz.ch/ZEBRA	Bayesian, Hybrid
KCORRECT	Blanton & Roweis 2007 http://cosmo.nyu.edu/blanton/kcorrect/	Model templates
Le Phare	Arnouts & Ilbert http://www.oamp.fr/people/arnouts/LE_PHARE.html	Template
EAZY	Brammer *et al.* 2008 http://www.astro.yale.edu/eazy/	Template

We now analyze the result of running four different photo-z algorithms on the same data sample of ~5400 luminous red galaxies from the 2dF and SDSS LRG and QSO (2SLAQ) survey. We use a training set method, ANNZ, a template maximum likelihood method, HYPERZ with both synthetic (BC) and empirical (CWW) templates, and a Bayesian method ZEBRA which, in addition to its use of Bayesian priors, can be thought of as a hybrid template and empirical code. This is because a training set of galaxies (with available spectroscopic redshifts) are used to *repair* the templates until a suitable match is obtained between the spectroscopic and photometric redshifts.

We can easily compare the results from the different codes, as the spectroscopic redshifts are available for all these objects. As an example, a plot of the spectroscopic redshift versus the photometric redshift obtained using the neural network code, ANNZ, is shown in Figure 13.3. In a similar way, ANNZ was used to create a catalogue of LRGs with a million photometric redshifts (MegaZ-LRG; Collister *et al.* 2007).

In Figure 13.4 we plot the one-sigma scatter on the spectroscopic redshift in each photo-z bin as a function of the photometric redshift. We also plot the bias in each photometric redshift bin, defined simply as the mean difference between the photometric and spectroscopic redshifts in each bin. The training method ANNZ is almost free from bias due to the presence of a complete and representative training set across the entire redshift range. However, the one-sigma scatter is higher for this method in the lowest and highest photo-z bins due to a lack of sufficient

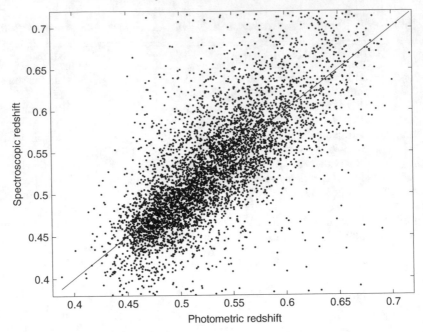

Fig. 13.3. Photometric redshift versus spectroscopic redshift for 5482 luminous red galaxies from the 2SLAQ survey. The photometric redshifts were derived using ANNz. (From Abdalla *et al.* 2008b)

numbers of training galaxies in these bins. The template method gives very different results based on whether empirical or synthetic templates are used. The synthetic templates seem to produce much better results in this case. The Bayesian method suffers from a large bias and scatter in the highest photometric redshift bins, but performs slightly better than the maximum likelihood method across the entire redshift range.

It is obvious that each photo-z method has its advantages and disadvantages, depending on the dataset and redshift regime in which it is being applied. Other empirical methods, such as support vector machines (e.g., Wadadekar 2005) and kernel regression (e.g., Wang *et al.* 2007), have also been used for photometric redshift estimation, but to our knowledge are yet to be made available as public codes.

13.8 The role of spectroscopic datasets

First, we should distinguish between spectroscopic sets for training methods (e.g., ANNz) and spectroscopic data for calibration of the photometric redshift distribution obtained by either template or training sets. Typically, a larger spectroscopic set is needed for calibration than for training. We also note that one needs

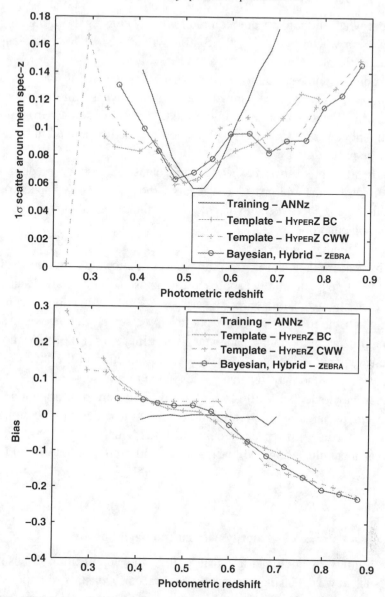

Fig. 13.4. The one-sigma scatter on the mean spectroscopic redshift in each photometric redshift bin (upper panel) and the bias (lower panel) as a function of photometric redshift obtained after running three different codes on a sample of 5482 luminous red galaxies from the 2SLAQ survey. The three types of codes compared are the training method ANNz, the maximum likelihood template method HyperZ, using both synthetic (BC) and empirical (CWW) templates, and the Bayesian hybrid method ZEBRA that uses a training set of galaxies with known spectroscopic redshifts to optimize templates. (From Abdalla *et al.* 2008b.)

to consider whether photo-z's are required for every single galaxy (e.g., to produce a catalogue) or whether an $N(z)$ distribution will suffice, as for weak lensing and baryon acoustic oscillation studies.

Let us consider three scenarios. Firstly, when no spectroscopic datasets are available, one has to use either the template method to derive a photo-z, or the training methods with synthetic data, based on a model for $N(z)$ derived from additional observational constraints. However, photometric redshifts derived in this way will probably be prone to biases.

Secondly, if there is a *perfect* spectroscopic dataset, which is a random subset of the photometric data, and provided it is large enough (10^4-10^5 spectra), we can use it both to get a photo-z catalogue via the training method *and* to calibrate $N(z)$ for cosmological probes.

Thirdly, if there is an *incomplete* spectroscopic dataset that is not representative of the photometric data, but the incompleteness is known as a function of magnitude and redshift, a quantitative approach can be used. For a detailed study of such biases in the photo-z estimate, see Banerji *et al.* (2008). One approach is to use a weighting scheme to reconstruct the photo-z $N(z)$ given the incomplete spectra, therefore generating a photo-z catalogue as in Lima *et al.* (2008). We note that if the spectroscopic training set covers only a small fraction of the sky, it may suffer 'cosmic variance' and may generate biases in the testing sets.

Another way to derive $N(z)$ is with the aid of clustering information. The idea is to measure the angular cross-correlation between galaxies with photometric redshifts and galaxies with spectroscopic redshifts. This, together with certain assumptions about clustering and biasing, can yield more accurate $N(z)$ (Newman 2008).

13.9 Synergy with cosmological probes

The main utility of photometric redshifts is in their application to cosmological probes, such as weak lensing tomography and baryon acoustic oscillations. In weak lensing tomography, the idea is to slice a lensing survey into photometric redshift bins and to analyze the cosmic shear of high signal-to-noise galaxy images in each photo-z bin, e.g., as discussed by Ma *et al.* (2006), Amara and Refregier (2007), and Abdalla *et al.* (2008a). Uncertainties in the photometric redshifts will therefore propagate into errors in the cosmological parameters.

A simple back-of-the-envelope calculation helps to explain the link between the dark energy parameters, the photometric redshift performance and the number of spectroscopic redshifts. Consider a distribution of sources selected from a photometric redshift bin that results in a more complicated (e.g., Gaussian) distribution with respect to true (spectroscopic) redshift, with mean redshift $\bar{z}$ and variance:

$$\mu_2 = \left\langle (z_{\mathrm{spec}} - \bar{z})^2 \right\rangle. \tag{13.7}$$

Assuming Poisson statistics we can predict the variance in the mean redshift $\bar{z}$ given N_{spec} spectroscopic redshifts associated with that photo-z bin:

$$(\delta z)^2 \equiv \mathrm{rms}^2(\bar{z}) = \frac{\mu_2}{N_s}. \tag{13.8}$$

We can now crudely model the uncertainty in deriving the constant dark energy parameter $w = P/\rho$ from weak lensing, if the only uncertainty is due to photo-z errors:

$$\left| \frac{\delta w}{w} \right| \approx 5 \frac{\delta z}{\bar{z}} = \frac{5}{\bar{z}} \sqrt{\frac{\mu_2}{N_s}}. \tag{13.9}$$

The pre-factor of approximately 5 can be estimated from detailed modelling of the weak lensing power spectrum. It can be justified qualitatively by examining the sensitivity of cosmological distance and the linear growth to variations in w (Peacock *et al.* 2006).

For example, for a desired fractional error of 1% on w, $\bar{z} = 1$ and $\mu_2 = 0.06$ (derived from mock catalogues and ANNz averaged over a range of proposed Euclid optical and NIR bands) we find that $N_s \approx 15\,000$ in each bin or, for say 10 bins, a total of $150\,000$ spectroscopic redshifts are required.

Ideally, we would like to know the probability distribution with redshift, $p_i(z)$, exactly from a spectroscopic survey, where the redshift is derived from the spectrum of each galaxy. In reality, deep wide-field surveys will only provide us with multi-band imaging data that will allow us to derive photometric redshifts based on templates and/or spectroscopic training sets. We can relate the probabilities for the true redshift z_{spec} and the photometric redshift z_{phot} by the Bayesian rule of conditional probability:

$$p(z_{\mathrm{spec}}, z_{\mathrm{phot}}) = p(z_{\mathrm{spec}}|z_{\mathrm{phot}})p(z_{\mathrm{phot}}) = p(z_{\mathrm{phot}}|z_{\mathrm{spec}})p(z_{\mathrm{spec}}). \tag{13.10}$$

Consider now a sharp cut in a photo-z bin i, i.e., we select only those galaxies in the range $z_{\mathrm{phot}}(i) < z_{\mathrm{phot}} < z_{\mathrm{phot}}(i+1)$. We can write the probability for the true redshift distribution resulting from the photo-z slice as

$$p_i(z_{\mathrm{spec}}) = \int_{z_{\mathrm{phot}}(i)}^{z_{\mathrm{phot}}(i+1)} p(z_{\mathrm{phot}}, z_{\mathrm{spec}}) \, \mathrm{d}z_{\mathrm{phot}}. \tag{13.11}$$

Typically $p_i(z_{\mathrm{spec}})$ would have a wide spread, *not* in the form of a Gaussian, as asymmetric tails are present due to the photo-z catastrophic errors.

We can parameterize $p_i(z_{spec})$ directly based on the projection of the photo-z scatter diagram $p_i(z_{spec}, z_{phot})$, derived based on a spectroscopic training set or mock catalogues. The uncertainties in the shapes of the probability distribution function also have to be taken into account, due to the finite number N_s of the spectroscopic redshifts per bin in the calibration set. One can then marginalize over both of these uncertainties. The dependence on the number of spectroscopic redshifts is really an indirect expression of the scatter in $\bar{z}$ and in μ_2.

Consider the specific example of the dark energy equation of state, commonly written as $w(a) = w_0 + (1 - a)w_a$. One can also define a_p, the pivot value at which the uncertainty in the constant part $w(a_p)$ is minimal. Accordingly, we define $w_p = w_0 + (1 - a_p)w_a$. Armed with this parameterization we can now apply the standard Fisher matrix formalism, and define the Figure of Merit (FoM) as

$$\text{FoM} = \frac{1}{\delta w_p \delta w_a},\tag{13.12}$$

where δw_p and δw_p are the 68% errors on w_p and w_a.[11]

One can then explore how the FoM varies subject to photo-z performance. For example, Abdalla *et al.* (2008a) have shown that the FoM for weak lensing can be improved by factors of 1.5–1.7 by adding NIR to the optical Euclid filter. Banerji *et al.* (2008) have illustrated how the addition of NIR colours from VISTA can improve the measurements of the galaxy power spectrum.

13.10 Discussion

Photo-z estimation is a growing field, which is of great importance for the next generation of wide-field imaging surveys. Bayesian ingredients in some of the methods have proven useful in reducing the number of outliers. The recent applications to cosmological studies have illustrated that the relevant question is not 'What is the best photo-z one can estimate?' but rather 'Is the photo-z accuracy sufficient to achieve a certain science goal?'

Future challenges are:

- The development of hybrid template/training methods.
- Finding reliable methods for assigning probability distribution functions and diagnostics of errors.
- Developing methods for incorporating incomplete spectroscopic calibration sets.
- Seeking new approaches for accurate photo-z for clusters of galaxies and hosts of supernovae type Ia.

---

[11]   This is the FoM as defined by the Dark Energy Task Force (DETF) in Albrecht *et al.* (2006).

Acknowledgements

We are grateful to Adam Amara, Peter Capak, Adrian Collister, Carlos Cunha, Eduardo Cypriano, Andrew Firth, Josh Frieman, Simon Lilly, Marcos Lima, Huan Lin, Hiroaki Oyaizu, Jason Rhodes and other members of the DES and Euclid photo-z working groups for their contribution to the work presented here and for helpful discussions.

References

Abdalla, F. B., Amara, A., Capak, O., Cypriano, E. S., Lahav, O. and Rhodes, J. (2008a). *Mon. Not. Roy. Astron. Soc.*, **387**, 969.

Abdalla, F. B., Banerji, M., Lahav, O. and Rashkov, V. (2008b). arXiv:0812.3831; *Mon. Not. Roy. Astron. Soc.*, in press.

Albrecht, A. *et al.* (2006). eprint arXiv:astro-ph/0609591.

Amara, A. and Refregier, A. (2007). *Mon. Not. Roy. Astron. Soc.*, **381**, 1018.

Andreon, S., Gargiulo, G., Longo, G., Tagliaferri, R. and Capuano, N. (2000). *Mon. Not. Roy. Astron. Soc.*, **319**, 700.

Babbedge, T. S. R. *et al.* (2004). *Mon. Not. Roy. Astron. Soc.*, **353**, 654.

Bailer-Jones, C. A. L., Irwin, M. and von Hippel, T. (1998). *Mon. Not. Roy. Astron. Soc.*, **298**, 361.

Banerji, M., Abdalla, F. B., Lahav, O. and Lin, H. (2008). *Mon. Not. Roy. Astron. Soc.*, **386**, 1219.

Baum, W. A. (1962). *Problems of Extragalactic Research*. IAU Symposium No. 15, 390.

Benitez, N. (2000). *Astrophys. J.*, **536**, 571.

Bertin, E. and Arnouts, S. (1996). *Astron. Astrophys. Supp.*, **117**, 393.

Bishop, C. M. (1995). *Neural Networks for Pattern Recognition*. Oxford: Oxford University Press.

Blanton, M. R. and Roweis, S. (2007). *Astron. J.*, **133**, 734.

Bolzonella, M., Miralles, J.-M. and Pelló, R. (2000). *Astron. Astrophys.*, **363**, 476.

Brammer, G. B., van Dokkum, P. G. and Coppi, P. (2008). *Astrophys. J.*, **686**, 1503.

Brunner, R. J., Szalay, A. S. and Connolly, A. J. (2000). *Astrophys. J.*, **541**, 527.

Bruzual, A. G. and Charlot, S. (1993). *Astrophys. J.*, **405**, 538.

Coleman, G. D., Wu, C.-C. and Weedman, D. W. (1980). *Astrophys. J. Supp.*, **43**, 393.

Collister, A. A. and Lahav, O. (2004). *Proc. Astron. Soc. Pac.*, **116**, 345.

Collister, A. A. *et al.* (2007). *Mon. Not. Roy. Astron. Soc.*, **375**, 68.

Connolly, A. J., Csabai, I., Szalay, A. S., Koo, D. C., Kron, R. G. and Munn, J. A. (1995). *Astron. J.*, **110**, 2655.

Connolly, A. J., Szalay, A. S. and Brunner, R. J. (1998). *Astrophys. J.*, **499**, L125.

Feldmann, R. *et al.* (2006). *Mon. Not. Roy. Astron. Soc.*, **372**, 565.

Firth, A. E., Lahav, O. and Somerville, R. S. (2003). *Mon. Not. Roy. Astron. Soc.*, **339**, 1195.

Jain, B., Connolly, A. and Takada, M. (2007). *J. Cosmol. Astropart. Phys.*, **03**, 013J.

Koo, D. C. (1999). In R. J. Weymann, L. J. Storrie-Lombardi, M. Sawicki and R. Brunner, eds., *Photometric Redshifts and High-Redshift Galaxies*. ASP Conference Series. Vol. 191. San Francisco: Astronomical Society of the Pacific.

Kurtz, M. J., Geller, M. J., Fabricant, D. G. and Wyatt, W. F. (2007). *Astron. J.*, **134**, 1360.

Lahav, O., Naim, A., Sodré Jr., L. and Storrie-Lombardi, M. C. (1996). *Mon. Not. Roy. Astron. Soc.*, **283**, 207.

Lima, M., Cunha, C. E., Oyaizu, H., Frieman, J., Lin, H. and Sheldon, E. S. (2008). *Mon. Not. Roy. Astron. Soc.*, **390**, 118.

Ma, Z., Hu, W. and Huterer, D. (2006). *Astrophys. J.*, **636**, 21.

Naim, A., Lahav, O., Sodré Jr., L., Storrie-Lombardi, M. C. (1995). *Mon. Not. Roy. Astron. Soc.*, **275**, 567.

Newman, J. (2008). arXiv:0805.1409.

Oyaizu, H., Lima, M., Cunha, C. E., Lin, H. and Frieman, J. (2007). arXiv:0711.0962.

Oyaizu, H., Lima, M., Cunha, C. E., Lin, H., Frieman, J., Sheldon, E. S. (2008). *Astrophys. J.*, **674**, 768.

Peacock, J. A. *et al.* (2006). astro-ph/0610906.

Sowards-Emmerd, D., Smith, J. A., McKay, T. A., Sheldon, E., Tucker, D. L. and Castander F. J. (2000). *Astron. J.*, **119**, 2598.

Stabenau, H. F., Connolly, A. and Jain, B. (2008). *Mon. Not. Roy. Astron. Soc.*, **387**, 1215.

Wadadekar, Y. (2005). *Proc. Astron. Soc. Pac.*, **117**, 79.

Wang, D., Zhang, Y. X., Liu, C. and Zhao, Y. H. (2007). *Mon. Not. Roy. Astron. Soc.*, **382**, 1601.

Weymann, R. J., Storrie-Lombardi, L., Sawicki, M. and Brunner, R. J., eds. (1999). *Photometric Redshifts and High-Redshift Galaxies*. ASP Conference Series. Vol. 191. San Francisco: Astronomical Society of the Pacific.

Wray, J. J. and Gunn, J. E. (2008). *Astrophys. J.*, **678**, 144.

Index

Printed in the United States
By Bookmasters